EcoMechatronics

Peter Hehenberger · Maki Habib · David Bradley
Editors

EcoMechatronics

Challenges for Evolution, Development and Sustainability

Springer

Editors
Peter Hehenberger
School of Engineering
University of Applied Sciences
Upper Austria
Wels, Austria

Maki Habib
Department of Mechanical Engineering
The American University in Cairo
New Cairo, Egypt

David Bradley
Abertay University
Dundee, UK

ISBN 978-3-031-07554-4 ISBN 978-3-031-07555-1 (eBook)
https://doi.org/10.1007/978-3-031-07555-1

This Springer imprint is published by the registered company Springer Nature Switzerland AG
The registered company address is: Gewerbestrasse 11, 6330 Cham, Switzerland

Foreword

Introduction

When Barack Obama was asked on a US talk show in 2020 what his greatest lesser-known achievement as President was, his answer was surprising: appliance standards. He said new rules for energy-efficient appliances equated to *the equivalent of taking 20 million automobiles off the road annually* in greenhouse gas emission reductions.

Particularly in the energy sector, incremental improvements are often overlooked by politicians in favour of completely novel technologies. In the run-up to the United Nations climate change conference COP26, US climate envoy John Kerry caused a stir by saying that 50% of the reductions that are required to get to Net Zero are going to come from technologies that *we don't yet have*. There is a tension between those who believe we can live sustainably through a combination of existing low-carbon technology and behavioural change (e.g., much less flying and meat consumption) and those, sometimes referred to as EcoModernists, who believe that new technologies like nuclear fusion and geoengineering will be able to enable prosperous lifestyles for all without compromise.

What is clear, however, is that every increase in efficiency or more sustainable resource use will make Net Zero more achievable and less expensive. Incremental improvements can also be achieved faster than by developing new, unproven technologies. As climate change is a cumulative problem, near-term action is more valuable in preventing dangerous levels of global warming than emissions cuts in future decades.

The climate system is complex and modelling it involves inherent uncertainty, so every small positive change we make now reduces the chances of reaching various ecological tipping points that could result in worst-case scenarios for humanity and the natural world.

Definitions of Sustainability

The ultimate aim of the global energy transition that is underway must be to provide a good standard of living for the growing billions of people in the world in a manner that is sustainable. Various definitions of *sustainable* abound, and are explored in *EcoMechatronics*, but the most famous is perhaps Brundland's *development that meets the needs of the present without compromising the ability of future generations to meet their own needs.*

It is often not even easy to determine which technologies are sustainable. Often *renewable* energy is conflated with *sustainable* energy, but this is not necessarily the case. Nuclear power in its current form is not strictly renewable, but with some moderate innovation in uranium mining and recycling of waste, its fuel could become essentially inexhaustible, meaning that it would be sustainable. Conversely, using biomass for energy is technically renewable but, without very strict control of where the source material comes from, it may not be sustainable. Similar concerns surround policies to promote the uptake of electric vehicles given the high embedded carbon emissions and resource extraction involved in manufacturing, in particular, lithium-ion batteries. Rethinking our approach to transport may be a better idea than replacing one inefficient mode of transport with another.

Predictions that humanity would run out of key resources, or of ecological collapse, have also come and gone, of course. Humans have proven themselves ingenious at extracting more resources from the ground and improving farming productivity. Today we produce enough food globally so that famines are almost unheard of, except when caused by political strife. Innovation in the oil and gas sector to expand to reserves of extractable fossil fuels also means that climate change will not be solved by peak oil.

This ingenuity would be something to celebrate if the risk to the world were not so great. Our planet is a complex system and positive feedback loops may result in the acceleration of climate change to the extent that we cannot innovate our way to sustainability without rapid emissions reductions and action to preserve resources.

Risks and Challenges

Some of the risks and challenges of delivering a Net Zero world are explored in *EcoMechatronics*. The rebound effect, or *Jevon's Paradox*, is the observed phenomenon of how making something cheaper or more available will mean that people use more of it. For example, although recent advances in energy-efficient lighting have reduced overall energy use, it has become so inexpensive that much more artificial lighting is used than ever before.

Technological lock-in is another risk. Decisions made today may put us on paths that make sustainability more difficult to achieve. In reaching Net Zero, carbon capture and storage (CCS) is considered crucial. But CCS is not one technology, it is a fleet of technology options, and none have yet been proven to be economic at the scale required to have an impact. Multi-billion-pound investments will need to be made that may lead us done an expensive and inefficient technology cul-de-sac.

Both the rebound effect and technological lock-in effects may also occur due to the uptake of autonomous vehicles. If the auto industry delivers on the potential of self-driving cars, the convenience may mean that more sustainable forms of transport become less popular.

Cause for Optimism

There is reason to be optimistic, however. Cost reductions of wind and solar power have exceeded all expectations. Battery electric vehicles are following the same trajectory, albeit a decade behind. We are also developing more resilient infrastructure that is able to withstand rising seas and more powerful storms. More controversially, solutions to global warming that involve geoengineering are being taken more seriously. The collective commitments made at COP26 may have gotten us to a point where limiting global warming to two degrees Celsius is achievable, but only if every country lives up to its promise. Geoengineering may be necessary if they don't.

One trend that is already happening, and which will continue to develop, is the increasing electrification of our energy system. Electric car sales are rising exponentially and eventually in a large proportion of the heating systems in buildings gas boilers will be replaced, at least in part, by heat pumps.

These, and other developments, open up opportunities for making increased use of big data and smart technology to improve the efficiency of buildings and transport systems, especially as they become more interconnected.

EcoMechatronics

The boundaries between engineering disciplines have been breaking down for many decades. EcoMechatronics is an important next step. It recognises the benefits of interdisciplinarity and highlights the potential of engineering to make giant leaps in progress by integrating big data and emerging fields like artificial intelligence.

By embedding sustainability in all steps of the engineering process, EcoMechatronics is a field that can make an enormous contribution to our race to live sustainably on Earth before we reach an environmental tipping point from which we cannot recover.

It is not just about improving efficiency; it is about improving the user experience. Basic electric vehicles have been on the road for a long time, but until the last decade have been considered unfashionable and impractical. What has happened in recent years is that people are increasingly trying out electric vehicles and finding out they are not only cleaner than those with an internal combustion engine, but also offer a better user experience for the driver.

The benefit of EcoMechatronics, therefore, lies not just in making our devices, building, and vehicles more efficient, but in making them more desirable to the consumer. We cannot rely on the environmentalist instincts of people to choose the more sustainable option out of their altruistic instincts.

Building decarbonisation is one good example. It is sometimes said that home-owners like their gas boilers and will be reluctant to give them up, but they have not yet been presented with a better option. New smart heating systems and ventilation management may eventually encourage a rapid uptake. Homes will be able to control temperature and humidity room-by-room and they will generate their own electricity and share it with their neighbours. The EcoMechatronics approach will also see engineered systems interact not just with each other, but also responding and adapting to factors like the real-time changes in weather and occupancy.

In this context, the Covid-19 pandemic has made ventilation a more important factor to consider, especially in workplaces and public transport. It will be up to engineers to come up with solutions that ensure we can breathe clean air indoors without expending huge quantities of energy on inefficient air-conditioning or excessive deployment of air-sanitation devices.

Skills

In 2021, the Institution of Mechanical Engineers (IMechE) polled our members on what they thought would be the key skills required for manufacturing engineers in the future. The top answer, from a list of 22 options, was *automation, robotics and mechatronics.*

As professional engineering institutions (PEIs), the IMechE has certain key roles. In certifying university degrees, we need to encourage academia to embed the concept of sustainability across the whole curriculum. Engineering graduates should enter the workforce with a philosophy of environmental protection engrained in their thinking across design, manufacturing, operation and end-of-life.

A second important role is continued professional development. The pace of technological change means that engineers and technicians who qualify today may need to retrain and/or upskill multiple times throughout their careers. Businesses, governments and educational establishments should collaborate to ensure the workforce has the right skills today and they should also proactively horizon scan to plan for what is required 10 or 20 years hence.

It is also important to embed a philosophy of sustainability as early as possible in our education systems. In the absence of dedicated engineering teaching in schools, the IMechE and other engineering organisations help to train primary school teachers so that they have confidence in talking to students about such topics. Volunteers from this initiative, *Primary Engineer*, were in attendance at COP26 in Glasgow in 2021 showcasing hands-on activities for young people to engage with emissions reduction, including a flat-pack wind turbine.

Conclusion

Although I began this foreword with the importance of incremental improvements in technology, the EcoMechatronics approach is a step change from merely thinking about reducing the adverse effects on the environment. Actively thinking about *waste*, in all its various forms, can be put to productive use to benefit the entire system. As Simpson et al. explains in their chapter, EcoMechatronics challenges engineers to go beyond the mindset of *doing less bad* and to instead reframe from problem in terms of *doing more good*.

Matthew Rooney
Head of Engineering Policy
Institution of Mechanical Engineers
London, UK

Matthew Rooney is a Chartered Mechanical Engineer and Head of Policy at the Institution of Mechanical Engineers (IMechE) in London. He holds an MPhil in Technology Policy from the University of Cambridge and an MEng in Mechanical and Manufacturing Engineering from the Queen's University of Belfast.

After spending six years as a mechanical engineer working on the design and manufacture of components for particle accelerators at the Harwell Science and Innovation Campus, Matt returned to university to study technology policy. The main focus of his research was on deployment strategies for new energy technologies, with case studies including carbon capture and storage and small modular reactors. Following this, he spent 18 months as an Energy Policy Research Fellow at the prominent think tank Policy Exchange in Westminster before joining IMechE in May 2018.

He is a British Science Association Media Fellow, having worked briefly as a science journalist with *Times Higher Education*. His other media work includes appearances on *BBC News*, *BBC World Service*, and the *Naked Scientists* radio programme and podcast.[1] He has had articles published in *BusinessGreen*,[2] *Professional Engineering*, and *The Engineer* and he has been quoted in the *Financial Times* and on the *BBC News* website.

He has authored or co-authored over ten reports on the topic of energy and technology policy.

[1] www.thenakedscientists.com/.

[2] www.businessgreen.com.

Contents

Applications

Education

Conclusion

Editors and Contributors

About the Editors

Dr. Peter Hehenberger is Professor for Integrated Product Development, Head of the *Smart Mechatronics Engineering* Research Group and Academic Director of the *Institute for Agile Transformation* at the School of Engineering, University of Applied Sciences Upper Austria, Wels/Austria.

Prior to this appointment, he was an Assistant Professor and Deputy Head at the Institute of Mechatronic Design and Production of the Johannes Kepler University of Linz (JKU), Austria and Deputy Research Area Coordinator for *Process Modelling and Mechatronic Design* at the Linz Center of Mechatronics (LCM). In 2015, he was invited as a Visiting Professor at the Universite de Technologie de Compiegne, Departement Genie des Systemes Mecaniques, France.

He completed his Diploma and Doctorate in Mechatronics from JKU in 2000 and 2004, respectively. In 2016, he received the post-doctoral lecturing qualification (venia docendi) for the subject area of Engineering Design. Core research interests include *Sustainability in Development and Production of Mechatronic Systems* and *Engineering Processes* along with selected topics on *Digital Transformation.*

He has published over 140 peer-reviewed papers in international journals and conference proceedings and has been a guest editor for three journal special issues. He serves as an Editorial Board Member for the IFAC journals *Mechatronics, Computer-Aided Design and Applications and International Journal of Product Lifecycle Management* (IPLM). In addition, he has authored or co-authored 6 books related to his research field.

Dr. Hehenberger has practical experience with industry and design projects and has been a key researcher and manager of several successful national and international research projects. He is active in the research community where he refers to his research area of *Mechatronic Design* as group member of IFIP WG5.1 *Global Product development for the whole life-cycle,* IFAC TC 4.2 *Mechatronic Systems,* SAE SETC *Small Powertrains and Energy Systems Technology Conference.* He is Austrian representative and Communications Publications & Events Officer for IFIP TC 5 *Information Technology Applications.*

Prof. Maki Habib obtained his Ph.D. in Intelligent and Autonomous Robotics from the University of Tsukuba, Japan, and now specialises in Robotics and Mechatronics. He was previously a research scientist at RIKEN, Japan, a senior researcher at RISO-Laboratories, Japan, and a visiting researcher at EPFL, Switzerland. He has also been a Visiting Expert at the Asian Development Bank (ADB) serving the Malaysian industry, an Associate Professor at UTM, Malaysia, and a Senior Manager at MCRIA, Malaysia. He was a leading senior research scientist with GMD-Japan where he led the Telecooperation Research Group and an Associate Professor with Monash University (Australia/Malaysia) to develop and lead the Mechatronics Engineering programme.

He was appointed as a Full Professor of Robotics and Mechatronics at Swinburne University (Australia/Malaysia), an invited Professor at KAIST, South Korea, and a Resident Professor at Saga University, Japan. Since September 2007, he has been a Full Professor of Robotics and Mechatronics at the American University in Cairo.

He was, and continues to be, a consultant and technical advisor to many international industrial companies and research consortiums, such as Toyota and ABB. He also has research collaborations with several international research groups associated with research centres and universities. His current research includes:

- Robotics for Humanitarian Demining Use.
- Autonomous Electric Vehicles.
- Nanorobotics for Biomedical Purposes.
- Robots for the Physically Impaired using Biological Signals (such as EEG and EMG).
- Industry 4.0, Internet of Things (IoT), and CPS.
- EcoMechatronics and Bioinspired Design.
- Intelligent Drone systems.
- Robotics and Industrial Automation.
- Tele-cooperation, Distributed Mixed Realities, and Collaborative Control.

Prof. Emeritus David Bradley has been involved in the design, development and teaching of mechatronics since the 1980s and this current work is the fourth of a series of mechatronics related-books that he has edited for Springer. He has been the author, lead author, co-author or contributing editor for 10 books, the majority of which have had mechatronics as their main subject. He has also published over 50 journal papers and 150 plus conference papers and book chapters.

Having taught at Lancaster University, Bangor University and Abertay University in the UK, he was a visiting fellow at the Technical University of Denmark and a research consultant at Sheffield University and Barnsley Hospital, and a research engineer, while on sabbatical leave, at New Zealand Electricity. His research interests have included engineering design, power system harmonics, construction robotics, eHealth & mHealth technologies and the modelling of hydro-power stations. He remains active in research as a consultant/advisor to research groups working in areas such as the design and operation of advanced prostheses, health care and the built environment and was recently a contributor and co-author to an IMechE policy paper produced for COP26.

He has also acted as a consultant on mechatronics course development in, among others, South Africa, Mauritius, Singapore and Colombia, and has been a visiting lecturer on mechatronics programmes in Colombia, Singapore, South Africa as well as at several locations in Europe. A co-founder of the UK Mechatronics Forum, later the Mechatronics Forum, he was also the originator of the Mechatronics Forum Conference series.

Contributors

Paolo Boscariol has been associate professor of Mechanics of Machines at the Department of Management and Engineering (DTG) of the University of Padova, Italy, since 2021. He received his Ph.D. in Industrial and Information Engineering from the University of Udine, Italy, in 2012, where he has been from 2008 to 2015. He is author of more than 90 publications in the field of mechatronics and robotics. He is an active member of IFToMM Italy, the Italian branch of IFToMM.

His research interest lies in the broad areas of modern robotics and mechatronics, being mainly focused on the development of motion design methods for industrial devices and robots. In particular, he has been developing solutions to improve the energetic efficiency of mechatronic devices, by exploiting new motion design rules and using redundancy. His current research activity is focused also on the energy-optimal design of servo-actuated systems. He has also investigated the development of robust motion design methods, through which the performance of uncertain system can be enhanced. In particular, he has focused on the development of indirect model-based method based on variational solutions and sensitivity shaping. In addition, he has also made numerous contributions to the topic of vibration reduction in mechanisms with pronounced flexibility, by investigating the issues of modelling, simulating, controlling flexible-link mechanisms.

Luis Gabriel Carmona is a Marie Curie Research Fellow working on resource efficiency, material/energy services and the circular economy at University of Cambridge. In 2020, Gabriel completed his Ph.D. in Sustainable Energy Systems at the University of Lisbon (Portugal). He focused on the stock-flow-service nexus to facilitate a more comprehensive analysis of resource production, consumption, and accumulation. Gabriel worked on several H2020 projects, where he assessed the sustainability of emerging technologies. He also has over fifteen years of experience working as a consultant and sustainability professional. He advised Latin American companies in the public transport, infrastructure, and hydrocarbon sectors. Gabriel obtained an Environmental Engineering degree at La Salle University (Colombia) and a master's degree in Energy Efficiency and Industrial Ecology at the University of Zaragoza (Spain).

Dr. Jun Chen is currently a Professor of Mechanical Engineering at Purdue University. He received his B.S. degree and M.S. degree in aerospace engineering from Beijing University of Aeronautics and Astronautics. He obtained his Ph.D. degree in mechanical engineering from Johns Hopkins University in 2005 and did postdoctoral research at Los Alamos National Laboratory. He joined the Purdue faculty of the School of Mechanical Engineering in 2008. Dr. Chen's research interests are in the general area of experimental and applied fluid dynamics, including advanced flow diagnostics techniques, benchmark experiments and modeling, and applied fluid dynamics. He also led the student team in Purdue Global Engineering Program to develop low-cost energy solutions for rural communities in developing countries. His teams won the 2016 YCOSST P3 AWARD presented by the AIChE Institute for Sustainability and the 2018 ASCE Sustainable Development Award. His research on renewable energy devices received funding support from the United States Department of Energy, Environmental Protection Agency (EPA), and industry sponsors. Dr. Chen is one of the co-founders of Poseidon Energy Inc., which develops products to harvest energy from river currents.

Dr. Chen is an associate editor of the ASME Journal of Fluids Engineering and the chair of the ASME Fluid Mechanics Technical Committee. He has organized and chaired many technical conferences for ASME. He is an active reviewer for more than 25 journals, panels of national agencies, and several international agencies. He is a recipient of the 2005 Outstanding Fluid Mechanics Paper Award of Measurement Science and Technology and the 2014 ASME Robert T. Knapp Award. He was elected to the ASME Fellow in 2020.

Dr. Jean-Yves Choley obtained a Masters degree in Mechanics and Automation and a Master of Advanced Studies in Automated Manufacturing from ENS Cachan, France (recently renamed ENS Paris-Saclay). He obtained a Ph.D. degree in Industrial Engineering in 2005 from Ecole Centrale Paris, France (recently renamed CentraleSupélec), related to GPS mechanical tolerancing and 3D metrology. He gained industry operational experience (1986–1988) in systems engineering and quality/reliability at Arianespace, Kourou (French Guyana), France, working both on ground facilities and satellites launchers *Ariane* 1 to 4.

He is Professor and Research Director of ISAE-Supméca, France, responsible for the IS2M (Ingénierie des Systèmes Mécatroniques et Multiphysiques) research team of Quartz laboratory and is in charge of research and teaching activities, focusing mainly on tools, languages and methodologies for the collaborative design of complex systems such as mechatronic systems, CPS (Cyber-Physical Systems) and CPPS (Cyber-Physical Production Systems). In the past 10 years, he has mainly focused on the Digital Twin concept, Model-Based Systems Engineering (MBSE) and Model-Based Safety Assessment (MBSA), together with multiphysics modeling for the collaborative design of safety critical systems, such as autonomous vehicle and Factory of the Future systems (Industry 4.0 and Industry 5.0).

Abbas Dehghani-Sanji a Ph.D. graduate from the University of Leeds, UK, is professor of Bio-Mechatronics and Medical Robotics in the School of Mechanical Engineering at the University of Leeds. His research interests include bio-robotics, intelligent control, sensors, and actuators using integrated system design approach where functional materials, sensors, actuators, and control are brought together to develop intelligent systems/devices. His research focus is mobility and intelligent control in robots and development of devices to support mobility in patients and the growing aging population as well as wearable robotic systems for enhancing human capabilities. His current research includes design and development of intelligent robotic exoskeletons, soft robotics, and artificial limbs. Professor Dehghani is the leader of Robotic Exoskeleton Research and has had in his team 11 Postdocs and 40 research students. He has extensive government and industrial funded research and over 150 journal and conference publications. He has been Director of ExoLab research and was the founding Director of the Centre for Mechatronics and Robotics at the University of Leeds. He is a member of the IMechE Mechatronics Forum Steering Committee.

Margherita Finamore with more than 20 years experience in the public sector on retrofitting cultural heritage, Margherita Finamore has been for many years investigating and guiding sustainability implementation, especially in the design and retrofitting of schools, school-gyms and social-housing.

Margherita is currently leader of research on sustainable construction with responsibility as project manager for public works at Pesaro Council where she is engaged in environmental sustainability and energy efficiency design and is the lead in developing the requirement assessments and technical specifications to achieve high values of sustainability and energy savings. Since joining the council in 1997 her work has focused primarily on sustainable cultural heritage, public buildings and retrofitting projects where she has been involved in international projects with colleagues from different cultures around the world.

Having graduated in Architecture from the University of Florence she has gained Masters in Environmental Sustainability and Energy Efficiency on Historical Buildings (University of Ferrara) and BioArchitecture (University of Bologna) and is a Technical Architect specialist in Historical and Cultural Environment with the Italian National Trust. She was selected to participate in elements of the European Federation of Agencies and Regions for Energy and Environment (FEDARENE) programme and is currently a Ph.D. student at Bradford University (UK) researching the circular economy in the built environment with a special focus on existing buildings. Her last project, the Antonio Brancati School in Pesaro in Italy received the 2020–21 International Green Solution Awards *Energy and Temperate Climates* prize at COP26 in Glasgow.

Elias Flening is a doctoral student at KTH Mechatronics. His interest lies in exploring organizational issues in engineering systems design in general, and in the project management of mechatronic systems development processes in particular. The topic area of his doctoral research follows this interest, investigated through exploratory and mostly qualitative research designs. Elias Flening is cross-disciplinarily dispositioned, loving the mixing of potent (and sometimes explosive) different worldviews in research.

He was previously a research assistant at Stockholm School of Economics in various research projects, mostly relating to activity-based office design and development. He also retains a strong professional and personal interest in Japan, having studied and lived there over two periods and travels there regularly.

Mojtaba Ghodsi is a Senior Lecturer in Instrumentation, and Measurement Systems in the School of Energy and Electronic Engineering at the University of Portsmouth, UK. He received his Ph.D. in Precision Engineering from the University of Tokyo, Japan in 2007 and continued his research as a JSPS postdoctoral fellow (2009) at the University of Tokyo. His main research interests are around smart materials and structures to develop actuators, sensors, energy harvesters, vibration control, NDT and mechatronic systems.

Roger Goodall is an Emeritus Professor in Control Systems Engineering in the Wolfson School at Loughborough University, UK. He has a long history in both industry and university of contributing to the subject of mechatronics, particularly with significant theoretical and practical contributions relating to railway technology. He was elected as a Fellow of the Royal Academy of Engineering in 2007. He has held a variety of external appointments, including a long-standing involvement with IFAC for which he is a former Vice President, and is now both a Fellow and an Advisor of the Federation.

Martin Edin Grimheden is associate professor in Mechatronics at KTH Royal Institute of Technology and has been publishing studies on mechatronics education since 2001. Research interests include learning in mechatronics in a broad sense, with studies on experimental learning, collaborative learning, globally distributed teams as well as product innovation.

He has been director of KTH B.Sc. program in Mechanical Engineering as well as M.Sc. programs in Engineering Design and Mechatronics and is currently Head of Department at KTH Machine Design as well as Head of Gender Equality, Heterogeneity and Equal Treatment at KTH School of Industrial Engineering and Management.

Alexander Hagenberger is a junior researcher at the Institute of Thermodynamics and Sustainable Propulsion Systems at Graz University of Technology. His research focuses on the simulation of small hybrid propulsion systems.

Dr. Tim Harrison is a Senior Research Associate working with the Control Systems research group in the Wolfson School of Mechanical, Electrical and Manufacturing Engineering at Loughborough University. He holds a Ph.D. in Aerospace Engineering from the University of Manchester. After working in aerospace systems engineering for several years, Tim returned to academia focussing on rail related systems and technologies. His research projects have an emphasis on developing solutions to real world problems in collaboration with rail industry bodies, improving service reliability and resilience and reducing environmental impact.

Dr. C. Greg Jensen is currently a Professor of Engineering Practice, Mechanical Engineering at Purdue University. He received his B.S. and M. Design and Manufacturing degrees from Brigham Young University. His Ph.D. in Mechanical Engineering was granted in 1993 from Purdue University, West Lafayette, IN. He rejoined the Mechanical Engineering faculty at Brigham Young University in 1993 and retired from teaching and research after 32 years of service in 2015.

In 2018, after a short self-employment and engineering consulting endeavour, he joined the Mechanical Engineering faculty at Purdue University as their Director of Senior (Capstone) Design. His interests and strengths over his long career have been closely aligned with the optimization of product and process design. Today, he annually oversees more than 100 senior design/build/test/iterate semester-long projects. He has authored more than 50 journal articles, 5 book chapters, and 70 conference papers. He is the author, and inventor on more than 40 design and manufacturing patents, with others in provisional and pending status. He has worked for and/or consulted with, among others, Boeing, GM, Ford, Lockheed Martin, Pratt & Whitney and Raytheon.

Kristina Kerwin is a design engineer and a researcher working at London South Bank University (LSBU). She graduated from LSBU in 2017 with a B.Sc. (Hons) in Engineering Product Design and continued to a master's degree at Kingston University while working part-time at a secondary school. As part of her master's degree, Kristina carried out a study titled *Studying the efficacy of teaching STEM subjects to primary school age students using LEGO robots* to demonstrate how the use of robotic devices can provide a more engaging and active learning experience for pupils. She presented her preliminary findings at Horizons in STEM Higher Education Conference 2019.

Youhua Li is a research Assistant in department of Design Manufacturing and Engineering Management at the University of Strathclyde, Glasgow, UK. After working in manufacturing company for about 10 years, she joined the department of DMEM, working on several mechatronic education, research and knowledge exchange projects, including areas of smart farming, agricultural robotics and smart factory costing. Her research interests include software validation for engineering and agricultural applications, and design for manufacture.

Uriel Martinez-Hernandez received his Ph.D. degree in Robotics and Tactile Perception from the University of Sheffield, Sheffield, U.K., in 2015, and his M.Sc. degree in Computer Science from the Centre for Research and Advanced Studies (CINVESTAV), Mexico City, Mexico in 2008. He was Research Fellow in the Institute for Design, Robotics and Optimisation (iDRO), the University of Leeds, and Research Associate in Sheffield Robotics Lab, the University of Sheffield.

Currently, he is a Lecturer (Assistant Professor) in Robotics and Autonomous Systems at the Department of Electronics and Electrical Engineering, Faculty of Engineering and Design, University of Bath. Uriel is leader of the Multimodal Interaction and Robot Active Perception (inte-R-action) group where he undertakes research and supervises work on human-robot interaction, autonomous robots, activity recognition methods, tactile sensors for robotics, vision and tactile perception, wearable haptic devices and wearable assistive robots. He is core member of the Centre for Autonomous Robotics (CENTAUR), and member of the Accountable, Responsible and Transparent AI (ART-AI) CDT.

Uriel has published over 40 scientific and technical papers in international journals and conferences, 3 book chapters and he has 1 patent. He has also worked in large multidisciplinary projects funded by EPSRC, The Royal Society, Cancer Research UK and EU with universities and research groups across the UK and Europe. His research interests include active touch and vision perception, human-robot interaction and collaboration, machine learning for autonomous robots, bioinspired cognitive architectures for robot control, wearable assistive robots, multimodal activity recognition and haptics and wearable immersive devices for telepresence and telerobotics.

Brendan McGlynn gained a Bachelor of Engineering degree in Electrical with a focus on Advanced Maths and Electronics at the University College Cork in Ireland and is currently completing a Bachelor of Commerce degree from University College Dublin. He has pursued a career in radar and telecommunications R&D, avionics, global data networks and main frame computer systems working for airlines in Europe, East Africa, Middle East, S.E. Asia, Far East, USA and the Pacific. He has also been a project manager and regional sales manager in telecommunications engineering in Europe, the Middle East and the USA. He is also CEO of an SME covering electronic sensors and robotics across applications in security, health and military environments. He has also been funded for 7 years as a lead partner on two EU Commission projects and is currently working on wearable sensors and innovative renewable energy. He is active in swimming, trekking, cycling and the propagation of native deciduous trees and sequoia conifers on his own land.

Faïda Mhenni holds a Ph.D. in Engineering Sciences from Ecole Centrale de Paris, France entitled *Integration of systems engineering and safety assessment* and a Ph.D. in Mechanical Engineering from Ecole Nationale d'Ingénieurs de Monastir (ENIM), Tunisia entitled *Effect of tolerances on the behavior of mechanisms*. She has a degree in electromechanical engineering from from Ecole Nationale d'Ingénieurs de Sfax (ENIS), Tunisia.

She is Associate Professor at ISAE-Supméca in France and a member of Quartz laboratory, IS2M (Ingénierie des Systèmes Mécatroniques et Multiphysiques) research team. Her main research domains include Model-Based Systems Engineering (MBSE) methods and tools, safety assessment and MBSA (Model-Based Safety Assessment), mechatronics and multiphysics, tolerancing and multi-domain models consistency.

Dr. Will Midgley is Senior Lecturer in Intelligent Mechatronics and Control Systems Engineering in the Wolfson School of Mechanical, Electrical and Manufacturing Engineering at Loughborough University. He holds a Ph.D. in engineering from the University of Cambridge, United Kingdom. His research interests are the application of mechatronics engineering to complex systems such as transportation with the aim of decarbonising these systems and reducing their environmental impact. Previous work has focussed on reducing the fuel usage of heavy goods vehicles and rail vehicles. He works closely with industry to implement his research and to ensure it has impact.

Morteza Mohammadzaheri is a Senior Lecturer in Mechatronics at Birmingham City University, UK on leave from his position as Assistant Professor of Dynamic Systems and Control in the Department of Mechanical and Industrial Engineering of Sultan Qaboos University, Oman. He received his PhD from the School of Mechanical Engineering, University of Adelaide, Australia in 2011.

Dr. Mohammadzaheri has published/presented more than 120 peer-reviewed articles in technical journals and conferences mostly on application of artificial intelligence in control, modelling and optimisation of engineering systems. He has recently worked on micro/nanopositioning with use of piezoelectric actuators and structural health monitoring thru intelligent vibration analysis.

Cong Niu has been awarded the B.E. with Honours. Degree of Electronic Engineering from the University of Central Lancashire, Preston, UK in 2014. He was then awarded with the M.Sc. in Embedded Digital System from the University of Sussex, Brighton, UK in 2015. He completed his Ph.D. in Mechatronics in department of Design Manufacturing and Engineering Management at the University of Strathclyde, Glasgow, UK in 2021. He is currently a research assistant in Department of Design, Manufacturing and Engineering Management (DMEM), University of Strathclyde, Glasgow, UK. His current research interests include robotics, autonomous systems, mechatronics, field and indoor path planning, modeling and simulation, unmanned ground and aerial vehicles. He is a Student Member of IEEE.

Prof. Crina Oltean-Dumbrava is an academic with over 40 years' teaching and research experience in the UK, France, Germany, Italy, Greece, and Romania and is the chair of Sustainable Built Environment at the University of Bradford in the UK. Research interests cover all aspects of sustainability within the context of the built environment including decision support and risk assessment, water and waste management, transport infrastructure, road noise mitigation, energy efficiency in buildings and the circular economy in construction.

Prof. Oltean-Dumbrava was appointed in 2013 to the European Committee for Standardisation CEN/TC 226/WG6 (Noise Reducing Devices) and in 2014 as the liaison officer between CEN/TC 226 (Road Equipment) and CEN/TC 350 (Sustainability of Construction Works). In 2017 she became a partner and expert for the UN's Sustainable Buildings and Construction Programm. In 2021 she was appointed as an output assessor for REF 2021 Sub-panel 12: Engineering and became the academic member of a team representing the construction industry in the ICE Carbon Champions programme.

Prof. Oltean-Dumbrava is a Past Chair of the Association of Civil Engineering Departments in UK and a JBM Board member, and a member of the JBM Accreditation Visiting Panel involved in the accreditation of civil engineering courses in the UK. She is also a visiting professor at the Technical University of Cluj-Napoca in

Romania, represents Romania on the European Council of Construction Economists (CEEC), is a member of the Multi-Criteria Aid for Decisions, European Working Group in France, of the International Association for Housing Science (IAHS) in the USA and of the International Society on Multiple Criteria Decision Making.

Prof. Oltean-Dumbrava has published some 140 articles and papers in books, journals and at conferences and is or has been Principal Investigator, Co-Investigator or Collaborator for wide range of research and teaching projects, including multinational projects.

Stanislao Patalano obtained a Master's Degree (cum laude) in Mechanical Engineering in 1997 at the Univ. of Naples Federico II (UNINA) and a Ph.D. in *Machine Design and Construction* in 2001. He was a researcher working on Design and Methods of Engineering Design from 1999 to 2010 and Associate Professor of Design and Methods of Engineering Design from 2010 to 2018. He has been a full professor since 2018.

His research interests include CAD methodologies and Knowledge-Based Engineering, Variation Analysis, Tolerance Design and Design for Assembly, Object Oriented Programming and Simulation, Mechatronic Systems Design. He has published more than 100 papers in International conference proceedings, in International journals and National journals.

He has been member and co-ordinator of research groups in research projects funded by national public agencies (MUR, Campania Region). From 2011 to 2021, he has been scientifical responsible for the research activities of UNINA-DII with two PON Projects and one H2020 project, in cooperation with Fiat Chrysler vehicles.

He is currently professor of Geometric Modelling and Virtual Prototyping and Modelling and Simulation of Mechatronic Systems at UNINA.

Dr. Robert Rayner is a Senior Lecturer in Automotive Engineering, University of Hertfordshire where he teaches and researches extensively in his areas of specialism which include mechatronic systems, robotics, and autonomous vehicles. He is an experienced engineering University academic who has worked at a number of UK Universities including the University of Bath, Imperial College London, and Kingston University London. He has also spent time working at a private school in St. Albans, Hertfordshire where he was the Deputy Head of Mathematics. He has also spent time in industry with companies such as Cummins Power Generation, Molins ITCM and Mercedes Benz. Dr. Rayner is very active in the Engineering Education world and has worked IMechE Education Officer for the Greater London Region for several years.

Dario Richiedei is full professor of Mechanics of Machines at the Department of Management and Engineering (DTG) of the University of Padova, Italy, since 2021.

He is, currently, author or co-author of more than 120 papers, published in International indexed journals (more than 60 papers), in proceedings of International conferences or as book chapters. He is a member of the editorial board of 3 International journals.

His research activities are developed within the framework of several research projects funded by public institutions and private companies and in collaborations with scholars from Italian and other universities. His current research interests are mainly in the following areas of research, and including both theoretical and experimental activities: multibody system dynamics (modeling, simulation, co-simulation, model reduction, model updating, state estimation); motion planning and vibration control in flexible systems; model-based design optimization of vibrating systems through inverse dynamic structural modification; energy optimal design of mechatronic systems; motion planning of industrial mechatronic and robotic systems (energy efficient trajectory planning, motion planning and control of cable driven robots).

Dr. Hans-Jürgen Schacht is scientific senior researcher at the Institute of Thermodynamics and Sustainable Propulsion Systems at Graz University of Technology. His research focuses on small hybrid propulsion systems.

Dr. Stephan Schmidt is research area vice head at the Institute of Thermodynamics and Sustainable Propulsion Systems at Graz University of Technology. His research focus comprises the simulation, design and experimental development of sustainable small propulsion and energy systems as well as the use of sustainable fuels.

Eddie Simpson career started in the Army taking him across the world with particularly challenging appointments and considerable responsibility at HQ Scotland and the United Nations. Eddie continues links to the Ministry of Defence in his role as the Convener of Tayforth Military Education Committee promoting defence studies.

Since entering Higher Education as a lecturer nearly 30 years ago Eddie has been Programme Leader on various courses including B.Sc. (Hons) Quantity Surveying, B.Sc. (Hons) Construction Management and M.Sc. Construction Management. He has also created, implemented and developed innovative educational courses and collaborations including a B.Sc. (Hons) Sustainable Architectural Design. Eddie's experience and reputation have led to appointments such as Higher Education representative (Scotland) for the built environment and Chairman of Validation Panels and Qualification Development Specialist. His professional qualifications include Chartered Environmentalist, Fellow of the Chartered Institute of Building and Senior Fellow of the Higher Education Academy.

Eddie's research interests and recent publications in the last 5 years include a broad range of issues covering Higher Education, procurement in the context of MoD projects, Stoicism as a philosophy and its alignment with environmental issues, and sustainability. Recent publications include the following indicative issues: innovation and failure in mechatronics design education, corporate environmental sustainability, how Stoics can have a diet that aligns with environmental facts, and what we can learn from Stoic philosophy and Freirean environmental pedagogies.

Payam Soltani is currently a Senior Lecturer in Mechanical Engineering at the School of Engineering and the Built Environment at Birmingham City University, UK. He obtained his Ph.D. in Mechanical Engineering on nonlinear modelling and structural dynamics of nanostructures in 2012. He then contributed to research projects in piezoelectric-based smart structures in collaboration with the Structural and Material Computational Mechanics (MSC) group in the BATir department at the University of Brussels (ULB) and the Space Structure and Systems Lab (S3L) at the University of Liege (ULg).

He then joined the Dynamic Research Group (DRG) at the University of Sheffield, and as the member of the Programme for Simulation Innovation (PSi) project he developed novel simulation tools for Jaguar Land Rover aiming at capturing the effects of material ageing in vehicle components. As the Associate Fellow of Institute for Future Transport and Cities (IFTC) at Coventry University, he continued his research in simulation of dynamically coupled smart structures (dynamic absorbers and energy harvesters) and 3D-printed structures. Developing modern simulation methodologies, science, and tools (digital modelling and data-informed, data-driven simulation) for complex mechanical systems such as nanostructures, 3D-printed structures, 3D-printing processes, and smart structures are the main cores of his current research activities and interests.

Alberto Trevisani has been a full professor of Mechanics of Machines at the Department of Management and Engineering (DTG) of the University of Padova since 2015. He is the President of the Degree Courses (B.Sc. & M.Sc.) of Product Innovation Engineering.

He is the author or co-author of more than 150 papers and regularly serves as invited peer reviewer for several indexed journals and has national and international founding. He has been the principal investigator of 6 research programs funded by public or private agencies and coordinator of 12 research and technology transfer projects with private companies.

His current research interests are in the fields of flexible-link multibody systems (model reduction, model updating, state estimation, motion and vibration control), design optimization of vibrating systems (inverse dynamic structural modification, model reduction), industrial manipulators (performance evaluation, energy efficient trajectory planning, collaborative robots) and cable driven robots (modelling, trajectory. planning, performance evaluation, recovery strategies in case of cable failures).

Nicholas Valentine holds an M.Sc. in Robotics and Automation and a B.Eng. in Mechanical Engineering from the University of Salford and is a Trials and Experimentation engineer for the UK Ministry of Defence. Nick has worked in a variety of organisations across defence and has lived in both the Middle East and Canada. He is also an active volunteer with the Institution of Mechanical Engineers (IMechE) and is currently one of the Vice Chairs of its Young Members Board. Through his volunteering with the IMechE, he founded CHAIN Events, an organisation that produces events for young and developing engineers around the world. He is also a keen Army Reservist in the British Army's Royal Engineers. In what little spare time he has, he enjoys tinkering with his 3D printer and visiting as many museums as he can.

Dr. Peter van Duijsen is Professor of Practice at IIT Guwahati, India and has worked with computer simulation and power electronics for nearly 35 years. After receiving his Masters in Electrical Engineering, and working at the Technical University of Delft, he founded Simulation Research and developed the simulation program CASPOC. In 2003 he was awarded a Ph.D. degree in the field of Modeling and Simulation of Power Electronic Systems and currently heads the research and development department at Simulation Research.

From 2013 until 2019 he was a part-time researcher at the Technical University of Delft in The Netherlands where the focus was on electric machine control, wireless power transfer for EV charging and DC grids for fast charging. Since 2008 he has been an invited lecturer at the Korean University of Technology and since 2020 Researcher at the DC-LAB (dc-lab.org) at the THUAS University of Applied Sciences in Delft, The Netherlands. He was appointed as Professor of Practice at the IIT-Guwahati-India in 2020.

Dr. van Duijsen co-authored several textbooks in the field of Power Electronics, Drives, Solar, Wind and electric vehicles and is involved in the development of educational training hardware in the field of renewable energy and holds a patent on the wireless charging of eBikes.

Ferdinando Vitolo is assistant professor of Technical Drawing and Concept Design of New Vehicles at the University of Naples Federico II. His major research interests are focused on System Engineering and Knowledge Based Engineering, Mechanical Design, Methodologies for advanced CAD modelling, Mechatronic systems, Object Oriented Programming.

He received his Master's degree in Mechanical Engineering (cum laude) from the University of Naples Federico II in 2012. From 2012 to 2014 he has collaborated with the University of Naples Federico II on the National Project: PON 01_1268, Digital Pattern Product Development; a pattern driven approach for industrial product design, simulation, and manufacturing. After his Ph.D. in Industrial Engineering, in 2017, he won post-doctoral fellowship for research on *Human-Robot collaboration—Modelling and design of collaborative workplace and devices.*

Dimitrios Vogiatzis is a junior researcher at the Institute of Thermodynamics and Sustainable Propulsion Systems at Graz University of Technology. His research focuses on the development of hybrid propulsion systems for hand-held power tools.

Dr. Christopher Ward is a Senior Lecturer in Control Systems Engineering in the Wolfson School at Loughborough University, UK. His work focuses on the application of control, mechatronics and conditions monitoring to a wide variety of systems, but with a core focus on the rail sector. This work has a practical motivation and is driven by a desire to create a railway that is fit for the coming challenges of climate change, energy usage and customer demands. He holds a Ph.D. in automotive powertrain control from the University of Liverpool, is a chartered engineer and member of the IMechE.

Dr. Roger White is the founder, president & CSO of RC2 Inc, a research consultancy based in Ontario Canada. He has over 35 years experience in R&D, marketing, and entrepreneurship in optronic sensing, robotics, and AI focussed largely on the development of novel analytical systems for use in real-time macromolecular characterisation. After completing a Ph.D. in Materials Science from Bath University in 1985, he worked as a Research Engineer on thin film optics & distributed laser sensing for a variety of military and civilian application areas before moving first into technical marketing/product management and then joining a specialist personal robotics start up as European Product Manager for Laboratory Automation & Assisted Living.

In 1992, he established a specialist analytical instrumentation distribution, contract analysis and R&D company and was responsible for managing and participating in a range of collaborative research projects funded by UK & EU bodies as well as developing a low cost protein aggregation analyser based on guided wave laser light scattering. In 2006, he relocated to Ontario where he created RC2 Inc. as a design and development consultancy. Key projects undertaken to date have resulted in the spin out of an innovative low cost micro-volume protein analyser that can be operated using a smartphone, and a multi-channel fibre-optic UV detector capable of monitoring 12 process separations in parallel.

He has also been involved in a number of business development initiatives, including as a prime mover in a Welsh industry/academia collaborative project funded by the EU to develop an Optronic Innovation & Business Centre in North Wales. He is currently working on a Transatlantic project to assist EU and North American SMEs in developing IIoT[3]/AI and health related mechatronic products & services.

Kai Whiting is based at UCLouvain in Belgium. He did his Ph.D. in Sustainable Energy Systems at the University of Lisbon, Portugal and completed a Masters in Renewable Energy at the University of Zaragoza, Spain and a Masters in Environment and Business at Abertay University, Scotland. He worked on UNIDO's 2013 World Small Hydropower Report as a member of the International Centre on Small Hydropower's research and editorial team. He was a member of the civil society table and technical support group for Extractives Industries Transparency Initiative Colombia.

Kai's research interests are in sustainable resource use, particularly from a service perspective. He is currently working on tracing energy and materials from source to wellbeing. He also researches the integration of energy, material and ecosystem services. He writes and frequently appears on podcasts to discuss the application of Stoic philosophy to matters linked to sustainable development. He is the co-author of *Being Better: Stoicism for a World Worth Living in.*

Xiu-Tian Yan is Professor in Mechatronic Systems Technology and the Director of Space Mechatronic Systems Technology Laboratory (SMeSTech), based in the Department of Design, Manufacture and Engineering Management of the University of Strathclyde. Through over 30 years' research investigation, he has developed a portfolio of research topics in mechatronic system design, including the development

[3] Industrial Internet of Things.

and application of *mechatronic design process models*, modelling and simulation of these systems, and techniques in prototyping these systems for validation. He has managed many projects in his research fields funded by the UK government funding bodies, European Commission, international funding bodies and industrial partners. He was the Chairman of Robotics and Mechatronics Technical Professional Network of the Institution of Engineering and Technology.

He received his Ph.D. from Loughborough University of Technology. He is a Chartered Engineer and a Fellow of the IET and a Fellow of the IMechE. His research interests include mechatronic systems design, especially in recent years in multi-perspective mechatronic system modelling, design and simulation, computer support mechatronic systems design using AI techniques, knowledge intensive product modelling and simulation, and their applications in robotics for agriculture, manufacturing and space applications. He has organized and chaired four international conferences. He has published over 260 technical papers in major international journals and conferences and edited or wrote 6 books in the fields.

Dr. Haiyan H. Zhang is currently a Professor of Mechanical Engineering Technology in the School of Engineering Technology at Purdue University, West Lafayette, IN. He is also the founding director of Center for Technology Development (CTD) at Purdue University. He received his BS and MS Mechanical Engineering degrees from Shanghai Jiao Tong University, and his Ph.D. in Mechanical Engineering in 1996 from University of Michigan, Ann Arbor, MI. Dr. Zhang's research focuses on Analytical Mechatronics and Multidisciplinary Product Design.

Dr. Zhang is an expert of analytical mechatronics governing design and control of smart devices/smart machines, which can produce the thinking-based motions via the intelligence controlled, reconfigurable multidisciplinary mechanisms for high product value through augmented functionality.

Before Dr. Zhang joined Purdue University, he was a senior specialist of automatic transmission engineering in automotive industry, and designed electro-hydraulic control systems and sensors of automatic transmissions. His Multidisciplinary Design Laboratory (MDL), functions as a ThinkBox for Ph.D. students and FabLab for MS students in Mechatronic Systems, and is dedicated to the innovation of multidisciplinary products through the co-simulation and has been awarded with technical patents. Dr. Zhang also has experience in HEV Drivetrains, Digital Hydraulics, and large flow VFS Development for vehicle powertrains.

EcoMechatronics: Concepts, Objectives and Outcomes

Peter Hehenberger, Maki K. Habib, and David Bradley

Abstract The concept of sustainability in the application and operation of engineering systems is of increasing significance in relation to the requirements for the mitigation of climate change along with achieving global targets for the reduction of emissions of all types of greenhouse gasses. As a major contributor to the design and operation of many current systems, mechatronics has a great deal to contribute to achieving many of the climate-oriented goals that have been identified. This has in turn led to the development of the concept of EcoMechatronics in which the approaches to the design and operation of mechatronic systems which have evolved over the years are modified to encompass consideration of the issues of sustainability and environmental impact.

Introduction

Mechatronics [1–3] is an interdisciplinary field of engineering science centred on the synergy, at the design level, of multidimensional and multifaceted knowledge across a range of domains, facilitated by tools and methodologies that support the implementation of smart and sustainable systems and manufacturing processes of any scale, while promoting cleaner futures that reduce ecological impact while improving sustainability and quality of life. This implies adapting the functionality of mechatronics technologies with the aim of realizing a sustainable global society. Such an approach may be termed as EcoMechatronics.

The vision of EcoMechatronics is therefore one of aiming to improve human behaviour and quality of life while moving towards more sustainable lifestyles. In

P. Hehenberger (✉)
University of Applied Sciences Upper Austria, Wels, Austria
e-mail: peter.hehenberger@fh-wels.at

M. K. Habib
The American University in Cairo, New Cairo, Egypt

D. Bradley
Abertay University, Dundee, UK

P. Hehenberger et al. (eds.), *EcoMechatronics*,
https://doi.org/10.1007/978-3-031-07555-1_1

1 NO POVERTY	2 ZERO HUNGER	3 GOOD HEALTH & WELLBEING	4 QUALITY EDUCATION
5 GENDER EQUALITY	6 CLEAN WATER & SANITATION	7 AFFORDABLE & CLEAN ENERGY	8 DECENT WORK & ECONOMIC GROWTH
9 INDUSTRY, INNOVATION & INFRASTRUCTURE	10 REDUCED INEQUALITIES	11 SUSTAINABLE CITIES & COMMUNITIES	12 RESPONSIBLE PRODUCTION & CONSUMPTION
13 CLIMATE ACTION	14 LIFE BELOW WATER	15 LIFE ON LAND	16 PEACE, JUSTICE & STRONG INSTITUTIONS
	17 PARTNERSHIPS FOR THE GOALS		

Fig. 1 UN sustainable development goals [4]

doing so, it has the ability to effectively address the interdisciplinary challenges that such a shift presents with a potential to support the synergy of experience and ways of working, technologies and knowledge to optimize the design and development processes, and hence to maximize functionality while providing efficient and economical ways of utilizing resources, maximizing added value and limiting environmental and social impact. There is also a need to put forward effective goals which aim to support the development of a wide range of technologies, processes, products, systems and machines that can contribute to a reduction of dependence on the combustion of fossil fuels while lowering associated carbon footprints by reducing energy consumption and waste.

Such developments contribute to improved productivity, better environmental performance, the introduction of more energy-efficient machinery, reduced energy costs and reduced environmental impact along with more effective management of waste and recycling with less material usage and waste, enhanced environmental sustainability and safety with high user comfort. Ecologically sustainable design also represents an environmentally aware and motivated design process that integrates ecologically sustainable feature requirements in line with the United Nations Sustainable Development Goals of Fig. 1 [4] to improve environmental performance and optimize economic outcomes through the development of effective smart technologies and design processes along with near-zero waste manufacturing systems and strategies.

In order to achieve such outcomes and to produce an ecologically sustainable society, there is a need to develop novel design strategies which embed sustainability at their core. Over the years, a wide range of eco-design techniques and methods have been developed to evaluate and improve the environmental performance of products and their associated production processes, and these in their turn support the evolution of the EcoMechatronics concept.

Synergy of Digital, Sustainable and Agile Transformations for EcoMechatronics

The importance of design and development methodologies to support the whole life cycle of technical systems such as mechatronic systems, cyber-physical systems and others has become increasingly significant in recent years. The main drivers for these developments [5, 6] are the digitalization of systems such as lies at the heart of the concepts of Industry 4.0 and Smart Manufacturing which along with the guiding principles of processes such as Agile Management and Agile Production integrate with aspects of sustainability such as Green Production and the Circular Economy to drive forward the sustainability agenda. EcoMechatronics as a design paradigm then supports the goal of achieving an economical, ecological and social efficiency of products and processes as suggested in Fig. 2.

In that context, EcoMechatronics builds on previous insights into the development of the mechatronic concept, originating in part with the publication in 2010 of *Mechatronics in Action* [7] which looked at the ways in which the basic concepts of mechatronics were being applied across a range of application domains from manufacturing to vehicles. This was followed in 2016 by *Mechatronic Futures* [8] which considered how these underlying concepts were evolving both the meet new application challenges as well as accommodate new technologies such as those of The Cloud. Then in 2020, *Reinventing Mechatronics* [9] aimed to understand how mechatronics practitioners were looking forward to new applications areas such as agriculture and approaches accordingly and adapting their thinking and design processes accordingly. *EcoMechatronics* then takes this forward with a specific focus on how mechatronics can contribute towards achieving a sustainable society. It must also be noted that education is an important aspect of mechatronics. Each of the texts referred to has introduced chapters on various aspects of mechatronics education for all ages.

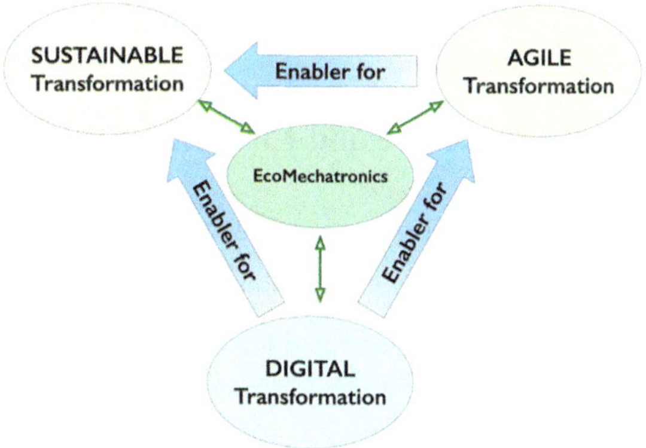

Fig. 2 Main drivers for EcoMechatronics

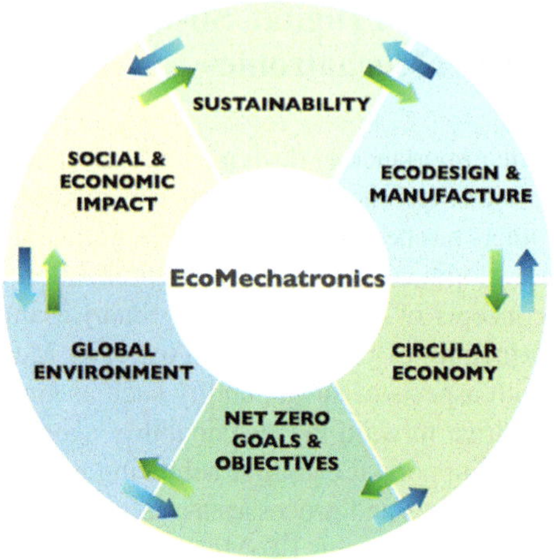

Fig. 3 EcoMechatronics structures

The EcoMechatronics-Paradigm

The concept of sustainability in the application and operation of engineering systems is becoming increasingly significant in relation to the requirement to mitigate climate change and achieve global targets for the reduction of the emissions of all types of greenhouse gasses.

As a major contributor to the design and operation of many current systems, mechatronics has a great deal to contribute to achieving many of the climate-oriented goals that have been identified. This has in turn led to the development of the concept of EcoMechatronics in which the approaches to the design and operation of mechatronic systems which have evolved over the years are modified to encompass consideration of the issues of sustainability and environmental impact. Figures 3 and 4 then present some of the key elements of EcoMechatronics and its evolution related to applications.

Goals, Structure and Content of the Book

In order to achieve the goals, the book is structured around both a discussion of some of the underlying issues associated with such a shift in thinking, for instance, in understanding the nature of sustainability and its implications and challenges for engineering design along with issues such as sensor technologies, artificial intelligence, machine ethics and privacy which impact on the way in which systems are designed and operated.

It also seeks to introduce through the technical chapters some of the areas in which current research is beginning to evolve into what might be considered first-generation EcoMechatronics systems in areas such as rail transport, agriculture and

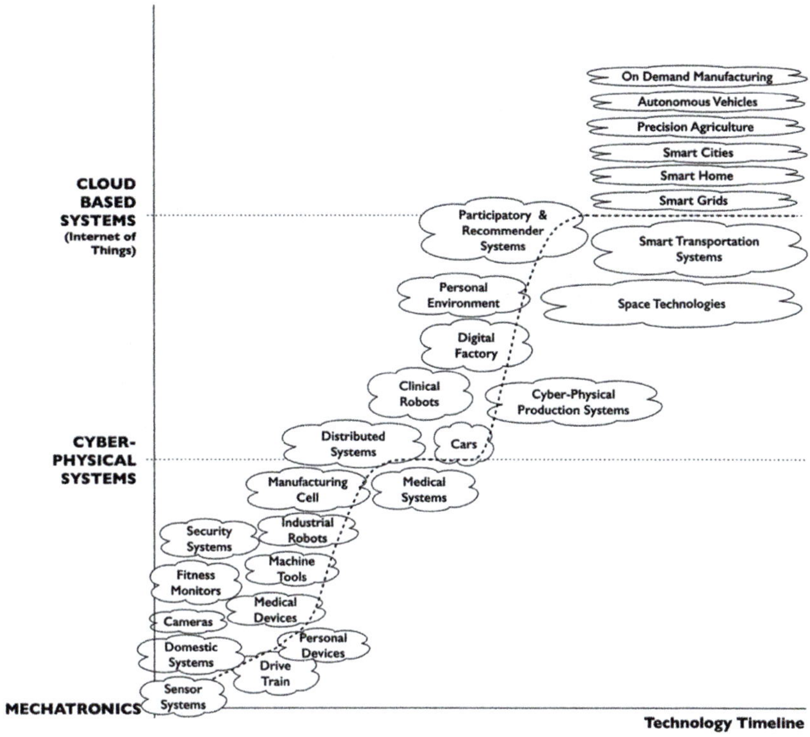

Fig. 4 The evolution of mechatronics & EcoMechatronics in relation to applications

design methods. However, it is not possible in a book of this size to encompass all approaches to achieving EcoMechatronic solutions, not to cover all aspects or applications of each of these efforts. Thus, there are a number of issues and topics that can only be considered in outline and these are presented here in the introduction and include.

4th Industrial Revolution—Referring to Fig. 5, we now sit in the middle of the 4th Industrial Revolution, one characterized by enhanced automation, the use of digital technologies and integrated communication. In this environment, manufacturing is structured around robotics and advanced machine tools, which are mechatronic in

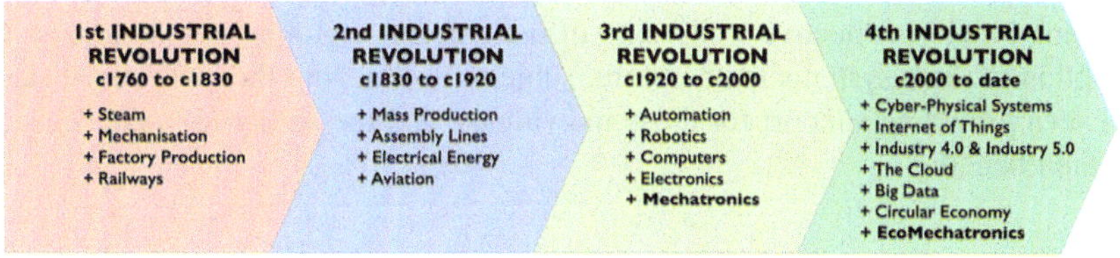

Fig. 5 Industrial revolutions

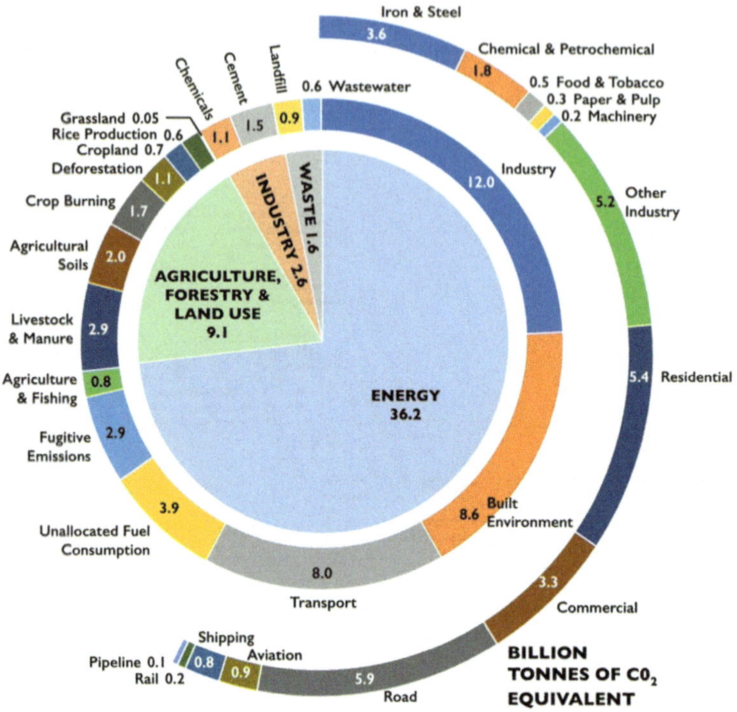

Fig. 6 Global greenhouse gas emissions by sector in 2016 (Our World in Data [10])

nature. By adopting an EcoMechatronic approach to design, these can be made more effective and efficient, consuming less energy and producing less waste.

Artificial Intelligence and Machine Learning—Artificial Intelligence (AI) and Machine learning have not as yet effectively demonstrated or fulfilled their promise or potential across a wide range of mechatronics-based applications. However, developments in several areas including, among others, Cyber-Physical Systems, Health Care, Manufacturing Technology and Robotics, each of which is structured around mechatronic components, systems and sub-systems, are beginning to demonstrate potential in respect of applications.

Autonomy—Systems are becoming increasingly autonomous in relation to aspects of their operation, relying on embedded software to make decisions that can impact the individual or collective well-being of individuals or groups of individuals. The adoption of such systems is, therefore, subject to achieving the necessary balance between providing support for those individuals, and the ability to make decisions on their behalf.

Built Environment—Referring to Fig. 6 it can be seen that in 2016 the built environment accounted for some 8.6 Billion tonnes CO_2 equivalent (~17.5%) of global greenhouse gas emissions without taking account of the embedded carbon in their production. Modern building design incorporates a wide range of mechatronic elements in relation to operational features such as energy efficiency and use and ventilation, as well as integrating within the concept of the urban Digital Twin.

Data—Mechatronics systems generate large volumes of data related to their operation which, when integrated with data from other sources provides a powerful resource to understand impact and outcomes. The processing of this data must however be subject to due and appropriate consideration of issues such as privacy, both of the individual and the organization.

Digital Twins—The concept of the Digital Twin evolved in relation to the operation and management of complex systems such as aero-engines and has now been extended to encompass the operation of buildings [11] and environmental issues such as flood management and control [12]. Again, many of the embedded systems are mechatronics and may also be supported by their own Digital Twin, resulting in a layered hierarchy.

Dynamics of Change—If the goal of achieving Net Zero greenhouse gas emissions by 2050 is to be achieved, then the first steps must be based on the effective deployment of existing technologies. The adoption of an EcoMechatronics approach to design and implementation has an important role to play in this.

Education—The educational requirements for mechatronics are incorporated within the wider concept of the STEM (Science, Technology, Engineering and Mathematics) approach. This means that it is important to engage students from the earliest possible age in the underlying concepts supporting sustainable development and the application of EcoMechtronics and others approaches to achieving this.

Information Management—As has been stated, mechatronic systems generate large volumes of data which can then be analysed and interpreted to provide feedback as to effective outcomes, and hence to adjust system operation accordingly.

Privacy and Ethics—Increased autonomy of systems is presenting challenges in respect of user privacy as well as raising issues as to the ethical implementation of such systems.

Society—The aim of moving towards a more sustainable approach to the design, development and implementation of a wide range of systems is to achieve a sustainable framework across all levels of society. Societal impact therefore needs to be increasingly incorporated into the design processes for EcoMechatronics.

Transport Systems—As can be seen from Fig. 6 in 2016 transport contributed some 8 Billion tonnes CO_2 equivalent (~16.2%) of global greenhouse gas emissions. Many of the systems involved in modern transport, from road to rail and aircraft are mechatronic in nature and EcoMechatronic thinking is therefore positioned to have a major impact on the reduction and management of these emissions.

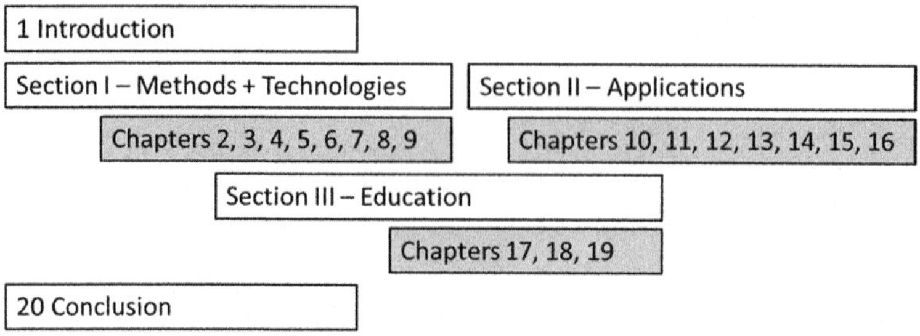

Fig. 7 Structure and content of the book

The book is structured around a series of chapters from invited authors, each of whom is an expert in a particular area of EcoMechatronics. In each case, the authors were challenged to establish the current state of the art using their own research or professional expertise as the starting point and then to try to isolate and identify those key areas in which significant development is needed or likely to take place in coming years. These chapters are organized into three specific sections as follows (see also Fig. 7):

- Methods and Technologies—Presents and describes the methodological background for EcoMechatronics.
- Applications—Presents selected applications for EcoMechatronics.
- Education—Presents aspects of the integration of the EcoMechatronics paradigm in education programmes.

The book thus provides the reader with an insight into the issues and challenges associated with EcoMechatronics and its role in achieving a sustainable future.

References

1. Harashima F, Tomizuka M, Fukuda T (1996) Mechatronics-what is it, why, and how? IEEE/ASME Trans Mechatron 1(1):1–4
2. De Silva CW (2005) Mechatronics–an integrated approach. CRC Press London, New York & Washington DC
3. Habib MK (2007) Mechatronics a unifying interdisciplinary and intelligent engineering science paradigm. IEEE Ind Electron Mag 1(2):12–24
4. United Nations (2015) The 2030 agenda for sustainable development, 17 sustainable development goals (SDGs). sdgs.un.org/goals. Last Accessed 31 Mar 2022
5. Merschak S, Hehenberger P (2019) Ecodesign methods for mechatronic systems: a literature review and classification. 20th International conference on research and education in mechatronics (REM). Wels, Austria
6. Hehenberger P, Vogel-Heuser B, Bradley D, Eynard B, Tomiyama T, Achiche S (2016) Design, modelling, simulation and integration of cyber physical systems: methods and applications. Comput Ind 82:273–289. https://doi.org/10.1016/j.compind.2016.05.006. Last Accessed 31 Mar 2022. (Elsevier)

7. Bradley D, Russell D (2010) Mechatronics in action: case studies in mechatronics-applications and education. Springer International. ISBN-13 978-1849960793
8. Hehenberger P, Bradley D (2016) Mechatronic futures–challenges and solutions for future mechatronic systems and designers. Springer International. ISBN 13 978-3319321547
9. Yan X-T, Bradley D, Russell D, Moore P (2020) Reinventing mechatronics, developing future directions for mechatronics. Springer International. ISBN 13 978-3030291303
10. Our World in Data. ourworldindata.org. Last Accessed 9 Sept 2021
11. Dembski F, Wössner U, Letzgus M, Ruddat M, Yamu C (2020) Urban digital twins for smart cities and citizens: the case study of Herrenberg. Ger Sustain 12(6):2307
12. digitaltwinhub.co.uk/credo-film/#credoapp. Last Accessed 9 Sept 2021

Methods + Technologies

Re-envisioning Innovation for Sustainability

Eddie Simpson, Kai Whiting, and Luis Gabriel Carmona

Abstract The chapter explores the concept of sustainability and its associated challenges within a mechatronics context. The primary aim is to promote critical thinking and innovation within an EcoMechatronics framing, i.e., via an application of mechatronics that at its core works with nature (rather than against it). Innovative technology per se is neither a force for good nor bad. However, EcoMechatronics permits a broader approach to innovation by considering socioenvironmental aspects as instrumental to improved performance, not as mere 'add ons'. Throughout the chapter, the authors tease out how innovation for sustainability can be so much more than merely doing 'less bad' (e.g. fewer emissions or less waste) when we are capable of doing 'more good', for example, by improving the soil or capturing and regenerating energy that would otherwise have dissipated. While the authors do compare two potential braking systems within an EcoMechatronics framing, their goal is not to provide a tick box methodology or a one-size-fits-all answer but instead to provoke deeper questioning and reflection that enables mechatronics practitioners to sustainably innovate according to their own specific set of circumstances and challenges.

Introduction

The 1987 WCED report [1], often referred to as The Brundtland Report, is a seminal investigation into the necessity for global action to ensure that resource use does not adversely affect the environment to the detriment of future generations. A key concept within the report is that of '*sustainability*'. Since its publication, various actors in a wide range of professional disciplines and industry sectors have attempted to shape

E. Simpson (✉)
Abertay University, Dundee, UK
e-mail: e.simpson@abertay.ac.uk

K. Whiting
Université Catholique de Louvain, Ottignies-Louvain-la-Neuve, Belgium

L. G. Carmona
University of Cambridge, Cambridge, UK

P. Hehenberger et al. (eds.), *EcoMechatronics*,
https://doi.org/10.1007/978-3-031-07555-1_2

sustainable development, in accordance with the norms, values and practices of their specialism or field. This chapter explores the concept of sustainability, innovation and their respective challenges within a mechatronics context. However, we do not just consider how mechatronics can be applied to environmental problems but rather highlight the added value of re-conceptualising mechatronics applications into an EcoMechatronics framing. The latter is an integration of knowledge and innovation across various engineering and computing fields in order to positively contribute to the environment. This means that EcoMechatronic design and practice goes beyond even a proactive anticipation of environmental harms and their mitigation because it leads those in the mechatronics field to design for improved functioning of the ecosystem.

The notion of sustainability and how it leads to the expansion of mechatronics into EcoMechatronics is explored through an in-depth philosophical discussion on what is often considered to be '*green*', and why a much more nuanced positioned is needed if we are to truly innovate for sustainability. In particular, the authors consider (1) the extent to which relative efficiency gains are beneficial given that they are known to often trigger increases in absolute resource consumption; (2) whether the circular economy can exist given the physical limits of recoverability, as dictated by the Second Law of Thermodynamics; (3) whether recycling is sustainable and (4) the limitations of biomimicry and other '*green*' solutions. The purpose of this critical examination is not to criticise efforts thus far to achieve sustainable goals nor to suggest that none of these aforementioned practices has been beneficial. Rather it is to give mechatronics practitioners the questions they need to be able to ask to establish what is required of them if they are to direct innovation, and by extension society, towards a more sustainable path in the future.

What is Sustainability?

The Brundtland Report [1] puts forward the notion of sustainability as '*development that meets the needs of the present without compromising the ability of future generations to meet their own needs*'. In this context, sustainability encompasses the following principles:

- The planet's ecology must be conserved, as the loss of plant and animal species is likely to impact the health and wellbeing of future human generations.
- Each generation has a moral obligation to conserve nature and maintain, and where possible improve the planet's biochemical cycles.

These principles are rooted in a holistic understanding of planet Earth as a multi-faceted and intricately interconnected complex system, whose health depends on the successful operation of all its parts and the relationships between them. Innovative technology per se is neither a force for good nor bad. However, restricting what is seen as '*good*' to the optimisation of the socioeconomic system or to what is favoured by a select group of human individuals, as opposed to what is good for all of humanity and non-humanity, has often generated unforeseen system feedback mechanisms to the detriment of the whole.

An increasing awareness of environmental problems has led companies to embark on sustainability drives focused on reducing specific aspects of environmental harm. For example, in response to environmental damage, the aviation sector has set sustainability targets linked to a reduction in CO_2 emissions. These targets can be achieved by reducing fuel consumption in absolute terms or by improving overall aviation fuel efficiency per air mile travelled, which is a reduction in fuel use in relative terms [2]. There is also a move towards new generations of aircraft using batteries, fuel cells or hydrogen [3–5] as alternatives to fossil fuels and commercial options for such aircraft are becoming increasingly viable. There are, for example, concepts for hydrogen-fuelled airliners being put forward by Airbus which combine hydrogen-burning gas turbines with hydrogen fuel cells that generate sufficient electrical storage capacity to carry up to 200 passengers for a predicted range of 2,000 nautical miles [6]. However, we are still some way from introducing these kinds of systems into commercial passenger operations. In the meantime, improving conventional fuel efficiency remains the principal strategy for sustainability.

Improved fuel efficiency offers considerable cost savings [7] and results in lower levels of air pollution per air mile travelled [8]. However, net environmental benefits will only occur if there is no increase in the number of air miles flown. In other words, any carbon emission reductions per air mile will only result in a better environment if cost savings *discourage* or *stabilise* air travel rather than promote additional travel. However, if a company can make more profit per mile flown through fuel savings, are fuel efficiency improvements likely to incentivise or reduce the market offer for flights? If cheaper fuel costs are passed onto the consumer because an airline wants to increase its market competitiveness, will this encourage the consumer to travel further or more often, as it is now cheaper to do so? This highlights the fact that even with the best intentions, it can be extremely difficult to argue that resource and/or carbon-intensive industries can ever have green credentials or meaningful contributions to a transition towards a more sustainable economy.

The paradoxical relationship between fuel savings in relative terms and increased consumption in absolute terms was documented in the nineteenth century by Stanley Jevons [9]. Jevons identified how technological innovations such as the steam engine and better iron smelting methods increased resource efficiency, which paradoxically increased resource consumption because cheaper commodities and energy prices heightened production demand [10, 11]. This paradox continues into the present day as shown in Fig. 1, which highlights how an increase in UK fuel efficiency in cars has, at least in part, enabled a higher overall fuel consumption nationally. Such consumption patterns accelerate changes in travel preferences, consequently in travel infrastructure, and thus a societal preference for cars leads to more roads

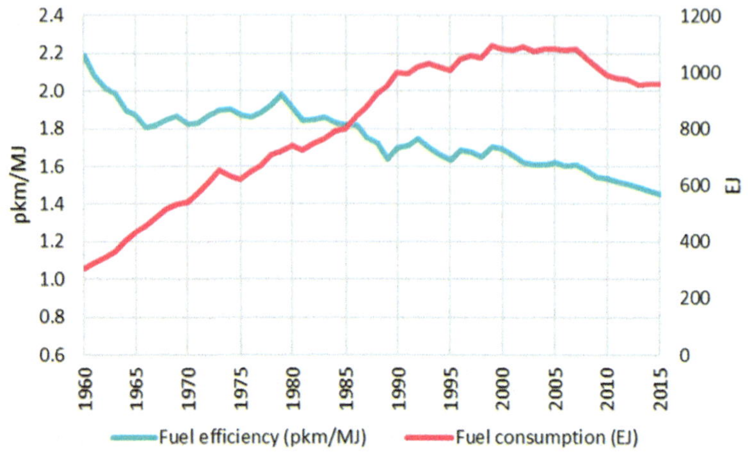

Fig. 1 An example of the Jevon Paradox: UK car fuel efficiency versus total fuel consumption. *Source* data (Rodrigues et al. [13])

and less public transport, creating and maintaining unsustainable consumer lock-ins. Here, the term '*lock-in*' refers to the phenomenon that occurs when our present decisions tie both present and future generations into behavioural patterns that are not easy to eradicate. Negative lock-ins are then those that diminish the quality of the environment rather than improve it [12].

Understanding the attitudes and behaviours of businesses and society, as explained through the Jevons' Paradox and the concept of consumer lock-ins, is key if mechatronic designs are to work with nature and support sustainable development. The notion of a *Circular Economy*, whereby the quantity of resources used and waste/emissions created are reduced, is being promoted as one way of thinking sustainably. This is due to its emphasis on the ability to reuse, repair, refurbish, maintain and recycle materials [14–16]. Extending material life expectancy through a cyclical approach that re-integrates waste material back into the socioeconomic system, rather than letting it leave via incineration or landfill, reduces the extraction and consumption of virgin materials and frees up land that would be otherwise designated to, or detrimentally affected, by waste management operations. The sustainability of the circular economy rests on the idea that humanity can self-contain its demand within a closed-loop system because eventually there will be a peak in resource extraction, meaning that society will no longer consume more than what has already been extracted.

The problem is that the Second Law of Thermodynamics dictates that with each cycle both the *quantity and quality* of resources degrade. In turn, this means that significantly, and successively more, energy and raw materials will be required to recover end-of-life material so that it is in a useable state. The Second Law also ensures that it is *physically impossible* to recover all material. No matter how careful one is or how efficient our technology is, some material will inevitably leave the system. No amount of innovation can change this fact and the best we can hope to obtain is a spiral rather than a circular economy [17]. Physical limits aside, there are also sociopolitical obstacles hindering the development of cyclic/spiral consumption patterns. For instance, to what extent can a circular economy operate alongside political influence and interference, which encourages rather than constrains economic growth?

This does not mean that the circular economy holds no value [18]. Indeed, its sustainable outlook on a complex problem has moved society in the right direction. Amongst other things, its strategies have led to a reduction of some end-of-life issues because of its promotion of a more sustainable design stage. Appropriate designs can effectively remove various downstream problems before they occur [17, 19]. They can also begin to mitigate some of the unforeseen damages, such as the plastic fibrous releases from clothes, which contaminate the oceans [20–22].

Decisions at the early stages of a product's life cycle also have the potential to either create or avoid trade-offs [23]. These include those linked to resource efficiency and product recyclability, reparability and life expectancy. Policy development and legislation can also support circular economy initiatives. For example, through the '*right to repair*' and a ban on planned obsolescence, as enacted by the French Law *n°2020-105* [24], society can create a more sustainable path forward. This legislation grew out of consumer dissatisfaction with planned obsolescence and the concerted efforts of the French consumer group *Halte à l'Obsolescence Programmée.*[1] This was built on the 2015 French Law *no. 2015-992* on *Energy Transition for Green Growth*[2] [25], which made planned obsolescence an offence.

In 2018, the Italian Competition Authority, *Autorità Garante Della Concorrenza E Del Mercato*, fined two hi-tech companies for forcing users to accept software that slowed their devices, leaving them with little alternative but to buy newer '*faster*' devices when their older ones were, if not for the updates, fully functioning [26]. Laws and legal proceedings underpinned by circular economy principles certainly hold merit, but how effective are they against the many commercial pressures to manufacture new products? How will designers influence the re-education of society regarding the role of technological innovation and lifestyle choices? To what extent can *national* laws address the *global* socioenvironmental cost of throwing away items long before they have reached the end-of-life stage?

Another complex problem linked to product longevity is that replacing a boiler that is perfectly functional but not very energy efficient with a newer one that is highly energy efficient is only a sustainable solution to environmental harm *if* we restrict our scope to carbon emissions. However, if one focuses on the environmental harm caused by waste generation, then replacing the boiler is completely unsustainable. The notion of sustainability, in its truer sense, is a holistic one that requires careful consideration of not only how nature operates but how it *flourishes*. The best strategies, including designs and ways of working, are those that harmonise rather

[1] HOP—Stop Planned Obsolescence.

[2] Energy Transition Law.

than antagonise biochemical cycles. They support rather than encroach upon plant and animal populations and their habits. This is the kind of innovation we propose in the transition towards a more sustainable approach to mechatronics design and function.

Nature, Inspiration and Learning

Nature operates efficiently because everything is useful and there is no material waste. Whilst the answer to '*what is nature*?' goes beyond the confines of this chapter, it is a system from which we have much to learn. Humans form part of this system and, as an integrated part of it, what we do impacts the rest, sometimes for better and sometimes for worse. As part of the system, we are also reliant upon its offerings and subject to its feedback mechanisms, including those that we accidentally, or purposefully, influence or manipulate. Whilst a relatively straightforward notion to grasp, it appears that we humans often fail to recognise that we are very much a product of nature and that nature is affected by, and evolves alongside, us. In this respect, we can choose to contribute positively or negatively to the health of the whole. Given that we are all intricately linked and thrive, or fail, together, it makes sense that we work with, rather than against, nature.

Studying nature provides both a source of inspiration and a repository of knowledge, which we can use to improve the world around us [27]. Natural cycles always tend towards equilibrium, the lowest energy state possible, and they operate in closed material loops. In nature, the inter-dependencies among organisms operate on the basis that the needs of one organism are met in some form or another by the constant rejuvenation and re-packaging of resources that have previously been used by another. Some forward-thinking innovators have begun to re-envision resources in this way via what has become known as '*Cradle to Cradle*' thinking [28]. This concept challenges individuals and corporations to go beyond the mindset of '*doing less bad*'[3] and invites them to reframe the problem in terms of '*doing more good*'. Fundamentally, it leads to the setting of positive targets such as creating compostable cups that can provide nutrients and improve soil structure. These kinds of targets encourage designs and practices that seek to avoid waste altogether by reconceptualising it and using it to improve the functioning of the ecosystem. Once we begin to think and act in this way, it becomes difficult to accept waste as an inevitable necessary consequence of production. Instead, we begin to understand it for what it is in many instances, namely a *design* failure and a lack of imagination [29, 30]. This paradigm shift is built upon asking more challenging questions such as why settle for trying to reduce the amount of toxins we emit to the environment if we can simply avoid them altogether? Why assume there is a trade-off between designing for aesthetics and

[3] For instance by producing less carbon emissions or solid waste.

sustainability? Why only substitute materials for cheaper options, if we can substitute them for better ones? Asking ourselves these kinds of questions opens up the possibility of re-purposing, re-integration and upcycling rather than merely thinking about how we might rescue resources from landfills via recycling [31].

The problem with recycling is that it results in material degradation and is not an effective method for preventing landfill in the long term. It also does little to increase resource utilization. Even recycled steel and paper which are, arguably, two of the biggest success stories have restricted uses because the virgin material is inevitably of better quality.

Taking the engineering example of a retaining wall, we can compare two different sustainable design proposals. In option 1, the approach may be to adopt concrete but reduce its environmental impact during the extraction, processing and production stages, by proactively reducing energy intensity and, consequently, CO_2 emissions, which are considerable [32, 33]. Yet producing cement as just one material for concrete, even with the best available technology, is still energy-intensive and emits various pollutants during processing [34, 35]. While improvements in cement manufacturing thus far have been significant [36], it still results in wastewater effluents, solid waste (e.g., cement kiln dust), particulate matter and considerable freshwater consumption [37]. Concrete is also difficult to recycle with each re-use resulting in material degradation and a reduction in consistency and strength. Some improvements have been made via microwave ovens. Certainly, the 0.93–2.86 $kgCO_2$/t aggregate is considerably lower than the kerosene furnace/rubbing or rotatory kiln/sorting process that generates 69.53–98.01 $kgCO_2$/t aggregate [38]. The microwave oven route also provides a higher quality recycled aggregate which increases potential uses and price [39]. This is certainly progress that the closed-loop process is made, but is there a better option?

In option 2, there is a vegetated retaining wall, which offers a structural design option that makes use of upcycled plastic that also supports plant and insect life, because it forms an extended part of their habitat. Vegetated retaining walls are not a new idea, and there are various examples going back more than 2,000 years, which made use of local and readily available materials such as reeds, willow and branches for reinforcement [40, 41]. All these materials are biodegradable and offer similar performance characteristics to newly developed innovative geosynthetic materials such as Flex MSE [42], which are much less biodegradable [43, 44].

Perhaps, an even better design option is to mimic nature rather than trying to become '*more sustainable*'? Biomimicry can give us models to work from which involve, for example, the observing of birds and their nest making activities, as a means to appreciate the physical limits of what is achievable in sustainable design and practice [45–47].

The Beijing National Stadium, designed by Swiss architects Herzog & de Meuron, is often described as the '*bird's nest*' stadium. This description comes from the appearance of the steel structure with interwoven twigs resembling an upturned nest [48]. The amount of steel required was approximately half of that under a typical stadium design and infill materials between the structural elements for insulation and protection from the elements decreased structural loading, via the ethylene tetrafluoroethylene (ETFE) panels [49]. The problem with ETFE lies with its recyclability in comparison to other potential options given that it is a plastic-based material. There were also some missed opportunities for doing more good when it comes to covering the stadium's needs; for instance, solar energy covers only the stadium's underground parking requirements [50].

This raises the question of just how far can we realistically adopt biomimicry as a specific source of eco-design inspiration? Even as we learn from nature and align our behaviour and technology more closely with natural systems, there will inevitably be constraints and compromises at each step of the life cycle, even if the goal is to design in such a way that the outcome contributes more to the environment than it takes out [51].

Sustainability Assessment and Life Cycle

To solve some of the issues mentioned above, one can undertake sustainability assessments to objectively determine whether the solution proposed is better than other reasonable alternatives. One particularly useful tool is life cycle assessment (LCA). According to Curran [52], the purpose of an LCA is to identify, qualify and quantify the environmental impacts generated within a given scope. An LCA thus permits the establishment and the quantification of the environmental impact inventory associated with a given product or service. It is used to state and compare environmental impacts from the cradle to the grave, meaning from extraction to disposal [53]. For an LCA assessment, one calculates that which enters the system (in terms of raw materials and energy) and that which leaves (i.e. air, water and residual solid emissions). These flows are then summed and subsequently represent the total effect of that product or service on the environment. All informational inputs should be transparent and objective. That said, an LCA is not a completely objective method and typically involves some form of interpretation. The scoping exercise and the definition of boundaries are also subjective, albeit based on objective factors. Any assumptions or purposeful exclusion from a case study should be stated as such [54]. For example, does the scope range from gate-to-gate (solely production processes), well to wheel (fossil fuel extraction to mechanical drive) or cradle to grave?

A well-considered LCA will enable a company or individual to calculate the sustainability of a product at each step across multiple environmental aspects such as material recovery, energy efficiency, carbon emissions and waste generation. To use an example, one can consider whether the upcycling of an end-of-life machining knife to a new wood turning tool is sustainable beyond the mere fact of delaying disposal and removing the need to mine for the metal associated with the wood turning tool. To do this, one could calculate the carbon footprint of the upcycled tool with an identical one made via conventional production. It may well be that the carbon emissions released during the grinding of the machining knife to provide the metal for the wood cutting tool is just as high, if not higher than the production of metal extracted from an ore and manufactured into a metal sheet. Of course, carbon emissions are just one aspect, which is why the weighting of all impacts is important. If one wishes to prioritise metal recoverability, then the LCA will indicate a positive result because the metal is upcycled rather than downcycled scrap. If one considers energy to be of primary importance, it may be that the upcycling has a negative result. Hence the subjectivity of an LCA. It may also be that the environmental considerations are secondary to social issues such as mining encroachment on indigenous land or child labour in mines. In such a situation, the above LCA would need to be expanded into a life cycle sustainability assessment (LCSA), which incorporates life cycle costing (LCC) and social life cycle assessment (sLCA). LCC takes into account the economic costs of the raw materials, equipment, energy and labour costs associated with product processing. An sLCA evaluates worker safety, delocalisation and migration, child labour practices, corruption issues and respect for indigenous rights [55]. One way to reduce the probability of these issues is through tracing and reducing the complexity of the supply chain. The LCA/LCSA helps with this objective because it encourages both transparency and accountability.

EcoMechatronics

Before exploring the nature of EcoMechatronics and how we can use it re-envision innovation, a brief analysis of mechatronics and its purpose is required. Mechatronics is an interdisciplinary engineering subject that integrates aspects of mechanical engineering, industrial engineering, production engineering, civil engineering electronics and computer systems with advanced controls for the designing, construction, and operation of '*intelligent*' (but not necessarily smart) products and processes [56]. The term smart has come to mean, in practice, the capability of being interconnected via the internet, including situations where there is no embedded capacity for data interpretation or analysis; thus '*smart*' tends to imply connectivity as part of the Internet of Things (IoT) and is a misnomer in that it does not actually mean that in every case there is an embedded capacity to make decisions on received data.

One area where mechatronics is extremely prevalent is in the transport sector. In 2010, 30–35% of a new, higher class, passenger car's economic value was linked to embedded electronics and mechatronics. This is unsurprising given that such a car will typically contain 2.5 km of cables, approximately 40 sensors, 100–150 electromotors and 45–75 microcontrollers of various types [57]. Mechatronic elements have also been increasingly incorporated into vehicles in response to environmental regulations, particularly in respect of emissions and air quality standards [58]. There has also been an increasing awareness that engineers must consider performance factors relating to customer satisfaction and economic cost reduction strategies [59].

Mechatronics, as a discipline, is also significant when one considers the potential for autonomous vehicles to influence not just the economy but how we think about road travel. An early mechatronics system introduced into vehicles was the anti-lock brake system (ABS) in the 1970s, enabled through developments in integrated electronics and microcomputers [60]. Since then, there have been various developments including traction control systems and, more recently, adaptive cruise control and automatic motor stop systems. The latter saves on fuel consumption by shutting the engine down when the car is operating but at a standstill.

Mechatronic elements are also favoured relative to conventional options as they tend to be lighter than the mechanical components they replace, leading to lower fuel consumption and lower power levels needed to obtain the same level of technical performance. However, as we have made clear in the previous sections, using mechatronics to reduce vehicle weight or within cost-cutting options does *not* automatically, or necessarily, lead to a sustainable policy or practice. This is especially the case if mechatronics solutions result in downcycling and an increased tendency towards landfill disposal, instead of recuperation, re-use, recycling and upcycling. In fact, and as we have already suggested, shifts towards mechatronics solutions that solely focus on energy savings at the cost of material efficiency or which fail to consider a more holistic evaluation of system efficiency can exacerbate rather than mitigate environmental issues.

This is where an EcoMechatronic approach comes in, whereby the utility of mechatronics is not just concerned with higher system performance in and of itself, or merely in relation to economic factors, but in terms of how the system and its parts impact the environment. Consequently, our conceptual framing of EcoMechatronics is as an interdisciplinary engineering subject that integrates knowledge and innovation across various engineering and computing fields, including environmental engineering and earth system science to ensure that the design, construction and operation of '*intelligent*' products and processes positively contribute to the environment and all those living beings that rely on optimally functioning natural processes (see Fig. 2).

Braking System—An EcoMechatronic Case Study

We will here examine an EcoMechatronics approach via two braking systems. The design of braking systems has typically been in terms of system performance linked to the benefits accrued in reducing vehicle weight and vibrations while improving braking effectiveness, system safety and reliability at a lower cost [61]. The environmental impact of the system's materials is not a conventional performance characteristic and is therefore typically ignored or of secondary importance. An EcoMechatronics approach means that the sustainability of the materials used in braking systems becomes a core criterion, rather than an added bonus. Moubarak et al. [62] provide an excellent EcoMechatronics example via their redesign of a brake system, which couples a magnetic brake to a regular brake. Their aim was to reduce the friction that causes copper to detach from the brake pad following its wear and tear.

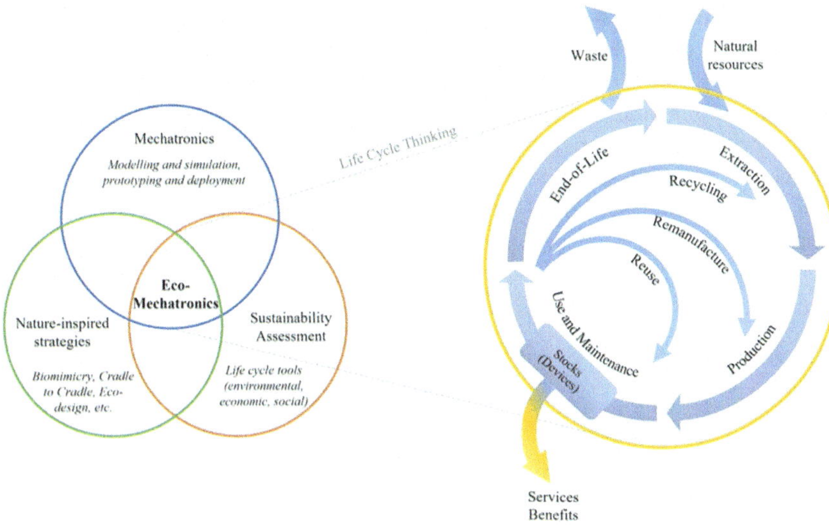

Fig. 2 A conceptual model for the EcoMechatronics approach

Elevated copper levels are toxic to aquatic life, causing mortality and morbidity and detrimentally affecting fish reproduction and growth. According to the US Environmental Protection Agency [63], brake pads are a significant source of the problem. This is because they are responsible for 24–192 g of copper being released into the environment, following road surface runoff, over the lifetime of the average car [62]. This problem is compounded by the fact that car ownership and use are increasingly prevalent. The global car fleet, for example, expanded from about 51 million in 1950 to over 800 million in 2007 [64]. The prototype developed by Moubarak et al. [62] reduced copper release by up to a third. The reduced friction also increased the brake longevity and consequently reduced waste disposal. However, the design by Moubarak et al. is still an example of doing '*less bad*' rather than '*more good*', in the sense it merely reduces copper contamination rather than innovates for positive environmental impacts.

In comparison, and as an example of a braking example that attempts to '*do good*', we can examine regenerative braking systems (RBS). The principle behind this system is straightforward. In the process of reducing vehicle speed, the RBS transforms kinetic energy into electrical energy, storing the electrical energy recovered in accumulators or batteries for later use [65]. This reduces the need for further inputs as some of the energy that would have dissipated is captured. It also reduces emissions either directly, through the reduced combustion of vehicle fuel, or indirectly due to the lower quantities of fuel consumed during the electricity generation process.

In one road test study, the incorporation of RBS significantly reduced CO_2 exhaust emissions from hybrid vehicles. CO_2 reduced from 140 g/km of CO_2 to 108.4 g/km of CO_2 or from 44.8 to 67 g/km, depending on the driving mode [66]. There are other positive benefits. Laboratory tests show up to a 10–30% decrease in electrical energy consumption from the electric/hybrid vehicle battery when RBS is utilised [67, 68]. RBS has also been shown to reduce braking friction, which as explained above reduces environmental contamination [69] in addition to noise while increasing the longevity of the component parts [68]. This is exactly the kind of innovative design and operation that those who wish to work in EcoMechatronics need to envision when designing for a sustainable future, precisely because it does not stop at merely doing '*less bad*'.

Conclusions

The chapter explored the concept of sustainability and its associated challenges within a mechatronics context. The primary aim was to promote critical thinking and innovation within an EcoMechatronics framing, namely an application of mechatronics that at its core works with nature rather than against it, to foster a better environment.

The conceptual framing of EcoMechatronics is as an interdisciplinary engineering subject that integrates knowledge and innovation across various engineering and computing fields in order to positively contribute to the environment. Innovative technology per se is neither a force for good nor bad. However, given that we are all intricately linked and thrive, or fail, together, it makes sense that we work with rather than against nature and use it as a source of inspiration, and a repository of knowledge, to improve the world around us. In nature, everything is useful and there is no material waste. Some forward-thinking innovators have begun to re-envision resources in this way via what has become known as *Cradle to Cradle* thinking. This concept challenges individuals and corporations to go beyond the mindset of '*doing less bad*'; for instance by producing less carbon emissions or solid waste and invites them to reframe the problem in terms of '*doing more good*'. This reframing encourages designs and practices that not only avoid waste altogether but actively use the '*waste*' that cannot be avoided to improve ecosystem function.

EcoMechatronics offers the conceptual framing needed to ask questions that will lead engineers and corporate leaders to align their targets and measure product performance against multiple socioenvironmental considerations with the end goal being sustainable development. Key questions might include:

- How do we as a company engineer our products so that we do more good, rather than less bad?
- What would such policies and designs look like in practice?
- What data do we need to collect and metrics do we need to invoke to ensure that our product (or service) contributes to the environment rather than merely mitigates our impacts?

- How does one go beyond environmental rhetoric and execute actionable plans with clear milestones?
- To what extent can we upcycle rather than recycle or avoid landfill?

And finally, how and to what extent can mechatronics solutions advance sustainable innovation and support critical thinking for the benefit of both present and future generations?

Acknowledgements Kai Whiting acknowledges the financial support by the *Fonds de la Recherche Scientifique*—FNRS under Grant No CR40001149. Luis Gabriel Carmona acknowledges the financial support of the European Union's Horizon 2020 research and innovation programme under the Marie Sklodowska-Curie grant agreement no. 101027892.

References

1. WCED (1987) Our common future. United Nations World Commission on Environment and Development (WCED), Rio de Janeiro. https://sustainabledevelopment.un.org/content/documents/5987our-common-future.pdf. Last Accessed 11 Dec 2020
2. Yilmaz N, Atmanli A (2017) Sustainable alternative fuels in aviation. Energy 140:1378–1386
3. Ng W, Patil M, Datt A (2021) Hydrogen fuel cell and battery hybrid architecture for range extension of electric VTOL (eVTOL) aircraft. J Am Helicopter Soc 66(1):1–13
4. Romeo G, Borello F, Correa G, Cestino E (2013) ENFICA-FC: design of transport aircraft powered by fuel cell & flight test of zero emission 2-seater aircraft powered by fuel cells fueled by hydrogen. Intl J Hydrog Energy 38(1):469–479
5. Misra A (2018) Energy storage for electrified aircraft: the need for better batteries, fuel cells, and supercapacitors. IEEE Electrification Mag 6(3):54–61
6. Kramer D (2020) Hydrogen-powered aircraft may be getting a lift. Phys Today 73(12):27–31
7. Singh J, Sharma SK, Srivastava R (2019) AHP-Entropy based priority assessment of factors to reduce aviation fuel consumption. Intl J Syst Assur Eng Manag 10(2):212–227
8. Brueckner J, Abreu C (2017) Airline fuel usage and carbon emissions: determining factors. J Air Transp Manag 62:10–17
9. Jevons WS (1866) The coal question; an inquiry concerning the progress of the nation, and the probable exhaustion of our coal-mines. Fortnightly 6(34):505–507
10. Lee JJ, Lukachko P, Waitz IA, Schafer A (2001) Historical and future trends in aircraft performance, cost, and emissions. Annu Rev Energy Env 26(1):167–200
11. Sorrell S (2009) Jevons' Paradox revisited: the evidence for backfire from improved energy efficiency. Energy Policy 37(4):1456–1469
12. Ivanova D, Vita G, Wood R, Laussele C, Dumitru A, Krause K, Macsinga I, Hertwich EG (2018) Carbon mitigation in domains of high consumer lock-in. Glob Environ Chang 52:117–130
13. Rodrigues B, Carmona LG, Whiting K, Sousa T (2021) Resource efficiency for UK cars from 1960 to 2015: from stocks and flows to service provision. Environmental Development (in review)
14. Rajala R, Hakanen E, Mattila J, Seppälä T, Westerlun M (2018) How do intelligent goods shape closed-loop systems? Calif Manage Rev 60(3):20–44
15. Jun HB, Kiritsis D, Xirouchakis P (2007) Research issues on closed-loop PLM. Comput Ind 58(8–9):855–868
16. Tiefenbeck V, Tasic V, Schoeb S, Staake T (2013) Mechatronics to drive environmental sustainability: measuring, visualizing and transforming consumer patterns on a large scale. In: IECON 2013 39th annual conference on IEEE industrial electronics society, pp 4768–4773

17. Valero A, Valero A (2019) Thermodynamic rarity and recyclability of raw materials in the energy transition: the need for an in-spiral economy. Entropy 21(9):873
18. Di Maio F, Rem PC, Baldé K, Polder M (2017) Measuring resource efficiency and circular economy: a market value approach. Resour Conserv Recycl 122:163–171
19. Ilic DD, Eriksson O, Ödlund L, Åberg M (2018) No zero burden assumption in a circular economy. J Clean Prod 182:352–362
20. De Falco F, Di Pace E, Cocca M, Avella M (2019) The contribution of washing processes of synthetic clothes to microplastic pollution. Sci Rep 9(1):1–11
21. Napper IE, Thompson RC (2016) Release of synthetic microplastic plastic fibres from domestic washing machines: effects of fabric type and washing conditions. Mar Pollut Bull 112(1–2):39–45
22. Almroth BMC, Åström L, Roslund S, Petersson H, Johansson M, Persson NK (2018) Quantifying shedding of synthetic fibers from textiles; a source of microplastics released into the environment. Environ Sci Pollut Res 25(2):1191–1199
23. Bradley D, Simpson E, Dawson D (2020) The challenge of complexity, COMEC occasional paper no 13. www.comec.org.uk/wp-content/uploads/2021/02/COMEC_Occasional_No13_Challenging-Complexity.pdf?x24141. Last Accessed 6 Dec 2021
24. Legifrance (2020) Bulletins Officiels des conventions collectives. www.legifrance.gouv.fr/jorf/id/JORFTEXT000041553759?r=yFtIxQRImr. Last Accessed 6 Dec 2021. (in French)
25. Legifrance (2020) Bulletins Officiels des conventions collectives. www.legifrance.gouv.fr/loda/id/JORFTEXT000031044385/. Last Accessed 6 Dec 2021. (in French)
26. Mondaq (2020) Italy: Samsung and Apple were fined for 'Planned Obsolescence'. www.mondaq.com/italy/antitrust-eu-competition-/751640/samsung-and-apple-were-fined-for-planned-obsolescence. Last Accessed 6 Nov 2021
27. Lurie-Luke E (2014) Product and technology innovation: what can biomimicry inspire? Biotechnol Adv 32(8):1494–1505
28. McDonough W, Braungart M (2010) Cradle to cradle: remaking the way we make things. North Point Press
29. Wise C, Pawlyn M, Braungart M (2013) Eco-engineering: living in a materials world. Nature 494(7436):172–175
30. Carmona LG, Whiting K, Carrasco A, Sousa T, Domingos T (2017) Material services with both eyes wide open. Sustainability 9(9):1508
31. McDonough W, Braungart M (1992) The hannover principles. William McDonough Architects. mcdonough.com/wp-content/uploads/2013/03/HP-20_email_121023.pdf. Last Accessed 6 Dec 2021
32. Schneider M, Romer M, Tschudin M, Bolio H (2011) Sustainable cement production-present and future. Cem Concr Res 41(7):642–650
33. Janssens-Maenhout G, Crippa M, Guizzardi D, Muntean M, Schaaf E, Dentener F, Bergamaschi P, Pagliari V, Olivier JG, Peters JA, Van Aardenne JA (2019) EDGAR v4. 3.2 global Atlas of the three major greenhouse gas emissions for the period 1970–2012. Earth Syst Sci Data 11(3):959–1002
34. Hasanbeigi A, Price L, Lin E (2012) Emerging energy-efficiency and CO_2 emission-reduction technologies for cement and concrete production: a technical review. Renew Sustain Energy Rev 16(8):6220–6238
35. Gursel AP, Masanet E, Horvath A, Stadel A (2014) Life-cycle inventory analysis of concrete production: a critical review. Cement Concr Compos 51:38–48
36. Sebaibi N, Boutouil M (2020) Reducing energy consumption of prefabricated building elements and lowering the environmental impact of concrete. Eng Struct 213:110594
37. Van den Heede P, De Belie N (2012) Environmental impact and life cycle assessment (LCA) of traditional and 'green' concretes: literature review and theoretical calculations. Cement Concr Compos 34(4):431–442
38. Quattrone M, Angulo SC, John VM (2014) Energy and CO_2 from high performance recycled aggregate production. Resour, Conserv Rrecycling 90:21–33

39. Noguchi T, Kitagaki R, Tsujin M (2011) Minimizing environmental impact and maximizing performance in concrete recycling. Struct Concr 12(1):36–46
40. Yang J, Tan FH, Tan A (2017) The ancient construction materials and methods: the Great Wall of China in Jinshanling as a case study. J Constr Eng Proj Manag 7(1):37–49
41. Chou NN, Yang KH, Barrett B, Wu HM, Liu TY (2020) Sustainable characteristics of reinforced soil structures–from ancient great walls to modern GRS walls. Transp Infrastruct Geotechnol 7(3):445–460
42. Scotscape 2020 Flex MSE living wall. www.scotscape.co.uk/services/flex-mse-vegetated-green-wall#Building-With-Flex-MSE. Last Accessed 6 Dec 2021
43. Cislaghi A, Sala P, Borgonovo G, Gandolfi C, Bischetti GB (2021) Towards more sustainable materials for geo-environmental engineering: the case of geogrids. Sustainability 13(5):2585
44. Oyenuga AA, Bhamidimarri R (2017) Upcycling ideas for sustainable construction and demolition waste management: challenges, opportunities and boundaries. Intl J Innov Res Sci Eng Technol 6(3):4066–4079
45. Mathews F (2011) Towards a deeper philosophy of biomimicry. Organ Environ 24(4):364–438
46. Heenan CB, Seymour RS (2011) Structural support, not insulation, is the primary driver for avian cup-shaped nest design. Proc R Soc B Biol Sci 278(1720):2924–2929
47. Dicks H (2016) The philosophy of biomimicry. Philos Technol 29(3):223–243
48. Ren X (2008) Architecture and nation building in the age of globalization: construction of the national stadium of Beijing for the 2008 Olympics. J Urban Aff 30(2):175–190
49. Rogers A, Yoon B, Malek C (2008) Beijing Olympic Stadium 2008 as biomimicry of a bird's nest. www.mcgill.ca/architecture/files/architecture/BiomimicrySSEFessay2007.pdf. Last Accessed 6 Dec 2021
50. Yuhuan G, Tian M, Peng C (2010) Contributions of Green Olympics to ecological environmental improvement in Beijing. In: 2010 International coneference on mechanic automation and control engineering. IEEE, pp 2051–2054
51. Braungart M, McDonough W, Bollinger A (2007) Cradle-to-cradle design: creating healthy emissions–a strategy for eco-effective product and system design. J Clean Prod 15(13–14):1337–1348
52. Curran MA (2008) Development of life cycle assessment methodology: a focus on co-product allocation. https://repub.eur.nl/pub/12679/PhD%20Thesis_Curran_May%202008.pdf. Last Accessed 6 Dec 2021
53. Anielski M, Wilson J (2010) Environmental footprinting for agriculture in alberta: literature review and analysis, environmental stewardship division of alberta agriculture and rural development. https://www1.agric.gov.ab.ca/$Department/deptdocs.nsf/ba3468a2a8681f69872569d60073fde1/0cd8b859692799e68725781c00674682/$FILE/Lit_Review.pdf. Accessed 4 June 2021
54. Stripple H, Erlandsson M (2004) Methods and possibilities for application of life cycle assessment in strategic environmental assessment of transport infrastructures. www.ivl.se/download/18.34244ba71728fcb3f3f6f6/1591704430908/B1661.pdf. Last Accessed 6 Dec 2021
55. Achten W, Barbeau-Baril J, Barros Telles Do Carmo B, Bolt P, Chandola V, Corona Bellostas B, Dadhish Y, Di Eusanio M, Di Cesare S, Di Noi C, Eisfeldt F (2020) Guidelines for social life cycle assessment of products and organizations. In: Guidelines for social life cycle assessment of products and organizations, p 138
56. Kopacek P, Ceccharelli M, Hajrizi E, Stapleton L (2006) Mechatronics education and international stability: the development of university-level education programmes in advanced engineering in Kosovo. IFAC Proc 39(23):1–7
57. Frede D, Khodabakhshian M, Malmquist D (2010) A state-of-the-art survey on vehicular mechatronics focusing on by-wire systems. www.diva-portal.org/smash/get/diva2:460614/FULLTEXT01.pdf. Last Accessed 6 Dec 2021
58. Ribbens WB (20120) Understanding automotive electronics, 7th edn. Waltham, Butterworth-Heinemann
59. Chiabert P, D'Antonio G, Inoyatkhodjaev J, Lombardi F, Ruffa S (2015) Improvement of powertrain mechatronics systems for lean automotive manufacturing. Procedia Cirp 33:53–58

60. Wellstead PE (1997) Analysis and redesign of an antilock brake system controller. IEEE Proc Control Theory Appl 144(5):413–426
61. Goodall RM, Kortüm W (2002) Mechatronic developments for railway vehicles of the future. Control Eng Pract 10(8):887–898
62. Moubarak JC, Ouellet P, Plourde D (2018) Innovative hybrid braking system engineering report. https://www.escholarship.mcgill.ca/downloads/0v838495q. Last Accessed 6 Dec 2021
63. US Environmental Protection Agency (2015) Memorandum of understanding on copper mitigation in watershed and waterways (United States). www.epa.gov/npdes/memorandum-understanding-copper-mitigation-watershed-and-waterways. Last Accessed 6 Dec 2021
64. Moriarty P, Honnery D (2010) Rise and fall of the carbon civilisation: resolving global environmental and resource problems. Springer, London
65. Chau KT (2014) Pure electric vehicles. In: Alternative fuels and advanced vehicle technologies for improved environmental performance, pp 655–684. Woodhead Publishing
66. Clarke P, Muneer T, Cullinane K (2010) Cutting vehicle emissions with regenerative braking. Transp Res Part D Transp Environ 15(3):160–167
67. Bielaczyc P, Sala R, Meinicke T (2021) Analysis of technical capabilities methodology and test results of a light-commercial vehicle conversion to battery electric powertrain. Energies 14(4):1119
68. Wager G, Whale J, Braunl T (2018) Performance evaluation of regenerative braking systems. Proc IMechE Part D J Automob Eng 232(10):1414–1427
69. Adamiec E, Jarosz-Krzemińska E, Wieszała R (2016) Heavy metals from non-exhaust vehicle emissions in urban and motorway road dusts. Environ Monit Assess 188(6):1–11

Mechatronic Applications in Respect of Sustainability and Climate Change

Brendan McGlynn

Abstract Sustainability and climate change are closely interlinked. Tipping points in climate change that were identified some 20 years ago have now been surpassed. Dramatic change is now inevitable. The responsibility for reducing the effects of climate change now lies with the world population itself and how they press for change at the national level. Sea levels are predicted to rise by up to 3 m by 2030 irrespective of climate change mitigating actions now being considered. Widespread weather changes are also now inevitable and are in progress. New tipping points are identified and will be upon us inside the next 15 years. All these changes herald crop failures, wildfires, land erosion, more severe winds and rainfall coupled with drought in other areas. Sustainability is under attack from many directions—famine, population migration, loss of employment, disease, transport and housing. EcoMechatronics offers solutions in mitigating climate change and its onward instability profile to disaster and has a strong role to play in inhibiting and reversing climate degradation. This macro approach to micro data points delivers data on land use and soil condition, crop health plus security. On the micro side, a wide range of Internet of Things (IoT) physical sensors is now available to support the smaller granularity world that improves modern lifestyles, sustainability and environments. These new and advanced sensors support the plans to improve lifestyle sustainability and mitigate against climate change. This chapter intends to show the beneficial effects that (a) the emergence of low cost, miniaturized sensors coupled with embedded processing and artificial intelligence delivers to some of the leading topics in sustainability and climate change, (b) the engineering and chemical advancements in renewable power, power storage, mobility power and (c) the removal of CO_2 from the atmosphere by substitution or active removal.

B. McGlynn (✉)
Scorpion Networks Ltd, Kildare, Ireland
e-mail: bmcglynn@scorpionnetworks.com

P. Hehenberger et al. (eds.), *EcoMechatronics*,
https://doi.org/10.1007/978-3-031-07555-1_3

Introduction

Climate change and sustainability have been in relative equilibrium since the end of the last Ice Age. Climate order was restored by nature over the past 10,000 years (The Holocene). The climate is now in a disorderly state having passed the equilibrium and degraded to the first tipping point of rapid change in recent years. To get back to Holocene equilibrium requires major investment and global commitment. It is inevitable that the world will degrade towards the next equilibrium point incurring major climate and sustainability failures in the process. International organisations and governments have not truly invested in the problem. In part, this may be because vested interests and some governments are negating the risks which are partially due to the climate change and sustainability problems not being irrevocable stated. This chapter outlines a process within EcoMechatronics that illustrates the current degrading situation and sets out milestones of equilibrium degradation until effective action reverses trends.

The Nobel Prize Summit *Our Planet, Our Future* held in April 2021 [1] brought together Nobel Prize laureates, scientists, policymakers, business leaders, and youth leaders to explore the question:

> What can be achieved in this decade to put the world on a path to a more sustainable, more prosperous future for all of humanity?

The science sessions clarified the new reality of the global sustainability challenges based on multiple lines of evidence to summarise the current state of knowledge, acknowledging that society and nature are deeply intertwined and embedded in the biosphere. The summit was hosted by the Nobel Foundation and organised by the US National Academy of Sciences (NAS) in partnership with the Potsdam Institute for Climate Impact Research and the Stockholm Resilience Centre and Beiger Institute [2]. The science sessions and seminars by Prof. Brian Walker of CSIRO Sustainability Ecosystems, Australia also provide a useful insight [3] into the roles of sustainability and resilience as paramount factors in climate change. Improvements in lifestyle sustainability and climate may then be founded on three pillars as follows:

- **Technology Investment**

Mitigation of blocking technology across a wide range of disciplines advances rapidly with the ever-increasing speed of development [4]. These advances deliver solutions for today's problems and improve prospects for our planet and its population. Sustainability is not one separate technology silo but is integrally connected with climate change and other silos. This chapter looks at the complex non-intuitive connective links between these silo domains.

- **Investment**

In its correct form, investment is critical in supporting technological advances. Government and private investment in the academic world to support climate change

R&D is not effective per se, and investment is also required to take the advances in R&D out into industry, and the market. The USA is a prime example of how the investment may be positively applied to R&D because of the private funding nature of its venture capital market. When governments are the prime investment vehicle using tax-based investment or grants, the outcomes suit the academic world far better than the enterprise world. History demonstrates clearly that it is the small enterprises that make the big advances [5, 6], but these enterprises are the least supported by governments. The corporate world is focussed on next quarter's results and profits. Governments are focussed on the next national budget and their strategies for re-election.

Centralised governments can propose a 20–50-year economic plan and work accordingly, but are less concerned with more immediate climate Change issues and quality of life. Until a bottom-up pressure from the general population emerges, governments will not react to climate change in a timely manner.

Without important organisations such as the Nobel Prize Seminars, the Paris Accord and the Club of Rome, enabling a bottom-up pressure will be an insufficient investment to prevent the climate from reaching the next tipping point. The first tipping point probably has been recently passed (Antarctic West Ice Sheet). Some scientists disagree with the global tipping point argument and say it is speculation. Others, for instance, Timothy Lenton, Professor of Climate Change and Earth System Science at the University of Exeter in the UK, argue that, despite the uncertainty, the stakes are so high that it is too risky to bet against. They say '*to err on the side of danger is not a responsible option*', and call for an international emergency response [7].

- **Mitigation of Blocking**

This will become an essential component of making substantial progress across climate change and sustainability. The question then becomes '*who are the blockers*'? The Silicon Valley model enables those with leading-edge research and development across technology and processes to be properly funded via venture capital. This avenue to investment is fuelled by the potential of the population to save funds and make investments. This is a low tax strategy.

In contrast within the European Union with its high tax culture, the enterprising population have fewer means to fund research and development. Funding is managed by governments and civil servants who operate directly opposite to the Silicon Valley model. Offering limited funding via 50–80% grants to enterprises when the governments get some 50% + back in taxes is a blocking mechanism.

Changing the blocking aspects of government policy is a daunting task but must be engaged with if lifestyle, climate change and sustainability action plans are to reach the goals as now set out by the UN, USA, Canada, the EU and to a smaller extent other global players. The COVID-19 actions at a global level illustrate the positive and negative actions that are embedded in the current global economy.

One could describe the global players involved as the good, the bad and the ugly. When populations were in immediate danger, investment flowed, governments cooperated, the corporate world set aside its asset interests for a short time and vaccine solutions were found. Climate change does not present immediate danger to many governments, but time has run out and climate change disasters are at their initiation stages. The bottom-up wave for change at a global level has begun within the youth and the informed. EcoMechatronics will be one of the prime delivery paths to reversing the trends and stopping the planet from reaching future degradation tipping points.

The climate change and sustainability curves in the base space cluster can be linked to other domain space curves of secondary climate change clusters [8]. In this way, each point on the base curve has a complex function linking it to a point on any one of the secondary climate change curves, thereby causing influences on the base curve parameters and shape (Fig. 1). This use of conformal geometric algebra enables a wider population audience to see exactly what the causes and effects are [9].This may be the key to having the general population understand, in a graphical way, what the consequence of current inaction holds.

It also holds the key to more complex modelling. Climate will drive any ecosystem towards equilibrium while further climate change will degrade the system unless recovery investment is applied. This is the first failure of governments and industry. Once a tipping point is reached, substantial investment can reverse the strong trend beyond the tipping point and back to equilibrium (X). Ineffective or no investment leads to a new and degraded equilibrium (Y). These cycles of instability and degradation of one system are intricately linked to other systems and feedback loops. The consequence of ineffective or no-action is ecosystem collapse to a degraded system from which recovery is substantially more difficult in the short to medium term of 10–100 years.

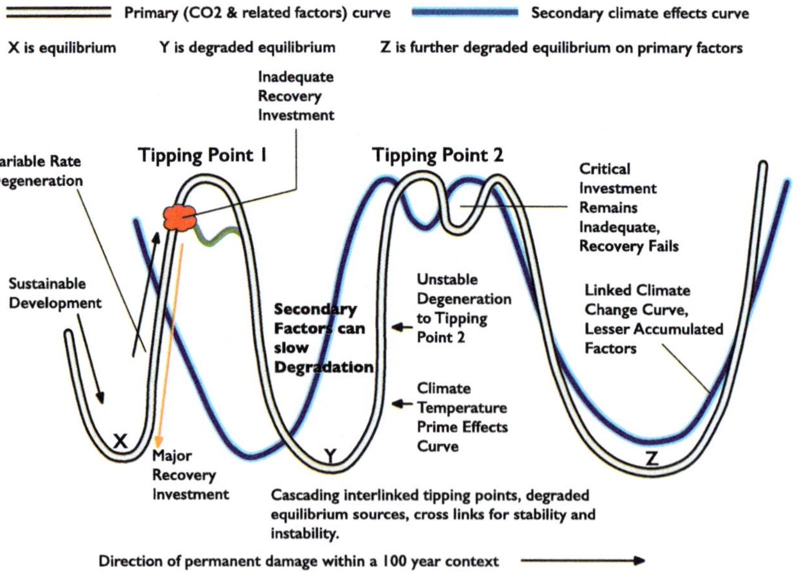

Fig. 1 Interaction of multiple climate change curves

The Technology of Sustainability

The path for technology has only in the past 7 years received its due focus as an important contributor to sustainability and climate change [10]. Major aspects of this are renewable energy, recycling technology, food production efficiency and the broad range of sensors that support these aspects at a granular level that can cover the planet's surface. Renewable energy is mainly seen as solar and wind. There are important contributions from hydro and pump storage which requires off-peak power and biomass. Future advances in nuclear fusion, expected within 30 years, will solve the base-load issue of power grids.

The variability of wind and solar power generation is a current block to the full replacement of the carbon cycle in coal, gas and oil-fired power stations. Energy storage technologies solve the block on eliminating fossil fuel power generation and in the past 5 years there have been major advances in the use of Li, LiP and LiFe battery designs, and recent new designs from R&D use hot liquid metals such as CaSb (calcium-antimony) and others that reduce the cost of Li-based batteries by 67% [11] and deliver enhanced features of fast charging and power delivery coupled to long life. These new batteries can support grid-level MW power storage. A recent advance in CaSb-type batteries removes the rare earth materials from the battery and are being deployed at the MW level in power grids. The geopolitical problems of Lithium supply can now be resolved due to the wide availability of both calcium and antimony across the planet. This battery revolution enables variable power from renewables to replace base loads power stations, enables a far more flexible power grid and facilitates the widespread adoption of the electric vehicle and fast home charging.

All these advances require AI in power supply management, sensors for all aspects of weather, new composite materials for new types of wind turbines, more granular solar and wind generation sites at local levels and the ability to locate and mine the antimony for MW levels at a global level of power storage. This is EcoMechatronics at its best. There is strong evidence that a bottom-up wave of innovation has left governments behind in the sustainability aspects of power.

Australia has encouraged this bottom-up approach. However, the EU continues to inhibit the renewable power strategy by using tax routes that offer no real incentives. Other nations have varying levels of government support between these two outlying strategies. Mechatronics is at the forefront of providing the required range of solutions to enable on-site power storage across the power grids, MW capacities at generation sites and multi-kilowatt capacities at the enterprise or home sites. This local power can fully charge an EV in under 15 min at home or at an EV charge station, thereby removing the primary range anxiety barrier to EVs. This locally stored power can also be extracted by the power grid to cover peaks but requires investment in edge-based AI.

Recycling materials carries substantial benefit when the technology involved is focussed on the initial product and the post-first life use by adaption, re-use and material disassembly coupled to re-assembly. Material resources are finite and having raw materials scarce due to political manipulations or depletion of natural resources raises the cost of these materials. A current case in point is Lithium as used in modern batteries. Lithium deposits on the Earth are very finite with one country holding over 70% of the current global supply. Geo-political issues now dominate this market. Mechatronics is now being applied successfully through material and electronic developments to bypass the lack of resources and replace Lithium with other materials easily sourced globally and new R&D to divert the battery market in a new sustainable direction. It is the enterprise that leads this field of moving to lower costing and easily obtained materials with positive consequences that reduce carbon emissions from vehicles and incorporates greater efficiency into power grids.

The world population is increasing leading to food supply stresses. Concurrently, climate change is disrupting food-producing lands in equatorial regions and altering the growing seasons in temperate climates. Mechatronics can play an important role in mitigating the effects of supply stress and climate consequences using satellite-based data, ground-level sensors (low-cost Internet of Things devices) and the data network to gather the information and generate actions to support food production. By linking sustainability and climate change through a wide range of technical solutions food production is going to deliver safer food, more food and reverse climate change effects such as global warming. Aspects of this positive process will be blocked by corporations and governments who have conflicting interests. It will fall to enterprise to overcome blockers and release the positive outcomes so desperately needed by large swathes of the global market (Fig. 2).

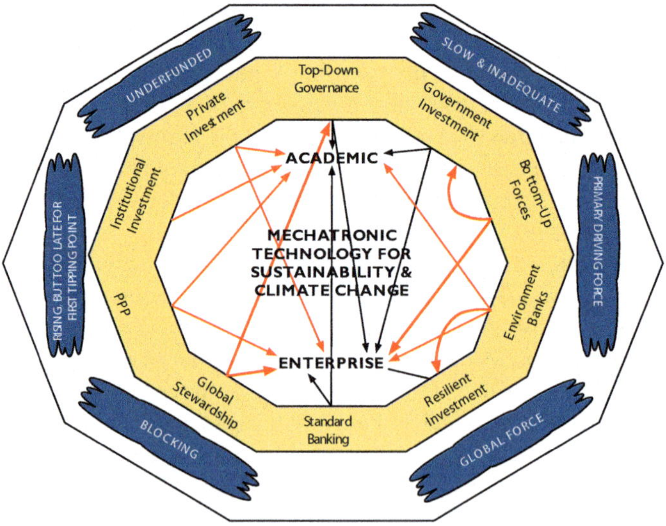

Fig. 2 Global reactions to climate change investment

Connectivity Web

Mechatronics technology is a core pillar for dealing with climate change and sustainability. It can deliver the solutions required across the full range of cause and effect, but it is only half the solution to limit climate change and raise sustainability. Bottom-up forces, which are actors of change that can trigger good investment strategies, as experienced in the younger generations have raised the impetus to effect change and re-think sustainability though the funding process remains fragmented and needs to be simplified. Governments have proven themselves to be unable to provide leadership or channel the investments. The conclusions drawn from Fig. 2 are that all actors need to establish a global stewardship domain that prevents investment blocking, sets targets to counter further tipping points and channels global investment to the most effective channels for academia and enterprise. The paths in the diagram need to be simplified and strengthened. The cost of inaction produces complex feedback loops in multiple eco-environments with dangerous results for the global population. Experts now agree that we have passed the first climate tipping point of Antarctic Ice Sheet melt and Atlantic Gulf Stream reduction. We must now prevent the next and consequentially more disruptive cascaded tipping point series.

EcoMechatronics—Technology of Climate Change

The application of all arms of mechatronics has already resulted in some progress on climate change, though EcoMechatronics remains hindered by a lack of investment in its core technologies and policy plans. However, the pace of climate action is insufficient to date and the first major tipping point has been passed. The Paris Climate Accord and COP process has highlighted a prevalence of talking and not doing. There is a wide expert body acknowledgement [1] that the world must expect consequences derived from passing the 1.5 °C target tipping point. The planet is not prepared for this shock (Figs. 3 and 4).

Power Generation

Advanced Electronics and Battery Chemistry

It is estimated that by 2050, 86% of electrical energy will be from renewables [13]. One of the most important elements of this process is power generation and usage. Previous means of generating power such as coal, oil and gas will be greatly reduced within 10 years. Carbon taxes may well aid this process but in the main it will be driven by the climate change ambitions of governments and their populations seeking attractive alternative generating solutions. Large-scale hydro-generating solutions

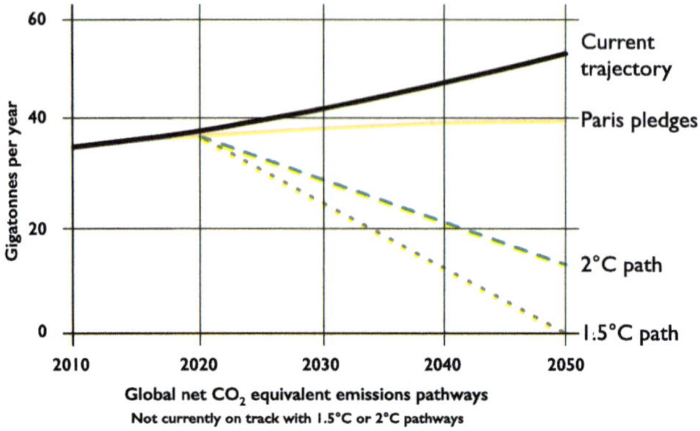

Fig. 3 CO_2 and climate temperature rise trajectories [12]

NET ZERO EXTERNAL BARRIERS

POLICY

- Major policy changes not yet applied to boost net zero plans.
- Government are taking a wait and see approach.
- No global carbon value yet set.

MARKET

- Limited low carbon offers on open market.
- Public awareness remains confused.

INVESTMENT

- The USA model is too short term
- New pathways that bypass government inaction are needed.
- A viable return on investment over periods of up to 30 years is required free from national interference.
- May be crypto-currency aligned.

TECHNOLOGY

- Energy technology is advancing and needs investment.
- Carbon sequestration is recently emerging.
- When implemented via bio-organism path is well known and increasingly applied.
- Technology of sustainability and resilience is only now beginning.

ENTERPRISE INTERNAL BARRIERS

OWNING A NET ZERO PLAN

- Middle management is reluctant to make changes required.
- Lack of employees training at all levels.
- Quarterly based corporate plans are a mismatch for net zero implementation.

LEADERSHIP SHORTFALL

- Few of the global large enterprises have a coherent policy that matches global initiatives such as COP and the Paris Accord.
- Academic leaders are clear with their message on Tipping Point and consequences.
- Carbon taxes by governments will not advance net zero plan by much.

ALIGN CLIMATE & SUSTAINABILITY PLANS

- Enterprises that embrace net zero plans are exposed to competitive disadvantages.
- The World Trade Organisation and major political blocks must ensure that no advantages pass to errant behaviour.
- A global penalty arrangement that is meaningful and impactful must be organised.
- Market forces are the optimum control tool to use.

CLEAR TARGETS & CONSEQUENCES

- Long term targets for the next 50 years are required with Tipping Points clearly defined.

Fig. 4 Barriers to progress on climate change

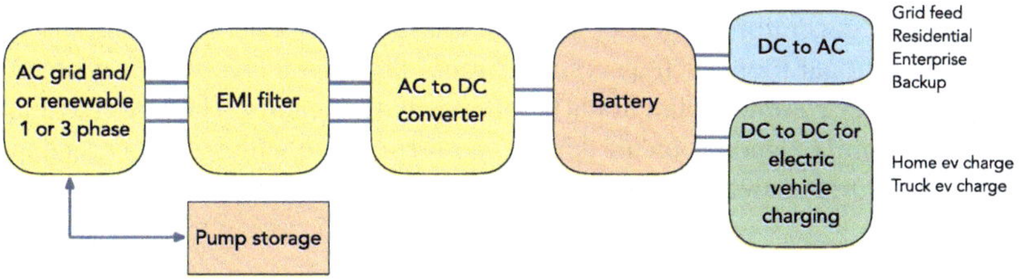

Fig. 5 Grid or renewable power flow to home batteries

are delivering carbon-free power but at the cost of disruption to land use and major earthworks. Hydro is an effective power solution for long-term base loads in some countries. However, the opportunity is limited to terrain profiles and the sufficiency of water flow (Fig. 5).

The nuclear industry has a chequered history on safety, and the economic cases for such power stations are often concealed by governments. Storage of radioactive materials is not necessarily presented to the public in true economic terms where some material by-products have radiating lives of 1000s of years. Relatively new power technologies that are becoming common with technological advances progressing rapidly include solar, wind, small hydro and biogas. Nuclear fusion is an estimated 30 years away from being economically viable but carries solutions and advances that nuclear fission cannot deliver.

Noncarbon-related power generation lowers CO_2 from the atmosphere via carbon sequestration, as an enterprise either by natural processes or hybrid solutions that use minimum carbon cycle work. This includes re-afforestation, protection of wetlands, seagrass planting and a move from fossil-fuelled heating to electric and electrical-powered sequestration. Reverse effects must be identified and countered. For instance, a by-product of fossil fuel use is SO_2 in the upper atmosphere which reflects sunlight. Removing fossil fuels from use also removes SO_2 and allows more heat onto the planet's surface. Side effects like this need to be accounted for in the predictions on climate change.

Green Power

Wind—The application of new generator and turbine designs

Mechatronics has been instrumental in the technological advances of wind power generation. The combined resources of mechanical, electrical, chemical and computational assets have delivered viable solutions with further advances on the horizon. Many engineering skills came together to design the common three-bladed high-tower wind turbines with power capacities ranging from 0.5 to 10 MW. Mechanical engineering drove the aerofoil designs, chemical engineering drove the move to composite materials for aerofoils, electrical engineering drove the development of more efficient generators, power converters and high magnetic field materials while new computational tools such as computational fluid dynamics (CFD) enabled the virtual testing of complex wind fields before costly fabrication. The power available, as per the formula, depends on air density, performance coefficient (tip speed, pitch angle), rotor surface area and the cube of the wind speed (Eq. 1).

$$p_w = \rho \cdot \frac{c_p \cdot (\lambda \cdot \delta)}{2} \cdot A_r \cdot V_w^3 \tag{1}$$

Computational fluid dynamics (CFD) tools indicate an advance in design that enables non-pylon/mast solutions with reduced capital expenditure and more acceptable conditions to nearby residents in the range of 500 kW–2 MW capacity. This trend supports the greening of electrical power and local ownership (smaller investors). However, the corporate influence of the power grids will need to be balanced against the interests of smaller more local ownership populations.

Solar

In the combined mechatronic domains of chemistry, quantum physics and electronics, the efficiency of solar panels has increased slowly over the past 20 years and now approaches 35% using hybrid organic–inorganic lead or tin halide-based material, as the light-harvesting active layer (Perovskite). Panels of 1 m^2 with 35% efficiency will produce 273 kWh/year based on average solar irradiance (Fig. 6). Low latitude sites in strong sunshine climates can generate 540 kWh/yr, while northern Europe sites can produce only 140 kWh/yr. This is coupled with efficiency improvements in the power inverter technologies using MOSFETs to replace silicon IGBT-based solutions with 99% inverter efficiency as possible. The solar power generated varies by the latitude of the site, the summer–winter sun height above the horizon, weather and the angle

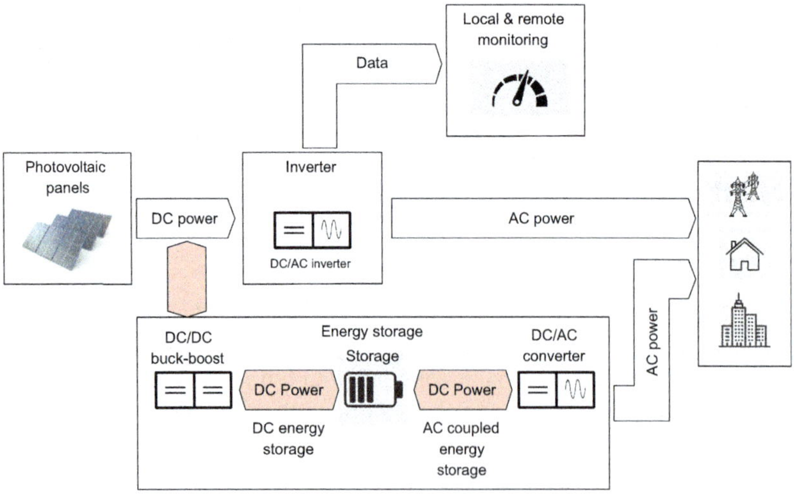

Fig. 6 Solar power generation

of the solar panel. This leads to a very variable renewable power in many global locations and dictates that a battery is required in the solution. Major improvements in battery efficiency and capacity have recently taken place and continue to improve on capacity, cost and use of easily sourced materials. CaSb and other new batteries open the potential for charging EV at home in 15 min for a 500 km range with more progress to come. This breakthrough will ensure the faster uptake of EVs with fast charging at home or on the road, thereby competing with fossil-fuelled vehicles. The effect on CO_2 emissions indicates that on average there will be a global reduction of 20% by eliminating much of the fossil-fuelled road vehicles.

Inverters (DC to AC) are a key component to bridge from power storage to power usage (microinverters for residential, single phase for residential and commercial and three phases for commercial and utilities). Power storage may be derived from solar (DC), wind (AC), pump storage (AC) or battery (DC). The battery removes the disadvantage of variable renewable power and capacities range from 200 kW (typical home) to MW (typical multiple MW storage for power networks). Tesla has launched the battery wall for homes which stores power from a variety of sources. Lithium, however, has a chequered history in fire dangers while the new hot metal batteries such as CaSb have no fire danger and can be mounted outside the home with as much capacity as is required for the home and EV charging [11, 14].

Hydrogen

Hydrogen as a transport fuel is another green fuel to replace fossil fuels. However, many hydrogen-producing plants use fossil fuel-derived power to generate hydrogen. Only when hydrogen is produced by renewable energy can this type of fuel be classified as green and renewable and the variable energy-producing solutions of solar and wind can usefully use excess power to generate hydrogen. Hydrogen needs to be stored with care and requires specialised distribution for vehicle use.

The emerging hot metal batteries are likely to be more attractive as a storage solution at the point of distribution. In this case, the power grid delivers the EV power to the distribution sites (re-charging stations) along with many other applications of electrical power. Current progress in hot metal batteries indicates 2 MW type static storage units of standard transport container size, and fast recharging capability for transport (trucks, cars, trains, shipping and short-haul aircraft), where EV battery tilting becomes a non-issue. Such hot metal batteries [11] will be deployed in the transport solutions themselves. These changes now under trial will negate the current hydrogen fuel trends.

Carbon sequestration—The application of bioremediation and technology

The Earth has been capturing carbon for some 2 billion years. In this way, the excess CO_2 in the atmosphere has been removed by the sea and plant life. This balance between the solar heating effect of CO_2 and runaway greenhouse effects has been naturally balanced (Fig. 7). If this had gone awry, then Earth may easily become Venus-like with no surface water and terrain temperatures of about 700 °C. Man has injected additional CO_2 into the Earth's atmosphere causing temperature increase and consequent climate change.

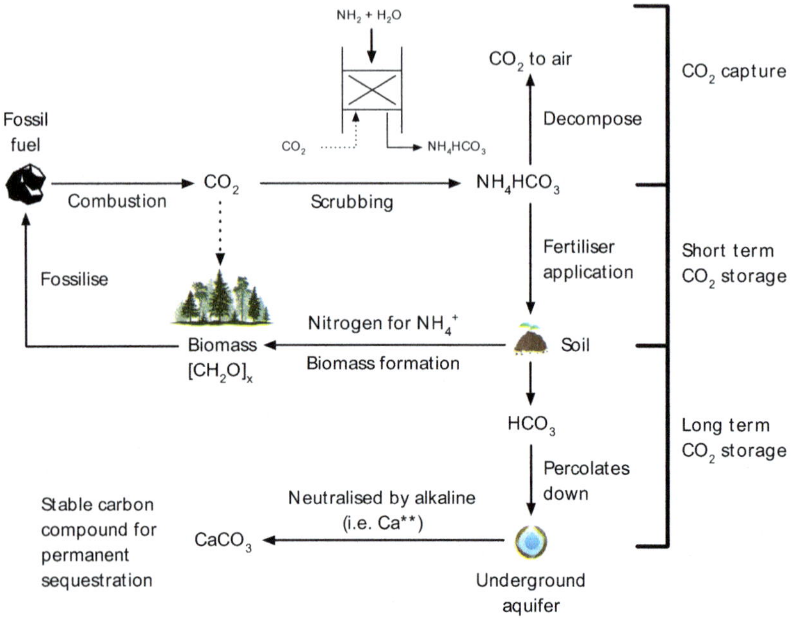

Fig. 7 Pathways to carbon sequestration [15, 16]

These changes may not be fully caused by man's activity and other cyclical natural effects may be involved. It now becomes one of the most important tasks of man to remove CO_2 from the atmosphere or suffer swings in the climate that have not been experienced since before the last Ice Age that ended 12,000 years ago or the earlier more severe climatic conditions of hothouse/warm house/snowball Earth.

Forests have been a major carbon sink when left to mature, decay and re-generate but burning forests breaks the cyclic sequestration and releases stored carbon into the atmosphere. The same applies to general vegetation on land and sea, in particular wetlands that generate peat, seagrass and kelp. Planting forests of coniferous trees where original forests were deciduous is a temporary solution. Coniferous forest is appropriate from temperate climates up to the tree line of the Arctic Circle but not at lower latitudes. Coniferous forests cause acidic conditions in the earth, rivers and streams.[1]

Technology, driven by excess electrical power, can scrub CO_2 directly from the atmosphere using an inefficient process requiring energy input—refer to the diagram below. The mechatronics approach of combining electrical, chemical, and mechanical systems has produced such solutions. However, it is likely, as populations grow, that excess electrical power will not be spare capacity in the future. One important step to take is to stop the clearing of forests in South America and Southeast Asia.

[1] The Sequoia tree is the most efficient carbon storage tree and can live for hundreds of years. The author has established a nursery based on cloning and seeds for general distribution.

The Global Population Needs Access to Verified Data

Satellite data and climate specialists can clearly present the problem that local works can damage global populations, however, politically orchestrated meetings and soundbites coupled with empty promises have left the global population sceptical. The information derived from the multi-source collected and audited data needs to be loaded into an EcoMechatronics engine free from political interference and false narrative. The global community, both young and old, must have confidence in an EcoMechatronics solution that is controlled by the many and not by the few. This may be sufficient to stop further climate destruction beyond the tipping points and present viable and enforceable options. This EcoMechatronics engine must have multiple segments with sufficient power, independent of governments, to identify the sources of the CO_2 and methane global metric tons/year emission problem, present the solutions and penalise those running counter to the interests of the global population. While EcoMechatronics can deliver the engineering, information technology and artificial intelligence solutions for such an engine, there is currently no organisation at a global level to carry out the task. A few have ambitions to be one such, but they must be free of political interference. COP26 highlighted the actions of the fossil fuel lobbyists in reducing the effectiveness of clear remedial solutions. The world needs a climate tzar unencumbered by narrow government restrictions. COP26 may have identified such individuals as one positive outcome.

The Strategy of Mitigating Blocking—How EcoMechatronics Can Affect Political and Civil Change

Much of the important developments and research in developed economies emerge from small enterprises and academic spin-out companies. The R&D efforts in large corporations are often confined by accounting principles (quarterly results etc.) and a lack of foresight resulting in slow solution-oriented progress. The small enterprise has no such limitations except being blocked by funding issues. Once a progressive concept emerges from a small enterprise, a category of first stage investors may emerge by presenting limited funding support in the knowledge that the small enterprise will need second stage funding. It is at this stage that small enterprise owners are diluted to give effective control to first and second stage investors. The scene is then set for early investors to sell successful solutions to larger corporations for the profit of the investors. The original small enterprise owners are often diluted to near zero in this mix. Mechatronics can offer a solution to this degradation of new solutions. Government departments managed by civil servants will fund the early investors but rarely the small enterprise itself to a meaningful level. The same government departments will, however, invest heavily in the corporate world to produce more employment and hence increase tax revenue.

Mechatronics is often seen as a technology-linked domain. If the best climate change and mitigation ideas coming from academic spinouts and small enterprises are to mature for the benefit of those taking the risks and the world, the impetus for these new technologies is blocked to varying degrees by being excluded from venture and other capital funding. Crowdfunding may be partially effective in resolving the lack of venture capital but requires better regulation and encouragement. A globally-based Mechatronics domain that is crowdfunded can be established. This will require well-regulated computerised systems, global standards and ethics engine, an accounting engine that exists outside of government interference but is transparent to all and ownership by those investing as enterprise owners and subsequent investors. In this way, the best solutions pertaining to sustainability and climate change can be shepherded to maturity.

Block chaining [17] offers a partial solution in that a wider spread of early investors is managed by enterprise-seeking funding. The investment is private, secure and global. Governments, banking and venture capital do not like block chaining because it can be fully independent and operated at a low cost. Those blocking progress in climate change technologies and sustainability are concerned by the in-depth and global reach that may in the future bypass the current vested interests in the use of fossil fuels. Whatever its faults, blockchain technology can manage the early investment for the greater global good. There are dangers in the globalisation of block chaining for the funding of new EcoMechatronics solutions, but the benefits of bypassing government and large corporate blocking far outweigh the disadvantages.

The Mechatronics of Food Supply

Climate changes that affect food production are already happening as the first tipping point is passed. This is derived from water shortage, heat, pest infestations, conflict because of water and food shortages, sea-level rises, floods and increasingly severe storms frequencies. A small percentage of farmed land is now compromised in equatorial regions and has become unsustainable. Mass migration of populations is the direct consequence, which in turn diminishes the food and the water safety buffer of populations living north and south of the equatorial regions. Mechatronics can deliver some of the solutions to this expanding problem, but global government policies are needed to deliver real solutions to reverse the trend. It remains possible that a global entity independent of governments and having an arsenal of EcoMechatronic solutions can enforce remedial actions and reverse the climate change trajectory. Sustainability follows remedial actions. The UN is not equipped to fill this global role because it is politically constrained.

Palm oil is one such issue. Rain forests in Brazil and Southeast Asia are being destroyed to make room for palm oil trees. Carbon sequestration is greatly reduced and fires are used to clear the land to inject CO_2 into the atmosphere. Palm oil itself remains a suspect food ingredient. However, corporate profits drive this process. Large-scale cattle breeding contributes CO_2 and methane to the atmosphere. Efforts to replace real meat and dairy products with synthetic alternatives are being made at a global level. Human health is at stake here and the long-term effects of these changes are undetermined. Compromises need to be made on the above and similar issues globally. This requires an independent global authority to manage the changes and compromises. A UN-type structure founded in science and EcoMechatronics with global government support is required, but as COP26 clearly demonstrated, this authority needs to be led by the world's most climate change affected countries and global youth equipped with accurate forecast tools relating to tipping points.

Satellite-based ground surveillance delivers data on crop health, surface mineralisation, pest infestation, crop water content, areas under different crops, forest expansion or depletion, nearshore conditions of seaweed, seagrass and coral reefs. However, the NASA Earth Observing System (in low earth orbit (LEO)) at 99-min interval, is too course to give good ground condition pixilation. An LEO system based on a *Rosae-type* constellation of 24+ satellites [18] using interferometry which interrogates 90% of the Earth's surface at 1 m^2 resolution has been proven to work mathematically, but funding is not available. This type of approach is likely to be the future of the Earth observation. Coupling the satellite data to ground sensors verifies the satellite data and augments it. AI can then analyse the data from all sources to give the following:

- Farming advice and forecasts, crop condition reports, water management.
- Global estimates of crop production per region.
- Location advice on suitable crop production areas as the climate changes.
- Food shortage predictions per region as an early warning of potential famine.

Food safety ensures less waste coupled with the protection of the food chain. Sections of EcoMechatronics are being applied in this respect. Next-generation sequencing is applied to ensure safety across complex supply chains with block chaining IT and AI supporting the protection of the food chain from farmer to fork. Improved management and safety contribute to less pollution and waste and the potential to reduce food shortages in regions experiencing the worst of climate change [19].

The Mechatronics of General Transportation and Batteries—Transportation Contributes an Estimated 32% to CO_2 Emissions: Technology Can Stop Most of This

Technology advances offer alternatives to the use of fossil fuels for transportation. The recent arrival of affordable e-cars (EV) has delivered the first step in moving the personal transport and some of the public transport to electrical drive. Tesla advances in electrical motors and drive trains, which indicates the potential for further advancements. The advances in battery technology currently enable affordable EVs to have a range of up to 400 km, based on Li-type battery material. More recent advances, about to emerge from R&D, facilitate up to 600 km on one charge with a 10-min quick charge, using battery material such as CaSb already referred to and other materials that are common and cheaper than Li. The CaSb-type [11] battery and future hot metal batteries while heavier than the Li types are ideally suited to heavy public transport (bus, train and heavy road vehicles) and shipping. These developments are based on new electrical motor designs, drive trains and fast re-charge stations using power storage based on local renewable power and slow bulk charge off national power grids. These match renewable energy production from solar and wind at a widely distributed level.

While mass airline traffic contributes some 1.9% of the CO_2 emissions associated with transport (at a total of 16.2% in 2016) [20], it is unlikely that we will see battery-driven aircraft above the 50-seat capacity and 200 km range for some 20 years. The advances in train transport based on the above battery developments coupled with the advances in magnetic levitation will enable high-speed rail transportation, fast re-charging and thereby greatly reduce the need for short-haul aircraft. The megawatt (MW) capabilities of the CaSb and other new hot metal batteries used for road vehicles, shipping and rail transport will enable renewable energy to further replace the CO_2 of fossil-fuelled power stations.

Global Impact on the Local—EcoMechatronics Helps Identify the Cause and Effect of the Climate Feedback Loops, Tipping Points and Other Complex Relationships

Multiple feedback loops occur in tiers (3D domains) with tipping points inherent in the system—refer to the 2021 Nobel Prize Conference material [21]. Timothy Lenton at Exeter University, England, and his team of researchers, had first warned in their landmark 2008 Proceedings of the National Academy of Sciences paper [22], about the risks of climate tipping points and that large-scale discontinuities in the climate system were considered likely only if global warming exceeded 5 °C above pre-industrial levels. In 2019, Lenton and six co-authors thought that a new study published in Nature in November 2019 [23], warned in language that is much starker than the Intergovernmental Panel on Climate Change (IPCC) forecasts, that risks are much more likely and much more imminent and that some tipping points may already have been breached. It is clear to most climate scientists that exceeding one tipping point in one sector is likely to trigger other tipping points in linked sectors. These links and feedback paths may not be initially evident. A disruptive cause and effect in a local ecology may trigger a tipping point in a seemingly unrelated ecology system, in another locality (Fig. 8) [24]. Human activity with its CO_2 and other emissions may contribute to the removal of further shallow Ice Age cycles but the disruptive cost to the equilibrium of the past 5,000 years is yet unknown.

A UN Environment Programme (UNEP) report [25] warns that unless global greenhouse gas emissions fall by 7.6% each year between 2020 and 2030, the world

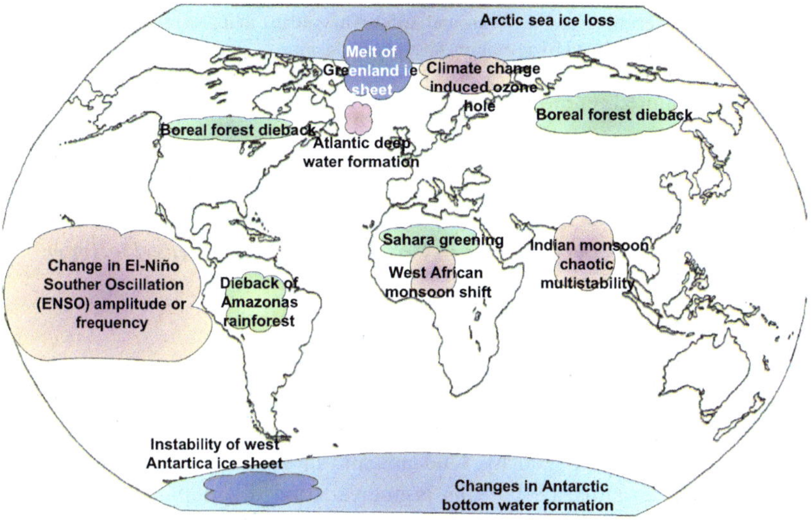

Fig. 8 Tipping points in Earth's climate

will miss the opportunity to get on track towards the 1.5 °C temperature goal of the Paris Agreement. Indeed, the report suggests that:

> There is alarming evidence that important tipping points, leading to irreversible changes in major ecosystems and the planetary climate system, may already have been reached or passed. Ecosystems as diverse as the Amazon rainforest and the Arctic tundra, may be approaching thresholds of dramatic change through warming and drying. Mountain glaciers are in alarming retreat and the downstream effects of reduced water supply in the driest months will have repercussions that transcend generations

In October 2018, the IPCC issued a special report [26] on the impacts of global warming of 1.5 °C, finding that limiting global warming to 1.5 °C would require rapid, far-reaching and unprecedented changes in all aspects of society. With clear benefits to people and natural ecosystems, the report found that limiting global warming to 1.5 °C compared to 2 °C could go hand in hand with ensuring a more sustainable and equitable society. While previous estimates focussed on estimating the damage if average temperatures were to rise by 2 °C, this report shows that many of the adverse impacts of climate change will come at the 1.5 °C mark. The report also highlights several climate change impacts that could be avoided by limiting global warming to 1.5 °C compared to 2 °C, or more. For instance, by 2100, global sea-level rise would be 10 cm lower with global warming of 1.5 °C compared with 2 °C. The likelihood of the Arctic Ocean being free of sea ice in summer would be once per century with global warming of 1.5 °C, compared with at least once per decade with 2 °C. Coral reefs would decline by 70–90% with global warming of 1.5 °C, whereas virtually all (>99%) would be lost by 2 °C. The report finds that limiting global warming to 1.5 °C would require:

> ... rapid and far-reaching transitions in land, energy, industry, buildings, transport, and cities. Global net human-caused emissions of CO_2 would need to fall by about 45% from 2010

levels by 2030, reaching 'net zero' around 2050. This means that any remaining emissions would need to be balanced by removing CO_2 from the air

expending limited renewable energy to do so [27].

Applying Advanced Sensors—Rapid Development in Sensor Technology Across All Domains of EcoMechatronics Delivers Solutions

Detailed data is required on all ecosystems at a finer granularity than is currently available. The IoT (Internet of Things) has delivered a wide range of sensory systems that augment the potential for fine-grained data across all aspects of the climate, sustainability, resilience, security, land use and more. Earth observation satellites can get to 1 m^2 granularity, but this is not yet funded. Coupled with Internet of Things type sensors and network communications EcoMechatronics can have the power to present the data after analysis such that all the populations can understand the interlocking complexity of the environment. This necessary approach ensures that the actors most active in all age groups can push from the bottom-up perspective. The top-down actors of governments and older institutions will take more notice of the bottom-up pressure because this cohort is also the cohort that elects them to office.

The Effect of the Foregoing Paths on Sustainability and Resilience—The Conferences of COP, Paris Accord, the Noble Prize Group and Stockholm Institute Indicates that Even if Governments Are Not Moving Fast Enough, Others Are! [8, 28]

The World Business Council for Sustainable Development [29] is a Swiss-based global organisation that aims to direct businesses towards a more sustainable future to reach net carbon zero. As the initial stage of its SOS 1.5 project, WBCSD has developed a business roadmap to help companies design and implement their journeys to achieve net-zero emissions before 2050. Developed with the Boston Consulting Group and WBCSD member companies across sectors, the plan is to provide key steps and actions which companies at any stage in the journey can follow to accelerate their journey to net-zero. The WBCSD is typical of several global [31, 32] organisations consulting with enterprises to get to carbon net-zero. Some of these global enterprises will do the minimum required while others under pressure from the bottom-up will do substantially more. Governments have the power to ensure large enterprises conform to a net-zero plan.

Conclusions

The lack of comprehensive data from global sensors has allowed governments to avoid their part in responsibility. The lack of understandable data that is presented for the global populations has ensured lethargy at a global level. These negative influences are now being addressed because of the clear and imminent dangers of climate change. The recognition of danger has been too late to prevent the first tipping point related to the 1.5 °C global average increase. The complex web that is EcoMechatronics can affect change mitigation and prevent further travel to the next tipping points with consequential damage to climate change and sustainability. Reversing past the last tipping point requires substantial global action and investment but the time delay to get back to an earlier equilibrium is complex and exceedingly difficult. The world must now face long-term climate change consequences irrespective of any future actions taken. EcoMechatronics is an important asset class that can stop the world getting to another tipping point and start a decade or more of recovery to get to the recently lost Holocene equilibrium. Nature is itself robust, and given time, investment and care from the next generation, we can regain our lost equilibrium.

For the past 3 million years, Earth's climate has been in an Icehouse state characterised by alternating glacial and interglacial periods. Modern humans evolved during this time, but greenhouse gas emissions and other human activities are now driving the planet towards the warm-house and hothouse climate states not seen since the Eocene epoch of 34 million years ago. During the early Eocene, there were no polar ice caps, and average global temperatures were 9–14 °C higher than today [30]. The youth of today inherits the climate change effects and its degraded sustainability and resilience but is now motivated and equipped to use bottom-up pressure to effect change. EcoMechatronics is there to be an essential part of the change process when governments are forced to act and its solutions and predictions, in being clear and unequivocable, will not be ignored.

Acknowledgements Many enterprises and academic sites were researched to give EcoMechatronics foundation to the above chapter. This includes solar, permanent magnet generators, batteries, IT, AI, satellite earth observation, wind turbines, pump storage and recent conferences on climate change, sustainability, resilience and the consequences of 1.5 °C. R&D material from Scorpion Networks Ltd. and its partners in EU Commission-funded consortia in FP7, Horizon 2020 and Horizon Europe 2021 has been incorporated.

References

1. www.nobelprize.org/events/nobel-prize-summit/2021. Last Accessed 14 Nov 2021
2. beijer.kva.se/news-item/urgent-plea-from-scientific-community/. Last Accessed 14 Nov 2021
3. www.stockholmresilience.org/. Last Accessed 14 Nov 2021
4. National academies of sciences, engineering, and medicine (2021) Progress, challenges, and opportunities for sustainability science. In: Proceedings of the workshop in brief, The National Academies Press, Washington, DC

5. Wu L, Wang D, Evans JA (2019) Large teams develop and small teams disrupt science and technology. Nature 566(7744):378–382
6. Collison P, Nielsen M (2018) Science is getting less bang for its buck, The Atlantic, 16 November. www.theatlantic.com/science/archive/2018/11/diminishing-returns-science/575665/. Last Accessed 8 Dec 2021
7. www.weforum.org/agenda/2019/12/climate-change-tipping-points-earth/. Last Accessed 14 Nov 2021
8. CGA complex equation links for climate change R&D at Scorpion Networks Ltd.
9. https://en.wikipedia.org/wiki/Conformal_geometric_algebra. Last Accessed 14 Nov 2021
10. www.youtube.com/watch?v=Gzhr9z32X9I&list=WL&index=2&t=696s. Last Accessed 14 Nov 2021
11. ambri.com/technology/
12. www.wbcsd.org/contentwbc/download/9499/144264/1. Last Accessed 14 Nov 2021
13. www.irena.org/publications/2019/Apr/Global-energy-transformation-A-roadmap-to-2050-2019Edition
14. www.infineon.com/cms/en/applications/industrial/solar-energy-systems. Last Accessed 14 Nov 2021
15. www.wku.edu. Last Accessed 14 Nov 2021
16. www.usgs.gov/faqs/what-carbon-sequestration?qt-news_science_products=0#qt-news_science_products. Last Accessed 14 Nov 2021
17. Hassija V, Chamola V, Zeadally S (2020) BitFund: a blockchain-based crowd funding platform for future smart and connected nation. Sustain Cities Soc 60:102145
18. patents.google.com/patent/US4602257A/en. Last Accessed 8 Dec
19. www.cremeglobal.com/predictive-microbiome. Last Accessed 14 Nov 2021
20. ourworldindata.org/ghg-emissions-by-sector. Last Accessed 14 Nov 2021
21. www.nobelprize.org/events/nob. Last Accessed 14 Nov 2021
22. Lenton TM, Held H, Kriegler E, Hall JW, Lucht W, Rahmstorf S, Schellnhuber HJ (2008) Tipping elements in the earth's climate system. Proc Natl Acad Sci 105(6):1786–1793
23. Lenton TM, Rockström J, Gaffney O, Rahmstorf S, Richardson K, Steffen W, chellnhuber HJ (2019) Climate tipping points—too risky to bet against. Nature 575:592–595. www.nature.com/articles/d41586-019-03595-0
24. /en.wikipedia.org/wiki/Tipping_points_in_the_climate_system
25. www.globalstewards.org/environmental-issues.htm. Last Accessed 14 Nov 2021
26. www.ipcc.ch/sr15/. Last Accessed 6 Dec 2021
27. netzeroclimate.org/. Last Accessed 6 Dec 2021
28. www.youtube.com/playlist?list=PLUWHlbMj0sOFGQQJIsNqck_zLlDCJWY9Z
29. www.wbcsd.org/Programs/Climate-and-Energy/Climate/SOS-1.5/Resources/SOS-1.5-The-road-to-a-resilient-zero-carbon-future. Last Accessed 14 Nov 2021
30. phys.org/news/2020-09-high-fidelity-earth-climate-history-current.html?_escaped_fragment_=&deviceType=desktop. Last Accessed 14 Nov 2021

Other Sources

Stockholm resilience centre TV, www.youtube.com/channel/UCG0QUKa0MTBzBKiW18AZfTg. Last Accessed 14 Nov 2021

Bierbaum R, University of Michigan and University of Maryland, towards sustainable futures: governance, inclusiveness and stewardship. www.youtube.com/watch?v=ZHaKqIFwBE8

EcoMechatronics and Bioinspired Design Ecology, Circular Economy, and Sustainability

Maki K. Habib

Abstract Bioinspiration is an interdisciplinary, creative, and innovative design approach that merges technology with biology by observing nature, properties, and biological systems. It aims to learn, discover, and capture essential principles and concepts inspired by nature, transform ideas, create new designs, and develop new technologies, techniques, algorithms, and systems. Biomimetic is a type of bioinspired design approach that aims to reproduce aspects of functions or properties of the biological systems by solving practical problems through functional analysis. Biomimicry represents a subcategory of the bioinspired design approach that considers nature an adaptable model and a source to learn from and find evolved and proven solutions, strategies, principles, valuable functions, and structures to meet sustainable development challenges in social, environmental, resources and biodiversity perspectives. Hence, for effective synergy between EcoMechatronics and bioinspired design approaches, it is vital to understand applications, ecology, environment, sustainability, and the management of ecosystems requirements through intensive analysis associated with the study and observation of natural systems. The goal is to have a conscious conclusion to emulate nature's best properties and functions. Therefore, it is essential to understand the conditions that allow such an approach to produce creative designs introducing new ideas, functions, and systems. Creative design inspired by nature is the way to support new innovative and environmentally friendly technological development. The development of smart materials is laying down the foundation to create new biomimetic-based technologies. However, the development in the field is facing difficulties by different technical constraints that continue to challenge and need to be addressed effectively.

M. K. Habib (✉)
The American University in Cairo, Cairo, Egypt
e-mail: maki@aucegypt.edu

P. Hehenberger et al. (eds.), *EcoMechatronics*,
https://doi.org/10.1007/978-3-031-07555-1_4

Introduction

Mechatronics is an interdisciplinary field of engineering science centered on the synergy of multidimensional knowledge. It focuses on developing intelligent and sustainable designs with adaptive behaviors and optimized performance. It aims to promote a prosperous and healthy future and a cleaner environment that reduces ecological impact while improving the sustainability and functionality of mechatronic technology to realize a sustainable global society. Such an approach within the Mechatronics field is termed EcoMechatronics. Thus, the visionary concept of EcoMechatronics aims to enhance human behavior and improve the interactive human and technology relation with the environment toward a sustainable lifestyle. EcoMechatronics can effectively address the interdisciplinary challenges with a potential synergy of experience and way of work, technologies, and knowledge to optimize the design development process and maximize functionalities. This approach contributes to developing new efficient, economical, energy-efficient, and environmentally friendly technologies. Also, it utilizes resources with added value practices while reducing the dependence on the combustion of fossil fuels and lowering the associated carbon footprints, reducing energy consumption, and reducing waste. The circular economy concept facilitates and manages these outcomes. To achieve such requirements and to have an ecologically sustained society, there is a need to develop novel design strategies.

Nature as a collective phenomenon of the physical world includes all plants, animals, the landscape, and all forces, features, and products of the Earth that exist independent of humans.[1] Nature always works to optimize its design strategies as evolutionary processes by removing those that are not surviving while revising and enhancing efficient and sustained outcomes. Such capability contributes to the operation of natural selection. The natural evolutionary process helps biological organisms, phenomena, and systems to succeed in solving problems encountered in their natural environments. Bioinspiration, biomimetics, biomimicry, bio-replication, and bionics with their overlapping boundaries are overgrowing investigative fields reflecting the inspiration of nature as a leading innovative design reference. These research directions attract researchers' interest and help derive inspiration and solutions to address and solve existing design problems and challenges [1]. Researchers in bioinspired design and ecodesign consider nature a sustainability model and use biological systems, organisms, environmental interactions, and properties as analogies when developing an environmentally sustained design. Thus biology is described as a model for sustainability [2]. However, researchers have yet to discover, learn from and apply sustainable design solutions within each living organism. Biological systems exhibit essential physical properties and have always served as an endless source of inspiration to humans.

Bioinspiration is a creative and innovative design approach that merges technology with biology by observing nature, properties, and biological systems. It aims to learn, discover, and capture essential ecological and sustainable principles

[1] Mariam-Webster Dictionary & Oxford English Dictionary.

and concepts inspired by nature, transform ideas, create new designs, and develop new technologies, techniques, algorithms, and systems. As a result, creative and innovative developers have investigated bioinspired design approaches with their overlapped areas to support further developments and applications. Using nature as a source supporting ideas and solutions inspiration, the bioinspiration design approach helps produce conscious abstraction of new principles for creative design and solutions. It can contribute to answering many of the challenges by focusing on ecology and sustainable development in different domains, directions, and needs. Numerous bioinspired applications domains are inspired by nature available in many real-life artificial fields, such as intelligent and autonomous robotics, optimization, information processing, architecture design, structures, materials, etc. [3–6] along with area-related applications, including transport, architecture, automotive, medical, aerospace, etc. [1, 7–10].

However, for effective synergy between EcoMechatronics and bioinspired design approaches, it is vital to understand applications, environment, and sustainable requirements through intensive analysis associated with the study and observation of natural systems. This approach would enhance the design process for a better explanation, evaluation, and generation of effective solutions supporting innovative ideas and design [3, 11]. Of course, there is no assurance that the bioinspired approach will lead to creative design outcomes [12]. However, bio-design has been the basis of innovative designs and is understood [13]. In addition, it is necessary to understand that solutions developed by nature are not always the optimal ones. Hence, bioinspired solutions should identify and address the required design criteria supported by the required attributes and technologies.

Of course, biology solves problems differently than engineering. However, such biological solutions offer insight into alternative strategies as a source of inspiration to develop efficient engineering systems reflecting EcoMechatronics features by expanding the knowledge and concept spaces to help visualize alternative and innovative design.

Biomimicry concerns many sectors of human activity: medicine, research, industry, the economy, architecture, urban planning, agriculture, ecosystem management, development, and so forth. Different levels of learning and idea abstractions are involved in successfully implementing biomimicry principles. For example, the innovative design approach may cover mobility, functionality, behaviors, thinking process, material, structures, and intelligent functions of biological creatures in nature [3, 14]. But, on the other hand, it is motivated by learning and adopting concepts, ideas, mechanisms, behaviors, and capabilities from nature, and using a potential scientific approach led to the effective development of new materials, shapes and structures, process, algorithms, systems, and other benefits.

Bioinspiration is an emerging fast-evolving design and development approach that adopts and uses processes, knowledge, and operational principles inspired by nature. As a result, designers can collaborate in a scientific context to draw their ideas from interdisciplinary and bioinspired sources of knowledge and ideas to address and face the challenges of developing effective alternate innovative design solutions supporting different levels of functionalities under different scenarios [15, 16].

This chapter introduces bioinspired design principles and presents the current engineering design methodology. The chapter also explains the role of biomimicry and ecologically sustainable design. It highlights the strengths ad weakness of design methodologies and practices that lead to consistent and sustainable bioinspired design. Further, it discusses the requirements for integrating an effective bioinspired design strategy featuring sustainability, ecosystem, and circular economy principles. Finally, it underlines the needs and the challenges for progress in this field.

Bioinspiration: Biomimicry, Circular Economy, and Sustainability

Bioinspiration

Current conventional technologies and methodologies are associated with constraints, which stimulates the demand to develop innovative and effective solutions that enhance functionality and performance while reducing or eliminating waste and reducing energy consumption. Such demands encourage creative thinking to create new solutions that drive research to consider nature as a critical motivation reference and source for new solutions. The biologically inspired design strategies that integrate context-specific solutions represent a practical approach to overcoming constraints.

Biomimetics is a term that originally came from bios which means life, and mimesis, which means imitate, and accordingly, the term implies imitation of life. In addition, the definition of biomimicry describes innovations inspired by nature [5, 17–20]. Therefore, Benyus [5] suggests considering nature as a reference design model (source of innovation as a standard reference), a measure (performance of new ideas), and a mentor to help understand the bioinspired concept of learning and knowledge building sustainable design and methodologies. Further, bioinspiration, biomimetics, biomimicry, bionics, ecodesign, and biognosis are other bioinspired developments motivated by nature. These techniques were used to capture new ideas, functions, and structures. This direction is facilitated by the superiority of biological systems' properties in terms of robustness, flexibility, agility, efficiency, and the ability to adapt continuously and improve performance while surviving in real life. Furthermore, biological systems also work in harmony to sustain the different ecosystem levels [5, 6, 19–22].

Biomimicry is an interdisciplinary design approach encompassing biology and engineering to describe a methodology and innovative process linking ecology and technology. This approach helps to merge biological principles and physical applications to find effective and environmentally friendly solutions adopted by nature through its evolutionary processes to address modern society and human life problems. Further, this approach helps abstract, emulate, and inspire systems, functions, and models from different species to improve life by developing innovative ideas, novel theories, efficient designs, and technologies. Hence, understanding biological systems and linking efforts between biologists and scientists present unique opportunities to develop potential concepts, ideas, methods, and paradigms that serve both sides in finding solutions and efficient design generations of intelligent tools, devices, products, and materials [18].

Biomimicry: Ecosystem and Circular Economy

Ecology studies the individual living organism within a population of organisms, the relation of these organisms to one another, the interaction of each organism with its local environment impacting it, and the interaction and adaptation of the organisms with that local environment [23]. Hence, it requires understanding all the components forming the ecology, including the environment's effect on a population of organisms. Ecology also studies the interacting populations of biological communities representing multiple ecosystems and their relations to compete or cooperate for resources. Further, ecology closely relates to physiology, behavior, genetics, and evolution. An ecosystem thus represents a specific biological community that shares a limited but loosely defined physical environment. The physical environment is essential in deciding the interacting living organisms and the non-living components available in an ecosystem. Hence, an ecosystem is a subset of ecology, and it can be of different types.

Biomimicry emulates the strategies available in real life of bringing together much of already existing features and functions in ecological innovation by helping to assess the way ecosystems work in nature and how they can contribute to solving human problems within the design context [24–26]. Ecosystems are adaptable, evolving, and resilient to disturbance. At the same time, they can create conditions that tend to support life continuity [27]. Such features help ecosystems to provide a model associated with a set of performance targets with their indicators. Correctly understanding different ecosystems can determine measurable targets, such as water catchment, emissions levels, energy production, carbon storage, and resource production. Ecosystem biomimicry enables tangible benchmarks to be devised over different periods in time.

A circular economy represents an integrative paradigm inspired by nature with the main focus on enhancing environmental sustainability and improving the economy with no concept of waste while reducing carbon emission. Instead, nature gives us the solution when every organism is considered a resource for the next higher level in the cycle of interdependence [28, 29]. The functionalities of circular economy aim to extend the value of materials and resources, reducing the need for further extraction of natural resources. Therefore, based on the concept of ecosystem, resource efficiency and environmental sustainability are key elements enabling circularity within human social activities and actors. However, this issue raises the need to change how to deal with the flow of material and energy at the system level while understanding the unsustainable practice of how current production systems, consumer behaviors, and waste disposal influence sustainability with the involvement of different interdependence actors. This understanding helps in transforming the linear models into circular ones. Interdependency between all elements of an ecosystem can be initiated by learning from nature to understand the cyclic behavior at a different level of shared purpose, intentions, logic, interaction, and values that enable interconnections among other levels or actors [30, 31]. It is vital to note that the outcome of the system level associated with an ecosystem will be higher than the outcome of the collective individual effects at each level making the ecosystem [30]. Therefore, the circular economy concept requires creative design, new intelligent technologies, and innovations to develop and implement cyclic technological loops focusing on energy and materials. In line with such requirements, the circular economy demands critical changes in management and business models and the sustainability directions toward advancing and fulfilling its goals by building circular systems with low environmental impact. Therefore, it is essential to assess the natural cyclic performance loops and compare them to human-made systems and methods for better understanding and further enhancement to secure resource supply and environmental sustainability.

Biomimicry and the circular economy are complementary approaches reflecting nature as a principal model. Both represent models that depend on ecosystems' operation, i.e., this operation is featured by waste-free, circular and autonomous processes. Biomimicry helps develop methodologies and systems for the circular economy by finding solutions to problems and transforming waste into resources. This can be achieved by drawing positive inspiration from one species that feeds on another species' waste. Biomimicry makes rethinking to make the current circular economy model possible. It allows researchers to conclude that the circular economy of nature works with flexible, complex, and open-ended loops. Hence, from the prospect of biomimetic thinking and development, the ecosystem is the principal model of inspiration. In addition, natural ecosystems inspire the circular economy, and the circular economy provides biomimicry with an economic context. However, inspired learning from biological processes should be ecosystemic and global, i.e., it should be done by considering the overall ecological functioning when seeking to emulate the innovations from different species. Further, the global view of biomimicry should ensure renewable energy and effective recycling.

The circular economy has emerged as an essential paradigm in environmental sustainability that challenges the linear model, which creates excessive waste and drives climate change and pollution while resource use is inefficient. The linear model is associated with severe environmental problems, subjecting the biosphere and its resource provider ecosystems to high risk [32]. Driven by the sustainable design requirements, it is possible to list three pillars that describe the circular economy principles: eliminate waste and pollution, circulate products and materials, and regenerate nature [33]. It decouples economic activity from the consumption of finite resources and critical global challenges ranging from climate change, waste, pollution, and biodiversity loss.

The circular economy demonstrates quality economic models inspired by nature to facilitate sustainable business opportunities where cyclical rather than linear processes dominate. It promotes sustainable development goals and intends to create economic and industrial structures that help reduce natural resource usage and waste by reconciling the economy with nature [28]. It is a resilient system that is good for business, people, and the environment. The shift and the transition toward a circular economy indicate the need to introduce new bioinspired ideas, innovative design, and efficient technologies [34].

The circular economy concept aims to achieve the desired principles related to redesigning, repurposing, remanufacturing, reusing, recycling, refurbishing, and recovering water, waste, materials, and nutrients to preserve natural resources. It lays out the necessary conditions that enable economic and social players to develop and adapt strategies to stimulate sustainable actions. However, the increase in complexity of sustainability means that managerial economics and traditional engineering alone cannot face the new challenges and fulfill the requirements toward environmentally efficient solutions [35]. Products and services in a circular economy are designed to be reused in biological or technical cycles to tackle global climate change, biodiversity loss, waste, and pollution. This framework reduces resource input and optimizes environmental, social, material, and economic values while maximizing waste prevention.

Biomimetic Design and Sustainability

The key idea of the biomimicry concept is that nature always functions on a principle of economy and optimal efficiency, minimal waste. Nature represents the best example of sustainability as it provides novel and original solutions with innovative features associated with change and adaptation to its environment and needs. Besides, it eliminates design strategies that cannot withstand changes in their habitats while optimizing design solutions over many revisions to achieve sustained outcomes. Therefore, bioinspiration and sustainability should work together as interdisciplinary approaches for novel effects that develop systems and devices with a zero-waste life cycle. Applying bioinspiration to stimulate innovation and facilitate learning from natural systems to emulate and create a reliable, intelligent design for sustainability with high performance saves energy, reduces materials and cost, redefines and eliminates waste, and stops environmental damages [36].

Therefore, to achieve a sustained design and development quality of innovative products and services, it is necessary to understand the role of the designers in the sustainable world and the necessity within the design process to balance the three key pillars representing social, ecological, and economic factors. This demand for sustainable practices throughout the design process and development to integrate these factors as a green solution into environmentally friendly behaviors while considering natural resources as a part of the design process [37–39].

A bioinspired sustainability framework has been initiated to regain the advantages of societal embeddedness in its environment through biological inspiration. Thus, the biomimicry philosophy is explicitly part of a global strategy of responsible development, concerned with establishing a viable balance between the resources offered by the planet and their exploitation.

Biomimicry: Ecology, Circular Economy, and Sustainability

Biomimetic innovations inspired by nature can motivate sustainable design solutions because biological systems are inherently sustainable [40]. However, the ISO standard formulates top-down criteria to classify a product as a biomimetic. This criterion has three integral components necessary to satisfy a product as biomimetic. These components are: identify the main biological principles as a source triggering function analysis, abstract a model from the biological system of the identified principles, and then transform the model into a practical design for possible implementation [41]. Hence emulating and abstracting the time-tested patterns and ideas from natural systems in the living nature will result in better sustainable features and performance. Thus, to strengthen and widen bioinspired sustainable solution initiatives, there is a need to establish a practical interdisciplinary framework connecting innovation inspired by the natural biological system with the sustainable theory to achieve the best way of realizing sustainability in practice.

It is essential to mention that not all bioinspired designs are biomimicry. However, sustainability is a vital factor differentiating biomimicry from other bioinspired design approaches, as it aims to meet sustainable development challenges in design development. Biomimicry emphasizes learning from and emulating nature forms and living systems with their technical transformations at all levels of abstraction. Understanding living systems' technological transformation may include functions, patterns, processes, architectures, and ecosystems that are necessary to develop regenerative engineering solutions to meet specific performance challenges and solve problems sustainably. Hence, not all bioinspired solutions are sustainable, but biomimicry solutions are. Thus, knowledge of ecosystem principles in ecology, biology, industrial ecology, ecological design, and biomimicry is vital to consider ad use to formulate a group of ecosystem principles to capture an interdisciplinary understanding of how an ecosystem functions. Biomimicry then helps integrate ecological sustainability into the design process. Such a development approach aims to balance the design by economic and environmental factors serving sustainability.

Human activities are becoming unsustainable and cause severe damage, especially in terms of global depletion of resources, destabilization of natural systems, and harm to its harm life on the planet. By taking biological systems as a model, it becomes possible to reconcile industrial activities and economic development while preserving the environment, resources, and biodiversity. The key idea of the bioinspired approach and biomimetic concepts is that nature always functions on a principle of economy and maximizing performance and efficiency without waste. Therefore, the vision of biomimicry is explicitly part of a global strategy of responsible development, concerned with establishing a viable balance between the resources offered by the planet and their exploitation while addressing human life, pollution, and climate problems.

Whether animals, plants, or other biological models, drawing inspiration from the living is a real possibility and is now alive. Biomimicry concerns many sectors of human activity: medicine, research, industry, the economy, architecture, urban planning, agriculture, ecosystem management, development, and so forth.

Biomimetic innovations inspired by nature can motivate sustainable design solutions because biological systems are inherently sustainable [40]. Hence, learning, emulating, and abstracting the time-tested patterns and ideas from natural systems in the living nature will result in better sustainable features and performance. Thus, to strengthen and widen bioinspired sustainable solution initiatives, there is a need to establish a practical interdisciplinary framework connecting innovation inspired by the natural biological system with the sustainable theory to achieve the best way of realizing sustainability in practice. Biomimicry has a vital role in the evolving conceptualization of sustainability and measures of success needed to integrate science-based targets and the interests of diverse stakeholders.

Biomimicry's distinctive feature involves studying and learning from the ecosystems, their structures, biological systems and their interaction among themselves, the interaction with their living environment, and from nature's model to emulate functional strategies from natural forms, processes, and ecosystems tested and refined through biological evolution. Biomimicry can be applied to solve technical and social challenges of any scale. Biomimicry is a principal design approach that helps to:

a. Understand and learn how life in nature works,
b. Integrate this understanding with human experience and knowledge,
c. Emulate selected potential and efficient functions, processes, and systems and stimulate new potential discoveries merging through emulation,
d. Use the outcome of the emulation to develop new and innovative life-friendly engineering science designs contributing to a healthy planet and leading to a sustainable evolution.

Biomimicry should consider allowing for positive contributions of bioinspired technological development. Thus, it is not only considering limiting impacts on waste and usage of resources, but it should have a positive effect by creating a condition for life to exist and advance. When learning from natural and biological systems, it is critical to understand and transfer the most advantageous property of life itself reflected by sustainability. The property of life itself leads to sustainability.

Engineering Design Methodologies

The fundamental principle of a design methodology is to help improve design processes to find the best possible solution for each design scenario. Engineering design methodologies invoke different modes of design thinking (divergence and convergence) and aim to fulfill the engineering design requirements of systems in alignment with the list of desired functions.

A design methodology establishes a set of rules that guide obtaining desired objectives that cannot be generated by chance [41–44]. Hence, the development of design methodologies is essential to help in ensuring control of the design process and opens the way to explore and investigate new design tools. In general, design methodologies are about principles that guide design practices. Therefore, each available design methodology should justify and explain the necessity of finding suitable solutions. Some of the standard and traditional technology design methodologies may include [45, 46]:

1. The top-down-design starts from a solution and works backward while refining the design process at each step;
2. The bottom-up-design starts from the foundation principles and works toward a solution;
3. The structured design identifies and maps available inputs and desired outputs using graphical representation;
4. The structured analysis design describes the hierarchy of the system structures and functions;
5. The data structured system design determines the system structure; and
6. The object-oriented design focuses on a system of interacting objects

A proper understanding of the cognitive abstraction process helps develop an efficient design methodology. Hence, assessing how each design methodology will produce new knowledge is fundamental, and the available tools supporting it. However, modifying or reusing already tested design concepts cannot meet new challenges in designing and solving new engineering and science problems that demand effective and agile solutions. A successful design approach should combine natural creativity with the systematic design methodologies associated with the availability of the necessary design and assessing tools [47].

Bioinspired Deseign Methodologies

The analogy is a powerful design tool and process that helps designers identify and develop creative and practical solutions. Considering analogy in the design processes initiates inferences and helps to understand problems that lead to innovative solutions in the design process. Design is analogous if sharing at least one property/function/behavior from other systems [36, 48, 49]. The bioinspiration design approach is regarded as an example that adopts analogy. However, the interdisciplinary bioinspired approach does not need to be influenced only by the design appearance or style. Design solutions in nature undergo an evolutionary process that makes them inherently superior to human solutions. Therefore, a bioinspired design is approached using different design paradigms [1, 3, 50–55].

a. Develop a design that starts by specifying an engineering problem to find solutions by searching for a suitable biological analogy or a phenomenon, such as organisms or ecosystems, that can address the given engineering problem. Finding such an analogy leads to identifying/abstracting relevant principles from a biological model adopted from nature and then transforming/translating these into scientific design knowledge that can be applied as real-life solutions to engineering design problems. This approach is classified as an indirect design approach, and also it is referred to as a top-down or problem-driven approach, or "*design looking to nature/biology*". Efforts in developing databases with examples from nature are normalizing and standardizing the design process to help solve technical problems.

b. Develop a design technique that needs to understand biological analogies and their applications better. This paradigm starts with an engineering solution and tries to identify examples in nature/biology that help extend opportunities to develop principles/features/functions that can contribute to finding creative or improved solutions to the solution. This approach is a bottom-up, "*biology influencing design*", or solution-driven bioinspired design. However, this approach is less systematic.
c. The sustainability-driven paradigm is classified as biomimicry. This paradigm focuses on the technical transformation of biological forms, processes, patterns, and systems that connect biomimicry to sustainability. Nature always, through evolution, creates functions and resources that are efficient to support ecological solutions aiming to address challenges caused by humans (mainly the environment).
d. The bioreplication design paradigm focuses on realizing or reproducing specific functions or structures of a biological system.
e. The last paradigm represents a hybrid bioinspired design approach that combines two or more paradigms above. It was recognized that designers who work with the bottom-up design approach also employ the techniques extracted from the top-down design approach, i.e., these two design approaches are both meaningful and necessary throughout the design process. These design approaches are not sequential processes but somewhat interrelated and require interdisciplinary collaboration between biology and technology to fulfill functional decomposition and functional optimization design needs. However, this represents a significant need, but it is difficult to achieve this in isolation between the required fields and demand to interface biology, art, engineering, system thinking, and materials.

Need for Effective Bioinspired Design and Tools

In general, to have effective design and tools that help the design processes, products, forms, and functions, there is a critical need to learn how to design and, at the same time, to learn the design process itself. This approach requires an environment that associates learning with real-world practice [56]. The bioinspired design approach represents an invaluable tool that influences modern science and engineering design, fostering innovative and creative ideas, products, and functions. The following are highlights of the main challenges and considerations in the biomimetics field.

1. It is essential to recognize that learning about nature differs from learning from nature. Nevertheless, the needs require to fulfill both.

2. It is crucial to generate good data sets, tools, and guidelines to systemize and understand the bioinspired design process and available approaches. As a result, the entire ecosystems are emulated as a base for the design.
3. It is essential to recognize that nature may evolve different biological mechanisms to solve real-world problems in different environments and ways.
4. Bioinspiration does not mean adopting the weaknesses of biology systems together with the strengths.
5. Bioinspiration education must play a significant role in developing efficient design thinking. Therefore, the bioinspired design should be an essential element in the education syllabus of all engineering disciplines.
6. There is a need to understand the cognitive mechanisms and develop tools and techniques that help to support bioinspiration as innovative design solutions to engineering problems. This direction requires understanding the different operations that constitute the bioinspired design process.
7. It is necessary to understand the different approaches to sustainability solutions present in biology before developing practical tools and bioinspired design methodologies.
8. It is essential to recognize the potential coupling and interaction between biology and information and micro/nano spaces to enable new innovative solutions.
9. It is necessary to have tools that help assess the environmental impacts of concepts during the early phases of product design to select suitable, sustainable concepts or ideas.
10. Bioinspired designs are not all sustainable by default. Thus, it requires considering the entire product life cycle to determine whether a particular bioinspired feature confers a sustainability advantage to the design process and its outcomes.
11. It is necessary to have tools and guides to assess the environmental impacts of concepts during the early phases of design to select adequate and sustainable concepts and ideas efficiently.
12. There is an urgent need to develop sustainable bioinspired design tools and methodologies that help reduce the environmental impact of sustainable engineering designs and have added value features,
13. There is a need for in-depth studies of the theoretical bases of ideation methods and the factors involved in executing bioinspired methodologies for different knowledge domains to support creative and critical design scenarios.
14. There is a need to understand the assembly of biological systems at the components and system levels through their interaction.
15. It is necessary to consider different levels of abstraction appropriate in understanding the making up of each biological system and functional interactions as the abstraction at individual levels may lead to the wrong design if it is taken out without considering all the interactive components of the biological system.
16. Biological adaptation to a given environment depends not only on the circumstances of the environment but also on its entire evolutionary history.

17. Seek the possibility of integrating elements from different biological systems. It is also necessary to combine natural/biological materials with the strength of artificial components.
18. It is necessary to note that nature can still give us confidence in the correctness of a result to enhance the survivability of living functions and systems by reacting to changes in the dynamic environment, such as growth, learning, repair, act, move, and so forth. Therefore, biomimetic products design should reflect such capabilities.
19. It is critical to understand the principles underlying adaptive behaviors of intelligent systems and biological emotions to help develop systems/machines that are intelligent with the capability to interact socially. Hence, this demands definite design principles, paradigms, and theoretical formulations.

Conclusions

Maturing conventional design approaches and methodologies are often associated with inadequate performance due to a range of constraints. However, such difficulties stimulate the demand to find innovative and creative solutions to maximize functionality and enhance performance while minimizing costs in energy and materials.

In general, biologically inspired solutions are cost-efficient, multifunctional, and environmentally friendly. However, using a biological metaphor to inspire new innovative design approaches does not necessarily imply that the bioinspired design approach is well understood. In addition, biological organisms that emerge from natural selection and evolutionary processes are multifunctional, and accordingly, they cannot be optimized independently for any individual function.

It is not always necessary to solve biological problems to provide insight into potential solutions; instead, it can be a valuable and challenging experience to inspire exciting research ideas. In addition, it is not easy finding a natural element to provide the required answer for innovative design.

The available bioinspired processes supporting creative design are not intensively used and verified. Therefore, these processes are exploratory, and they cannot guarantee success. On the other hand, these processes may provide innovative solutions that can obtain original solutions, representing advantages over others such as normative processes.

The experience in developing a real, practical, and effective bioinspired design methodology demands closer interdisciplinary collaboration. Such collaboration should be associated with the necessary development of documentation and databases as a design reference supporting design inspired by nature. This approach will also help normalize and standardize them to generate systematic, efficient, and creative design approaches.

The main challenge is to have a successful connection between the theory of biomimicry and engineering through circular economy, allowing it to apply in everyday life. In addition, the absence of evaluation tools resulted in unforeseen impacts on the bioinspired design strategies and methodologies.

The biomimicry-inspired design will not by default lead to sustainable outcomes. Thus, it is necessary to synergize biomimicry with a good understanding of nature and ecosystems to realize sustainable design outcomes. This approach demands new ways of thinking, establishing efficient practices, and having qualified practitioners with creative visions and innovations. Such synergy is necessary to develop sustainable solutions and innovative EcoMechatronics processes, forms, and ecosystems inspired by nature. Therefore, it is essential to increase interdisciplinary research activities where biologists, engineers, and computer scientists collaborate to develop, assess, and tune methods and tools applied in each bioinspired design paradigm.

Finally, biomimicry needs to have a clear agreed definition with better visibility. However, while bioinspiration can create innovative solutions to many problems, it should have an ethical framework to assure proper usage to help humanity.

References

1. Lenau TA, Metze A-L, Hesselberg T (2018) Paradigms for biologically inspired design. Proceedings SPIE 10593, Bioinspiration, Biomimetics, and Bioreplication VIII, 1059302. https://doi.org/10.1117/12.2296560
2. O'Rourke J, Seepersad CC (2015) Using biology as a model for sustainability: insights for Ecodesign and bioinspired design practitioners. In: Weber C, Husung S, Cantamessa M, Cascini G, Marjanovic D, Venkataraman S (eds), section design theory and research methodology, design processes, ICED, 2015, pp 187–196
3. Salgueredo CF (2013) Modeling biological inspiration for innovative NPD: lessons from Biomimetics. Proceedings of the 20th international product development management conference, Jun 2013, Paris-France, pp.21
4. Speck T, Speck O (2008) Process sequences in biomimetic research. WIT transactions on ecology and the environment, design and nature IV, Brebbia CA (ed), 114, 3–11, section design theory and research methodology, design processes, ICED, 2015, pp 187–196
5. Benyus J (2002) Biomimicry: innovation inspired by nature, harper perennial (September 17, 2002)
6. Habib MK, Nagata F (2019) Bioinspired design: creativity and sustainability. International conference on mechatronics education and research (REM 2019), 23–24 May 2019, Wels, Austria
7. Zari MP (2009) An architectural love of the living: bioinspired design in the pursuit of ecological regeneration and psychological well-being. In: Brebbia CA (ed), Sustainable development and planning IV, Southampton, England: Wessex Institute of Technology Publisher, pp 293–302
8. Tucker VA, Parrott CG (1970) Aerodynamics of gliding flight in a falcon and other birds. J Exp Biol 1970(52):2345–2367
9. Norberg UM, Rayner JM (1987) Ecological morphology and flight in bats (Mammalia; Chiroptera): wing adaptations, flight performance, foraging strategy and echolocation. Philos Trans Royal Soc B Biol Sci 316:335–427
10. Budholiya S, Bhat A, Aravind Raj S, Hameed Sultan MT, Shah Md AU, Basri AA (2021) State of the art review about bio-inspired design and applications: an aerospace perspective appl. Science 11(11):5054. https://doi.org/10.3390/app11115054

11. Vattam SS, Helms M, Goel A (2010) A content account of creative analogies in biologically inspired design. Artificial intelligence for engineering design, analysis and manufacturing 24(4):467–481
12. Gonçalves M, Cardoso C, Badke-Schaub P (2012) Find your inspiration: exploring different levels of abstraction in textual stimuli. Proceeding of the 2nd international conference on design creativity (ICDC2012), Cognition Section, 189–198, Glasgow UK.
13. Fu K, Moreno D, Yang M, Wood KL (2014) Bio-inspired design: an overview investigating open questions from the broader field of design-by-analogy. J Mech Design 136(11)
14. Bar Cohen Y (2003), Biologically inspired intelligent robotics. In: Proceedings of the SPIE smart structures conference, San Diego, CA, Paper 5051–02
15. Yurtkuran S, Kırli G, Taneli Y (2013) Learning from nature: biomimetic design in architectural education. 2nd cyprus international conference on educational research, (CY-ICER 2013). Procedia Soc Behav Sci 89:633–639
16. Neves JCL, Francke ICM (2012) Creative product design using Biomimetics. WIT Trans Ecol Environ 160(2012):149–155
17. Wang J, Chen W, Xiao X, Xu Y, Li C, Jia X, Max Q-H, Meng Q-H (2021) A survey of the development of biomimetic intelligence and robotics. Biomimetics Intell Robot 1.https://doi.org/10.1016/j.birob.2021.100001
18. Habib MK (2011) Biomimetics: innovations and robotics. Int J Mechatr Manuf Syst 4(2):2011
19. Habib MK, Watanabe K, Izumi K (2007a) Biomimetics intelligent robots and biological inspiration. International conference on control, instrumentation and mechatronics, CIM; Malaysia, pp 824–832
20. Habib MK, Watanabe K, Izumi K (2007b) Biomimetics robots: from bio-inspiration to implementation. The 33rd annual conference of the IEEE industrial electronics society IECON'2007, Taipei, Taiwan, pp 143–148
21. Knight A (2009) Hidden histories: the story of sustainable design. ProQuest. Accessed Feb 23 2022. http://www.csa.com/discoveryguides/design/review.pdf
22. Habib MK, Nagata F (2018) Biomimetics and the evolution of robotics and intelligent systems. Handbook of research on Biomimetics and biomedical robotics, Chapter 1, IGI Global, 2018, pp 1–25
23. Lakna (2017) Difference between ecology and ecosystem, June 2017. Accessed Feb 23. https://pediaa.com/difference-between-ecology-and-ecosystem/EcosystemandEcology
24. Pedersen Zari M (2015) (2015) Mimicking ecosystems for bio-inspired intelligent urban built environments. Intell Build Int 8(2):57–77. https://doi.org/10.1080/17508975.2015.1007910
25. Reed B (2007) Shifting from 'Sustainability' to regeneration. Build Res Inf 35(6):674–680
26. Svec P, Berkebile R, Todd JA (2012) Regen: toward a tool for regenerative thinking. Build Res Inf 40(1):81–94. https://doi.org/10.1080/09613218.2012.629112
27. Pedersen Zari M (2010) Biomimetic design for climate change adaptation and mitigation. Archit Sci Rev 53(2):172–183
28. Aarikka-Stenroos L, Ritala P, Thomas LDW (2021) Circular economy ecosystems: a typology, definitions, and implications. In: Teerikangas S, Onkila T, Koistinen K, Makela M (eds) Book: handbook of sustainability, Chapter 17, Agency, Edward Elgar Publishing, Feb. 23rd 2022. https://doi.org/10.4337/9781789906035.00024
29. Sharma PK, Madsen K (2021) Curricular framework for advancing circular economy. Lucart Professional-educational
30. Thomas LDW, Autio E, Gann DM (2014) Architectural leverage: putting platforms in context. Acad Manag Perspect 28(2):198–219
31. Adner R (2017) Ecosystem as structure: an actionable construct for strategy. J Manag 43(1):39–58
32. González de Molina M, Toledo VM (2014) The social metabolism—a socio-ecological theory of historical change. Springer, Berlin, Germany, vol 3. https://doi.org/10.1007/978-3319-06358-4
33. Ellen Macarthur Foundation. What is a Circular Economy? Accessed on Feb. 23rd, 2022. https://ellenmacarthurfoundation.org/topics/circular-economy-introduction/overview

34. Saffré B, Nicolas Buttin N (2019) Activate the circular economy: reconcile economy and nature. Kindle Edition
35. Stefanakis A, Nikolaou I (eds) (2022) Circular economy and sustainability, vol 2. Environmental Engineering. 1st edn, Elsevier, Feb 23rd, 2022
36. Visser W (1996) Two functions of analogical reasoning in design: a cognitive-psychology approach. Des Stud 17(4):417–434
37. Herrmann C, Schmidt C, Kurle D, Blume S, Thiede S (2014) Sustainability in manufacturing and factories of the future. Int J Precis Eng Manuf-Green Tech 1(4):283–292
38. Bonilla S, Silva H, Terra da Silva M, Franco Gonçalves R, Sacomano J (2018) Industry 4.0 and sustainability implications: a scenario-based analysis of the impacts and challenges. Sustainability 10(10):3740
39. Kiel D, Müller JM, Arnold C, Voigt K-I (2017) Sustainable industrial value creation: benefits and challenges of industry. Int J Innov Manag 21(8):1740015
40. Mead T, Jeanrenaud S (2017) The elephant in the room: Biomimetics and sustainability?. Bioinsp Biomimetic Nanobiomater 6(2):113–121
41. Moreno DP, Hernandez A, Yang MC, Otto KN, Holtta K, Linsey JS, Wood KL, Linden A (2014) Fundamental studies in design- by-analogy: a focus on domain-knowledge experts and applications to transactional design problems. Des Stud 35(3):232–272
42. Mirjalili S, Gandomi AH, Mirjalili SZ, Saremi S, Faris H, Mirjalili SM (2017) Salp swarm algorithm: a bio-inspired optimizer for engineering design problems. Adv Eng Softw 114:163–191
43. Coyle S, Majidi C, LeDuc P, Hsia KJ (2018) Bio-inspired soft robotics: material selection, actuation, and design. Extreme Mech Lett
44. Pan H, Jing X, Sun W, Gao H (2018) A bioinspired dynamics-based adaptive tracking control for nonlinear suspension systems. IEEE Trans Control Syst Technol 26(3):903–914
45. Pahl G, Beitz W, Feldhusen J, Groe H-K (2007) Engineering design, 3rd edn. Springer-Verlag, London, p 2007
46. Hubka V (1982) Principles of engineering design, Zürich, Heurista
47. Otto KN, Wood KL (2008) Product design: techniques in reverse engineering and new product development. Prentice-Hall, Upper Saddle River, N.J.
48. Qian L, Gero J (2009) Function behavior structure paths and their role in analogy-based design. AI EDAM, 289–312
49. Linsey J, Laux J, Clauss E, Wood K, Markman A (2007) Effects of analogous product representation on design-by-analogy. International conference on engineering design, ICED'07. Paris, France
50. El- RMA (2012) Biomimicry as a problem solving methodology in interior architecture. Procedia Soc Behav Sci 50:502–512
51. Helms M, Swaroop SV, Geol AK (2009) Biologically inspired design: process and product. Elsevier 606–622:2009
52. Helms M, Vattam SS, Goel AK (2009) Biologically inspired design: process and products. Design Stud 30:606–622
53. Roshko T (2010) The pedagogy of Bbio-design: methodology development. WIT Trans Ecol Environ 38. 2010 WIT Press
54. Hatchuel A, Salgueiredo CF (2018) Beyond analogy: a model of Bioinspiration for creative design. AI EDAM, Cambridge University Press (CUP), 30, pp159–170
55. Vincent JFV, Bogatyreva OA, Pahl A-K, Bogatyrev NR, Bowyer A (2005) Putting biology into TRIZ: a database of biological effects. Creativity Innov Manage 14:66–72
56. Schön DA (1988) Toward a marriage of artistry and applied science. The architectural design studio. J Architect Educ 41(4):4–10

A Holistic and Sustainable View on the Product Creation Process for Mechatronic Systems

Peter Hehenberger

Abstract Nowadays sustainability is an important factor in the design and production of mechatronic systems, especially the environmental aspects which will be mainly defined in the development phase and influence the whole later life cycle. This chapter gives an overview of the main approaches like life cycle assessment (LCA) or related key performance indicators (KPIs) and considers their relevance to mechatronic design. Therefore, some methods for the early design phases are presented based on the well-known approaches V-model and design structure matrix. The summary will be reflected with two significant use cases.

Introduction

In recent years, sustainability has obtained more importance in consideration of the life cycle phases of mechatronic systems. The design/engineering phase is very significant here as the main properties will be fixed and have a major influence on production, usage and recycling.

Mechatronic systems are typically composed of multiple sub-systems, modules and components which may include electrical, hydraulic and pneumatic drive systems, automation and automated elements deploying a wide range and types of sensors and actuators along with the necessary ancillary and structural components. Crucial for the success of such a product from a sustainability point of view is the holistic consideration of environmental behaviour of the integrated whole, as sustainable-specific requirements essentially always relate to the whole system.

As state of the art, a lot of methods exist for modelling and analysing these aspects in several domains (e.g. life cycle assessment (LCA), "Materials, Energy, Chemicals and Others" MECO-matrix, see [1]). The motivation for this chapter is to present a methodological approach for the specific usage of analysing mechatronic systems.

Over the last few decades, work in product development has changed dramatically. The demands of the market for shorter product life cycles (and the resulting shortening

P. Hehenberger (✉)
University of Applied Sciences Upper Austria, Wels, Austria
e-mail: peter.hehenberger@fh-wels.at

P. Hehenberger et al. (eds.), *EcoMechatronics*,
https://doi.org/10.1007/978-3-031-07555-1_5

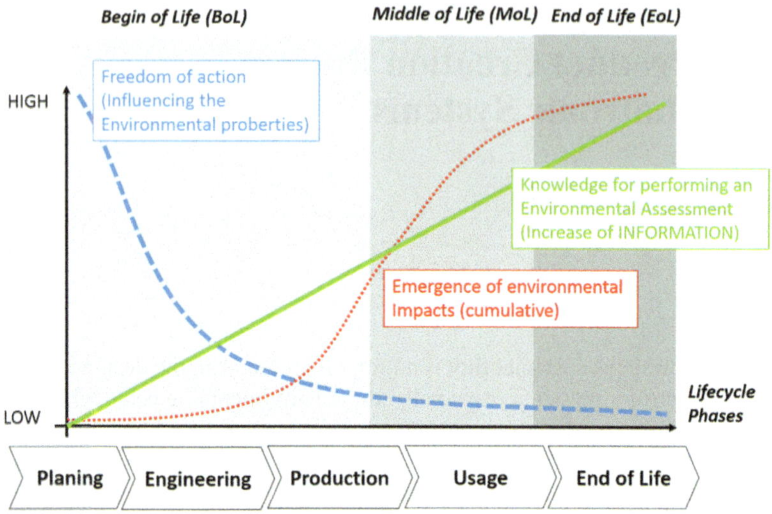

Fig. 1 Life cycle phases and their importance

of their product development times) pose ever new challenges to the product development process and for the improvement of the manufacturing processes of these products. At the same time, the ever-cheaper hardware and, in contrast, improvements in processor performance and memory capacities are making the computer an increasingly important companion in all phases of the design process.

In the development of new products, the majority of product costs and environmental aspects (see [2]) are determined in the planning and engineering phase. In contrast, most of the impacts are felt in the production, usage and end-of-life (see Fig. 1). The reduction of resource usage in these phases is achieved on the one hand by new technologies, selected materials, and energy flexibility of production, while on the other hand, there is potential for savings through the use of the most specific "Eco-design"-methods or "Design for Sustainability" [3, 4].

Background

Initially, consideration of the different terms and their relation to mechatronic product creation processes is presented.

Sustainability with specific focus on ecological aspects

The development of sustainable products is thus a central prerequisite for maintaining and increasing the competitiveness of manufacturing companies. Originally ecologically motivated, common definitions of sustainability placed ecological, economic

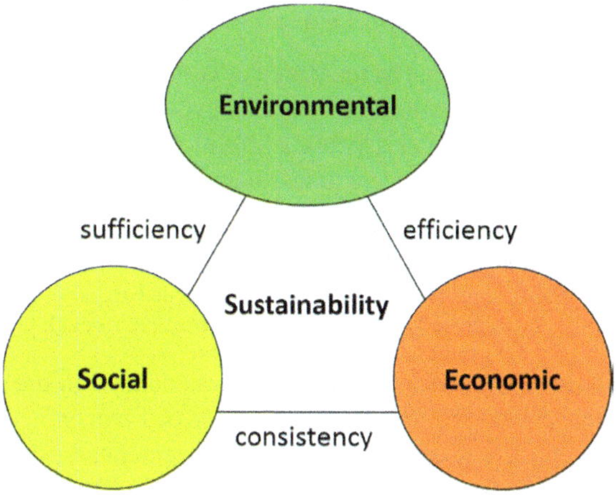

Fig. 2 Sustainability

and social aspects on an equal footing. This simple basic model is also called the "Magic Triangle of Sustainability" or the "Three Pillars Model" (see Fig. 2, [5, 6]).

- **Economic**: The economic dimension addresses the need for constant competitiveness, sustainable product development and, in particular, the need for companies to remain competitive.
- **Social**: The social dimension of sustainability takes into account aspects such as social justice, equal distribution of resources, health, qualification, education, human lifestyles and social resources.
- **Environmental (Ecological)**: The way of doing business has a significant impact on sustainable development.

Sustainable value creation must be ensured in terms of production, services and trade, for example through the implementation of a circular economy or an environmental management system. To ensure sustainability, the inclusion of various strategies is necessary [7].

- One of these is increasing efficiency, which aims to reduce material and energy use while maintaining the quality and quantity of products. This can be achieved through efficient processes, efficient products (e.g. low consumption), increasing the lifespan of products, their repairability and through optimised recycling.
- The sufficiency strategy targets human behaviour. The aim is to shape human behaviour in such a way that it is less harmful to the environment.
- The consistency strategy, on the other hand, aims to adapt the existing energy and material flows to the ecological system and thus avoid ecological damage in principle. This can be achieved, for example, through the use of renewable resources.

Life cycle assessment LCA as a selected method

Sustainability can be measured or verified by means of life cycle analysis, also called life cycle assessment (LCA). Life cycle assessment is thus a procedure for analysing the environmental aspects and effects of product systems. According to ISO EN 14,040 [8], a life cycle assessment can be defined as follows:

> LCA is a method for estimating the environmental aspects and product-specific potential environmental impacts associated with a product. The LCA study examines the environmental aspects and potential environmental impacts over the life cycle of a product (i.e. cradle to grave) from raw material extraction, through production, use and disposal.

In 1997, a structure for LCA was defined in the above ISO standard. The structure developed by SETAC (Society of Environmental Toxicology and Chemistry) was largely retained. Only the improvement assessment component was replaced by interpretation. The LCA according to ISO EN 14,040 [8] is divided into four phases, which are shown in Fig. 3.

Goal and scope definition

The first step in preparing a life cycle assessment is the definition of the objectives and the scope of the study. The definition of the objectives provides the basis for the scope of the study and is therefore crucial for the interpretation of the results. The depth and accuracy of the study are determined on the basis of the required study objectives.

In the course of defining the scope of the study, the following topics are described in more detail:

- The product system
- Technical system boundary

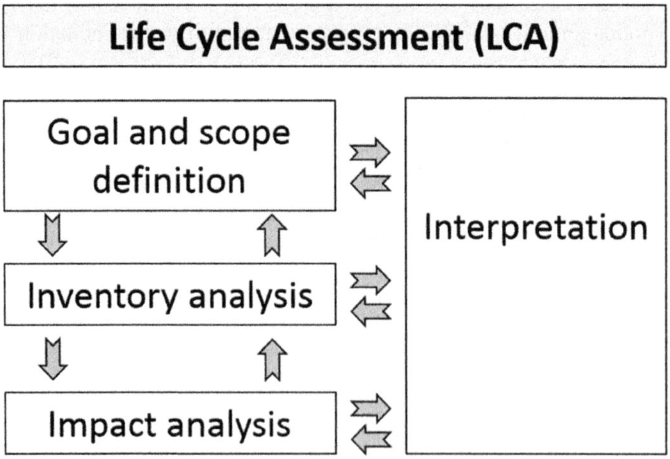

Fig. 3 LCA according to ISO EN 14,040 [8]

- Geographical system boundary
- Temporal system boundary
- The functional unit
- Data availability and depth of the study

Inventory Analysis

The next step is to draw up the life cycle inventory. This contains information on all physical input variables such as resources, materials or semifinished products and output variables such as emissions, waste or valuable products that occur in the product life cycle. The basis for many of these parameters is provided by databases that are either publicly accessible or generated by companies from historical values.

Starting with data collection, the following steps are important:

- **Flow diagram and reference quantities**:

  Building on the collected data from the phase of defining the objectives and the scope of the study, the product tree consisting of process modules is now elaborated in detail. The flow diagram consists of individual process modules, which represent the smallest unit in a life cycle inventory. A first pre-selection of which process modules are not considered was already carried out in the previous phase. In the next step, the individual process modules are broken down in more detail, whereby a distinction is made here between two variants (Multi-input processes or Multi-output processes). In a life cycle assessment, the data is related to the output of a process module, which means that the original data usually has to be converted into other units in order to create a uniform basis. For material products, the reference unit is very often the mass of the final product
- **Data collection and data quality**:

  Before starting the investigations and preparing the life cycle inventory, the impact categories and the requirements for data and data quality must be defined and selected. If, for example, only the impact area of climate change is to be investigated in the next step, the impact assessment, the life cycle inventory can be restricted to substances that have an influence on climate change. The necessary calculations of substance flows from the environment into the system under consideration and vice versa can be carried out on the basis of the data collected.

Impact Analysis

The impact assessment is necessary to describe the effect of the quantitatively recorded material flows on the environment. This makes it possible to assess all environmental impacts and to compare different impacts on the environment. The impact assessment consists of three main steps:

1. **Selection of impact categories**:

   The selection of impact categories has already been made in the previous steps, but should be reviewed again at this point and adjusted if necessary. Subsequently, the characterisation model and the impact indicator must be defined for each impact category. The cause–effect chain of the elementary flows determined in the life cycle inventory is described by the characterisation model. The appropriate impact indicator is then selected within this model. The impact indicator is environmentally relevant and represents the effects of inputs and outputs of the system on the impact endpoint.

2. **Classification**:

   In classification, an impact category is assigned to each elementary flow as determined in the life cycle inventory. Since several flows can contribute to one impact category, or vice versa, parallel as well as serial mechanisms must be considered.

3. **Characterisation**:

   In this step, the impact indicator values of the individual elementary flows assigned to a category are calculated by multiplying the result of the life cycle inventory by the characterisation factor of the substance. To obtain the result for the impact category, all elementary fluxes assigned to an impact category must first be converted into a common unit and then added together.

Interpretation

In the final step of a life cycle assessment, the information and findings obtained are summarised and interpreted. This should enable comprehensible statements to be made for the target group and provide support for decision-making. The evaluation is divided into three main sections:

1. **Identification of significant parameters**: The aim is to identify those outcome parameters for which there is a quantitative difference. This prevents misinterpretation.
2. **Assessment**: The assessment phase should serve to check one's own work. It should include a completeness, sensitivity and consistency check to support clear and understandable communication of the results.
3. **Conclusions, limitations and recommendations**: In the final part of the evaluation, conclusions are drawn, limitations defined and recommendations formulated for the target group addressed.

Key performance indicators for describing the environmental aspects

There exist many key performance indicators (KPIs) for environmental aspects. The most relevant mechatronic systems are presented here [1, 3, 9, 10]:

- **Carbon footprint**:

  The carbon footprint is probably the best-known method of life cycle assessment today and is used in numerous studies. Starting with simple products of daily life such as milk or bread, there are various online tools to calculate a personal carbon footprint. These are used as marketing tools for comparative studies of competing companies or to provide the wider population with a relatively simple and comprehensible proof as of how wastefully resources are sometimes used. But what exactly is the carbon footprint? Since there is a multitude of definitions, a definition that is valid for this work will be introduced here: *"The carbon footprint is a measure of the exclusive total amount of carbon dioxide emissions that is directly and indirectly caused by an activity or is accumulated over the life stages of a product"* [9, 10].

- **Material flow analysis**:

  Material flow analysis is understood as the input- and output-side balancing of ecologically relevant materials and energy flows. The term substance does not distinguish between chemical elements or compounds and more complex objects such as raw materials, semi-finished products or waste.
- **Global warming potential (greenhouse gas potential)**:

  According to the German Federal Ministry of Transport, Building and Urban Affairs, global warming potential (GWP) can be defined as follows: *"The global warming potential (GWP) is the potential contribution of a substance to the warming of the air layers near the ground, i.e. to the so-called greenhouse effect."* The ratio of the contribution of a substance relative to the global warming potential of the substance carbon dioxide is defined as the GWP value. The lower this GWP value, the lower the potential effect on global warming.
- **Ecological scarcity method**:

  The ecological scarcity method can be used in the course of preparing a life cycle assessment. Here, the weighting of environmental impacts of products or entire systems is examined. The result of the life cycle inventory is needed to carry it out. The result is the weighting of the data from the life cycle inventory, which has so-called environmental impact points as a unit. To calculate the desired eco-points, the life cycle inventory data are multiplied by the respective eco-factor. By adding up the individual eco-points, the total number of points is obtained and thus a measure of the environmental impact. This type of assessment can be used as a basis for decision-making.

The Linkage between Mechatronics, Cyber-Physical Systems (CPS), and Product-Service-Systems (PSS)

The word mechatronics was born over 40 years ago as an artificial word combining mechanics and electronics in Japan. With the advent of microelectronics and later, in particular, microprocessor technology, a further component of mechatronics was added in the form of information technology. The VDI guideline VDI2206 [11] describes mechatronics:

> ... as the synergetic interaction of the disciplines of mechanical engineering, electrical engineering and information technology in the design and manufacture of industrial products and in process design.

For many years now, information and communication technology has been the absolute central innovation driver (and unfortunately also cost driver) in originally mechanically dominated disciplines, such as the automotive industry. According to a recent study, the share of software development in the automotive sector will rise to 40% of total development costs in the next few years.

The latest development around mechatronic systems is that of intelligent systems, including so-called cyber-physical systems (CPS). Cyber-physical systems are systems that incorporate inherent intelligence and networking [12]. The term Internet-of-Things (IoT) should also be seen in this context; intelligence is here an important aspect in order to allow the networked systems to cooperate in a goal-oriented manner [13]. The development of CPS and software-intensive mechatronic systems is very complex compared to earlier, purely mechanical systems. Systems engineering is a possibility with an interdisciplinary approach to develop complex products [14]. The overall system is broken down into several smaller sub-systems, which can then be considered independent of each other. Systems engineering not only considers purely technical aspects but also those of project management.

Product-service systems (PSS) are understood as service offerings with significant shares of both tangible goods and services (hybrid service bundles). The services are not only provided at the beginning or on a few occasions (e.g. maintenance intervals) but represent an integral part of the PSS. They can, for example, be provided alongside the ordering, procurement, use and billing of the range of services. Examples of PSS in the area of consumer applications include rental bicycles and car-sharing systems that are booked, organized and billed via the internet.

In the area of industrial products, where the buyer or user is an industrial customer, there are now also numerous models for PSS. These include, for example, the production and drilling of boreholes according to their number and size as well as according to the material in which the holes are drilled. Also, for some users, it is no longer possible to buy aircraft engines, but only to lease them against billing for miles and time.

Customers benefit from the reduced capital investment and suppliers can use these business models to bind their customers more closely to them and gain important information from the use of their products.

- **Pure product**:

  Here, the product is owned by the customer and the customer is independently responsible for the results or repairs. For example, the purchase of a washing machine without additional services.
- **Pure service**:

  In this example, it is a pure service, such as a launderette or car rental. Here, the customer does not own the product, but benefits exclusively from the outcome of the service offered.

  Three types of PSS are as follows [15]:

- **Product-oriented PSS**:

  After selling the product to the customer, services such as warranty and maintenance are provided [15]. For example, the sale of a TV or a smartphone to an end customer. In the case of a product-oriented PSS, the classic sale of products is at the centre of the business model.

- **Use-oriented PSS**:

  Customer uses the product from the service provider but cannot own it [15]. For example, the taxi service and hotel accommodation. In the case of a use-oriented PSS, the product remains the property of the provider, and the use of the product is the key to the business model.
- **Result-oriented PSS**:

  Customer expects the outcome instead of the usage of the product retained by the company [15]. For example, one can ask for food from the catering company instead of renting utensils from a catering service and cooking food of their own. In the case of an outcome-oriented PSS, an agreement is reached between the customer and the provider on the goal to be achieved. The choice of product and approach is up to the provider, the results count and are rewarded.

  The influence on sustainability from PSS is presented by [16].

Viewpoints on Methodological Approaches for Mechatronics

This section presents some viewpoints for systematic processes to include environmental aspects in the design and production of mechatronic systems.

Life cycle phases of products and their production systems

For a holistic view of the sustainable properties of mechatronic systems, it is necessary to consider all life cycle phases and also the interaction between the product and the production system. The phases established are beginning-of-life (BOL), middle-of-life (MOL) and end-of-life (EOL). BOL phase contains the initial product design, its development, testing, production and initial marketing. The MOL stage consists of external logistics, use and support. The final phase is EOL, where the retired products are re-collected and re-manufactured for recovery [17].

From an engineering viewpoint, product lifecycle management (PLM) stages are directly related to the development, production and distribution of the product. Figure 4 shows a graphic representation of the PLM stages and processes from the engineering viewpoint. The five phases of the product life cycle can be described in detail as follows:

- **Planing**: The ideation phase happens when the product is just a concept in a person's mind. Aspects such as functions and requirements might be addressed.
- **Engineering**: Followed by the definition phase, where the idea is portrayed in detail.
- **Production**: Then there is the realization phase, where the product is built.
- **Usage**: Afterwards follows the use/support phase, where the customer uses the product. This stage is a rich source of information about the product. During this

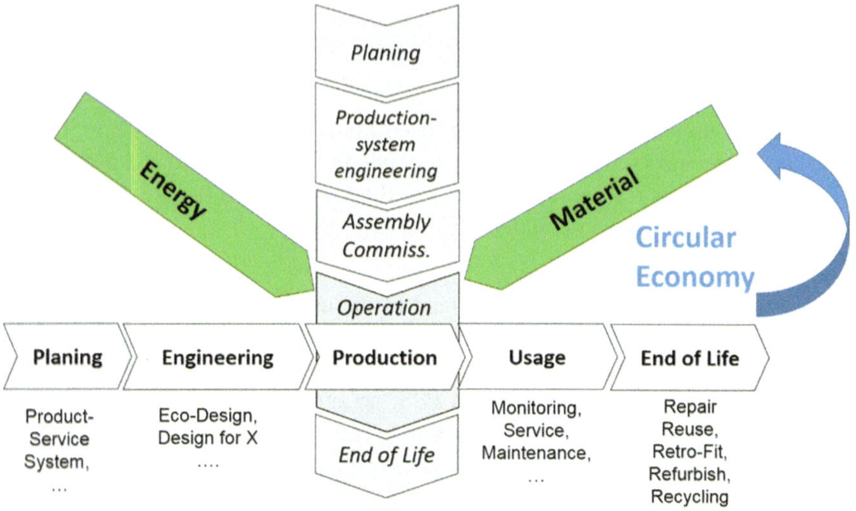

Fig. 4 Product—production system life cycle

phase, the performance of the product is observed. Based on that, it is decided if changes in the design have to be done.

- **End of Life**: The final phase is disposal/recycling/retirement. In this phase, the customer disposes of the product. Depending on the material and the manufacturing process, the product is recycled or deposited in a landfill.

The production system has the same phases and interacts with the other PLMs. In the production phase the resources are very relevant and can be classified as:

- **Energy**: This is needed for the machines operating the production process and also for the factory [18, 19].
- **Material**: This begins with the extraction of raw materials and also the recycling of materials is important, see Circular Economy [20].

Eco conceptual mechatronic design—modularisation

It is proven that there is a permanent change in the product development process. This transformation is influenced by changing market conditions and new product requirements. Product complexity increased due to a much greater product variety suitable for several markets and the constantly increasing availability and capability of electrical components and the associated embedded software (mechatronics) in a product [12].

The product development process is usually carried out as a project. The workflow consists of activities that build on each other according to a model-like, theoretical logic. The model is based on the procedure for solving complex problems and is therefore generally valid for product development. The concrete workflow, meaning the structural and procedural integration of these activities in a company, must, on the other hand, always be adapted to the specific development context which forms the basis for project management. The detailed planning of product development processes is complicated by iterations and the co-evolution of the problem and the solution associated with problem-solving. Competing objectives demand compromises and the open solution space, which allows a multitude of possible solutions, and recurrently requires a systematic evaluation of alternatives. These framework conditions lead to typical requirements for the design of product development processes. It needs to be possible that the initial objective can be readjusted regularly during development in accordance with the progress of knowledge. Iterations in the process must allow new findings to be taken into account. In order to keep iteration loops as short as possible, verification and validation of the fulfilment of requirements already need to be performed for partial results during development [21]. In a sustainable view, the concept decision (with analysis of the interaction between the involved disciplines) is an important stage before the structure will be fixed.

Modularity is an absolutely key issue in achieving sustainable products—the product architecture has a major influence on the life cycle of a product. Thus, modularity allows for achieving sustainability benefits in every life cycle [22].

- **Green modularisation methods**: Refers to methods for modularising products (methods for developing product architectures), taking into account the entire life cycle. Most scientific publications refer to this overarching aspect.
- **Design for X**: Design for X (DfX) methods are well-known and established methods for product development. An extension with regard to sustainability and environmental protection has been the subject of numerous scientific publications, especially in recent years. Design for end-of-life and design for disassembly contain guidelines for a broad application during product development. The aforementioned DfX methods often make use of the properties of modularity [23].
- **Manufacture**: The production of highly engineered goods is the pillar of prosperity in industrialised countries and modularisation offers a way to make production efficient. This can be done by optimising the utilisation of production lines, increasing the efficiency of work organisation, improving quality and reducing lead times. All these aspects can contribute to improving the eco-balance of a production.
- **Supply chains**: The supply chain of production is another lever for improving sustainability. In many industries, a long supply chain is necessary to produce the products. If the supply chain is converted to a well-functioning just-in-time procurement, for example, and the storage quantity is thus kept small, expensive and resource-consuming, intermediate storage can be dispensed with. Here, too, modularity helps to optimise the process.
- **Usage**: Refers to the phase in which the product is used. Modularity can offer several advantages in this context. Technically obsolete modules can be replaced, and maintenance and repairs can be carried out more easily (e.g. by exchanging modules).

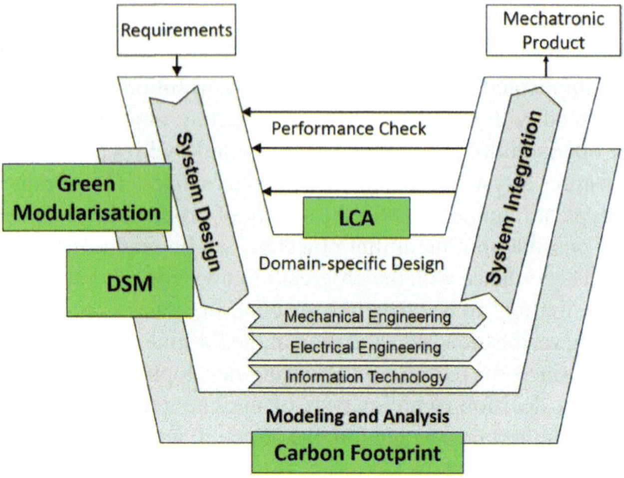

Fig. 5 EcoMechatronics V-Model

EcoMechatronics and V-Model

The design process that follows a V-model is one of the methodologies available in order to design a mechatronic system [11]. It is more popular nowadays and widely used in a lot of different processes. This procedure is also known as the integration of engineering across traditional boundaries. The design of a mechatronic system is divided into a hierarchical sequence. In the following, it is possible to see an example of a V-model, where the main mechatronic components have been integrated into the system. These components are the mechanical system, the electronic system and the information technology. Figure 5 is based on the VDI guideline 2206 [11], which uses the V-model as the macro level and shows in addition integration of modelling and analysing of environmental aspects at different stages.

As it has been explained in the previous section regarding the phases of the mechatronic system design, the particular sub-systems which form the overall mechatronic structure are decided after evaluation of the product specifications. The sub-systems are developed simultaneously, and afterwards, they are tested. The next step is to integrate them after which the execution of the combined system is analysed. If it is necessary to correct something, all the phases have to be completed again.

With so many variables it is easy to make an error in the process, especially considering that some of the variables are time-dependent. On each level, the simulation and modelling have to be done separately. Once the system model has been efficiently designed, the engineer or group of engineers who are in charge have to assemble the model to run the simulation. The less critical functions available, which do not have as much priority compared with other ones, are changed or not taken into account.

Figure 5 presents the extension of the well-known V-model with specific tasks related to the consideration of the ecological properties of the product. This leads to an EcoMechatronics design procedure. In the system design phase, the relevance of environmental aspects will be shown by using green modularisation methods (integrating the DSM). If some design parameters are already fixed, the analysis with LCA-based approaches can start.

Applications

In this section, two applications will be presented. One is related to the engineering phase and the other discusses a production-specific aspect.

Approach for evaluation in design phases

The identification of the main influencing components in a product system can be carried out using the following method framework. With little available LCA data, this allows the identification of the component that is likely to have the greatest impact on the LCA already in the product development phase. Once the component has been identified, this framework can be applied to it. The design structure matrix (DSM) is therefore often used as a tool [24, 25]. A DSM is a highly flexible network model method with many applications in engineering management and other fields. A DSM is also a tool used to describe the components comprising a system and their interactions, thus emphasising the architecture of the system. The architecture of a system is defined as the structure of a system comprising components, the relationships between them, and the design and development guidelines, altogether resulting in the functions and behaviours of the system. DSM is especially suitable for applications in the development of sophisticated and engineered systems and applications include health care management, public policy, natural sciences, social systems and financial systems [26]. A square N $\times$ N matrix is used to represent a DSM, mapping the interactions between the set of N system components. The rows and columns are labelled and ordered equally. The off-diagonal components show linkages between the ondiagonal components. DSM describes different architectures.

LCA-DSM Framework

The framework consists of the following steps (Fig. 6):

1. **System description**: The first step is to describe the product system in more detail. The design or structure and function are explained and the tasks of the individual components are explained. The description of the system from a scientific point of view includes naming the individual material and energy flows as well as defining the boundaries to the environment.
2. **List of components**: In the second step, the system is broken down into its individual components and presented in a list. Here, attention must be paid to the correct degree of decomposition. If the system is broken down too deeply down to the last screw, the view of the essential components will be lost. A too

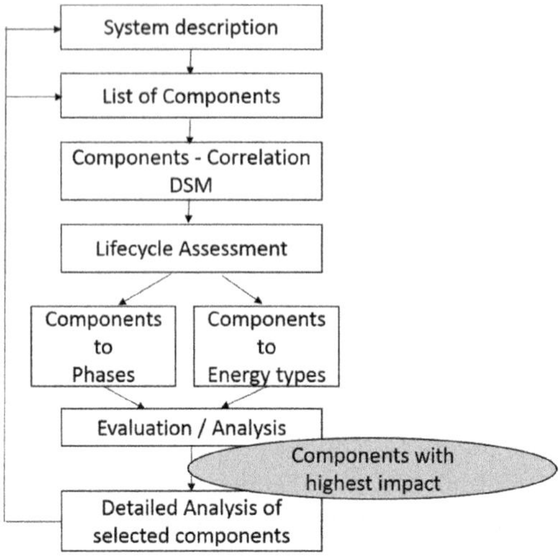

Fig. 6 LCA-DSM framework

coarse subdivision will not give clear information about the influences of the mentioned components in the further steps.

3. **Component correlations DSM**: Depending on the type of system to be considered, the mutual dependencies of the individual components are now presented in a design structure matrix. It makes sense not only to create a DSM for the component geometry, for example, but also to represent other dependencies.
4. **LCA**: In the LCA determination, all LCA data already collected or available are summarised. This includes the breakdown of the life cycle into the individual life phases such as production and distribution, use phase, maintenance if required and the phase of disposal or recycling. For the further steps, it would be ideal if concrete data on the energy consumption in the individual life phases of the product were available.
5. **Components to phases**: The assessment of the influence of the components over the life phases is carried out here. An assessment system suitable for the product system must be determined beforehand. The fewer data is available or the more unclear the influence of the components is, the smaller the subdivision of the evaluation scale is chosen. After analysing or assessing the individual influencing components, the level of influence is to be entered in a matrix-based representation. By simply adding up the assessment, an initial evaluation can be made of the phases.
6. **Components of energy types**: In this step, the individual forms of energy supply for the components in the individual life phases are considered. The basis for this is knowledge of the electricity mix, i.e. the division into non-renewable—fossil or nuclear and renewable—biomass, wind, solar, geothermal or hydro. Subsequently, in a matrix-based representation, the assessment carried out from the components to phases step can be multiplied by the percentage share of the individual energy forms.

7. **Evaluation/analysis**: In the evaluation/analysis, first a weighted evaluation of the evaluation steps components over phases and components over energy forms is carried out with factors from common evaluation methods. Possible evaluation methods are the cumulative energy input, Ecoindicator 99 or the ecological scarcity. The individual weighted evaluations are then analysed and the components with the greatest influence on the product life cycle assessment are determined.
8. **Detailed analysis of the components**: After identifying the main influencing components, this methodological framework can be applied to them in order to identify further optimisation potential.

Case study refrigerator: Considerations and results

In this section, the methods' framework will be presented by means of a case study. The aim is to identify the components that have the greatest influence on the energy balance in terms of production/distribution, operation, maintenance and disposal.

The principle of refrigeration is based on the physical principle that liquids that evaporate extract heat from their surroundings. If a liquid evaporates, its temperature remains the same, but it lowers the ambient temperature to liquid temperature. On the other hand, when the vapour is compressed, the ambient temperature increases and liquefaction of the vapour occurs. Refrigerators combine these two processes and carry them out with the so-called coolant in a closed circuit. Figure 7 shows the main components and their geometrical and cooling flow interactions.

The individual life phases of a refrigerator, namely production and distribution, operation, maintenance and disposal, were each assessed using assessment methods (e.g. Cumulative Energy Expenditure, Eco-Indicator 99 and Ecological Scarcity) over the individual life phases. A high significance has the production and operation with more than 90% share. The environmental impact of the components for maintenance and disposal can be neglected, as this type of equipment requires very little maintenance and most components are recycled and thus follow the cut-off approach.

G Geometry
R Refrigeration flow

	1	2	3	4	5	6	7	8	9	10	11	12	Production and Distribution	Operation	Simplified Evaluation
1 Enclosure		G	G	G		G	G	G	G	G	G	G	1	0	1
2 Door	G		G										1	0	1
3 Insulation	G	G											1	0	1
4 Compressor	G		G		G,R	G,R		R	G,R	G			2	2	**4**
5 Refrigerant				G,R		G, R	G	G,R	G,R	G			1	0	1
6 Condenser	G			G,R	G,R		G	R		G			1	0	1
7 Dryer	G				G	G		G		G			1	0	1
8 Capillaries	G			R	G,R	R	G		G,R	G			0	0	0
9 Evaporator	G			G,R	G,R			G,R		G			1	0	1
10 Piping	G			G	G	G	G	G	G				0	0	0
11 Thermostat	G												0	0	0
12 Electronics	G												1	1	**2**

Fig. 7 Main components of a refrigerator and presentation over phases

In this section, the design structure matrices are presented and evaluated across the phases. The influence of the components on the LCA in the individual phases is evaluated with a scale from 0 to 2 as shown in Fig. 7.

- 0: The component has no or negligible influence on the LCA.
- 1: The component has an influence on the environmental balance, but is not significantly responsible for a negative balance.
- 2: The component has a major influence on the life cycle assessment.

In the production and distribution phase, the compressor has the greatest influence on the life cycle assessment due to the high proportion of steel and copper and its complex design compared to the other components. The capillary, the piping and the thermostat, on the other hand, are negligible in terms of component size and the amount of materials used and are therefore rated 0. The remaining components have only a moderate impact on the product life cycle assessment and were given a rating of 1.

During operation, with the exception of the compressor and the electronics, no components have a negative impact on the life cycle assessment. Due to the conversion of electricity into cooling, the compressor represents the largest environmental impact variable in this phase and is therefore rated 2.

Here the DSM is evaluated in terms of the influences of the respective components in the individual phases. First, the evaluation is carried out using a simple summation and then a weighted evaluation is carried out using the cumulative energy input, the ecological scarcity method and the Eco-Indicator 99. The simple evaluation shows that the compressor has the greatest influence on the environmental impact and that the influence of all other components can be neglected. Furthermore, it can be seen that the production and distribution phase has a significant influence on the overall system.

Energy-flexible production systems

In the production processes of mechatronic systems, the usage of energy plays an important role (e.g. raw material and manufacturing processes for mechanical components, production of electronic components). Nevertheless, the relevance of supply chains and logistics also have to be considered due to their worldwide network, which is necessary for establishing the high-efficient production of mechatronic systems. These industrial production processes consume a large amount of resources, especially in the operation phase of machine tools. On the other side, renewable energy's share is increasing. In factories of the future, it is now relevant to have a holistic view of energy, which includes energy consumption, energy costs and the carbon footprint of the energy type used.

Currently, approaches to energy efficiency can be found in a number of research papers and projects. Since a large proportion of the energy consumed is used by manufacturing companies and electrical energy, these are often the target of an investigation. Often the approaches to energy saving can only be applied very specifically. Energy flexible measures are a synergetic measure for energy efficiency. Here, the focus is not on saving resources in terms of quantity, but on minimising associated target variables, such as costs or emissions. In order to be able to use flexibility, volatility of the target values is fundamental (see VDI5207 [27]). In order to use energetic flexibility, energetically relevant target values must be used. In production systems, a number of different energy sources are used whose target values are volatile in different ways. The volatility takes into account not only the period of change but also the relative change in value. With regard to the types of energy used industrially, electrical energy, in particular, stands out in terms of volatility. The volatility of electrical energy is influenced in particular by the day-ahead tradability on energy exchanges, as well as the varying availability of electricity from sustainable energy production.

Specifically, this includes the energy price on the wholesale market and the CO_2-equivalent emissions released. The CO_2 equivalent emissions are an indicator that takes into account a whole range of greenhouse gases and converts them to the global warming potential of CO_2. When calculating the CO_2 equivalent emissions for electrical energy, the type of energy production, in particular, is taken into account. While sustainably produced energy from wind power or photovoltaics, for example, has a low inherent CO_2 equivalent, larger amounts are released when energy is produced from fossil fuels. Due to fluctuations in demand, solar irradiation and wind, the composition of the electrical energy fluctuates. Furthermore, there is a continuous import and export across the defined control zones of the electrical energy grid, which creates additional volatility.

In order to account for the fluctuations that occur, there are a number of possibilities, which were defined in Fig. 8 by Grassl [28]. However, these differ in the degree of the operational or strategic time horizon. At the production process level, for example, the production parameters can be adjusted. This does not necessarily entail a change in the total manufacturing energy required, but may only change the manufacturing time. At the machine level, for example, energy can be stored. In production systems, storage can be used in particular by AGVs. Other applications

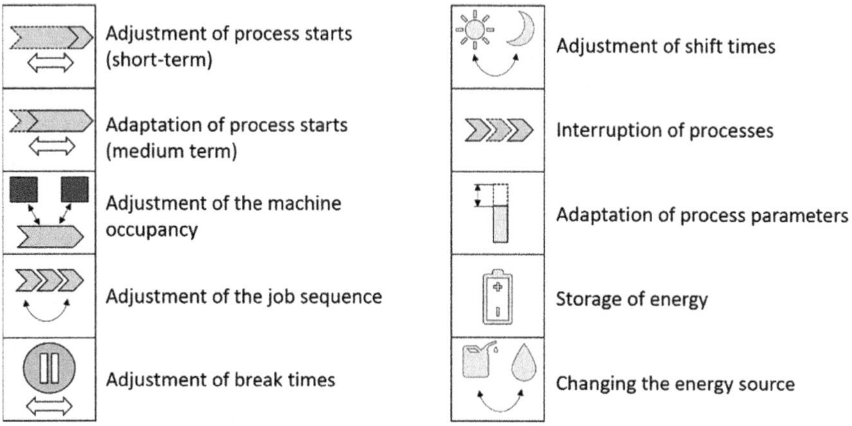

Fig. 8 Possibilities for managing energy-flexibility in production [28]

are based on the concepts of routing and scheduling, respectively, adjustment of the machine occupancy and adjustment of the job sequence. However, these take place at the level of production planning and control. Well-founded market data is only available one day in advance on the day-ahead market. For energy-flexible production, these restrictions must therefore also be taken into account.

Conclusions

There are many future research activities necessary to reach a holistic view. The main influence has the engineering phase, where different methods like LCA are used. In the production phase, it is the way from a lean factory over a lean digital factory to a green lean digital factory. The digitalisation helps to establish more transparency about energy and resource consumption during production. This goes in the strategic direction of a resource-efficient twin.

Acknowledgements The presented considerations and results were conducted within the research projects of the research group "SMART MECHATRONICS ENGINEERING" at University of Applied Sciences Upper Austria.

The author thanks all partners for valuable discussions within the H2020-project "EnerMan Energy-efficient manufacturing system Management" (grant agreement No 958478) and the Project "XLCA CO2 Life Cycle Analysis in early stage design phase" as part of the COMET-Project "Research Center for Low Carbon Special Powertrain", which is funded by the Federal Ministry for Transport, Innovation and Technology (BMVIT), the Federal Ministry for Digital and Economic Affairs (BMDW) and the Provinces of Upper Austria and Styria.

References

1. Klöpffer W, Grahl B (2014) Life cycle assessment (LCA): a guide to best practice. Wiley-VCH, Weinheim. https://doi.org/10.1002/9783527655625
2. Ehrlenspiel K, Kiewert A, Lindemann U (2007) Cost-efficient design. Springer, Berlin Heidelberg
3. Su D (ed) (2020) Sustainable product: development tools, methods and examples. Springer Nature Switzerland AG
4. Buchert T, Pförtner A, Stark R (2017) Target-driven sustainable product development. In: Stark R, Seliger G, Bonvoisin J (eds.) Sustainable manufacturing, sustainable production, life cycle engineering and management. Springer International Publishing, Cham, pp 129–146. https://doi.org/10.1007/978-3-319-48514-0_9
5. Morelli J (2011) Environmental sustainability: a definition for environmental professionals. J Environ Sustain 1(1):1–10. https://doi.org/10.14448/jes.01.0002
6. Merschak S, Hehenberger P (2019) Ecodesign methods for mechatronic systems: a literature review and classification. 20th international conference on research and education in mechatronics (REM), May 23–24, 2019, Wels, Austria
7. Hauschild MZ, Rosenbaum RK, Olsen SI (2018) Life cycle assessment theory and practice. Springer

8. ISO 14040 (2006) Environmental management—life cycle assessment—principles and framework. British Standards Institution, London
9. Pandey D, Agrawal M, Pandey JS (2011) Carbon footprint: current methods of estimation. Environ Monit Assess 178:135–160. https://doi.org/10.1007/s10661-010-1678-y
10. Wiedmann T, Minx J (2007) A definition of 'Carbon Footprint'. Pertsova CC ecological economics research trends: Chapter 1, pp 1–11, Nova Science Publishers, Hauppauge NY, USA, June 2007
11. VDI 2206 (2004) Design handbook 2206. Entwicklungsmethodik für mechatronische systeme/design methodology for mechatronic systems. VDI Publishing Group, Düsseldorf
12. Hehenberger P, Bradley D (2016) Mechatronic futures—challenges and solutions for future mechatronic systems and designers. Springer International Publishing
13. Hehenberger P, Bradley D, Dehghani A, Traxler P (2020) Mechatronic and cyber-physical systems within the domain of the internet of things. In: Stjepandić J, Wognum N, Verhagen W JC (eds) Systems engineering in research and industrial practice. ISBN 978-3-030-33311-9, Springer, London, Berlin, Heidelberg
14. Hehenberger P, Vogel-Heuser B, Bradley D, Eynard B, Tomiyama T, Achiche S (2016) Design, modelling, simulation and integration of cyber physical systems: methods and applications. Comput Indus 82:273–289. Elsevier.https://doi.org/10.1016/j.compind.2016.05.006
15. Bertoni A, Bertoni M, Panarotto M, Johansson C, Larsson TC (2016) Value-driven product service systems development: methods and industrial applications. CIRP J Manuf Sci Technol 15:42–55. https://doi.org/10.1016/j.cirpj.2016.04.008
16. Doualle B, Medini K, Boucher X, Brissaud D, Laforest V (2016) Design of sustainable product-service systems (PSS): towards an incremental stepwise assessment method. Procedia CIRP 48:152–157. https://doi.org/10.1016/j.procir.2016.04.074
17. Terzi S, Bouras A, Dutta D, Garetti M, Kiritsis D (2010) Product lifecycle management: from its history to its new role. IJPLM 4:360. https://doi.org/10.1504/IJPLM.2010.036489
18. Chiu M-C, Chu C-H (2012) Review of sustainable product design from life cycle perspectives. Int J Precis Eng Manuf 13:1259–1272. https://doi.org/10.1007/s12541-012-0169-1
19. Menghi R, Papetti A, Germani M, Marconi M (2019) Energy efficiency of manufacturing systems: a review of energy assessment methods and tools. J Cleaner Prod, 240
20. Chouinard U, Pigosso DCA, McAloone TC, Baron L, Achiche S (2019) Potential of circular economy implementation in the mechatronics industry: an exploratory research. J Cleaner Prod, 239.https://doi.org/10.1016/j.jclepro.2019.118014
21. Bender D, Gericke K (ed) (2020) Pahl/Beitz Konstruktionslehre, Methoden und Anwendung erfolgreicher Produktentwicklung. Springer, Berlin, Heidelberg
22. Krause D, Gebhardt N (2018) Methodische Entwicklung modularer Produktfamilien, Hohe Produktvielfalt beherrschbar entwickeln. Springer, Berlin, Heidelberg
23. Vajna S, Weber C, Zeman K, Hehenberger P, Gerhard D, Wartzack S (2018) CAx für Ingenieure—Eine praxisbezogene Einführung. Springer Vieweg, Berlin
24. Buzuku S, Kraslawski A (2017) Use of design structure matrix for analysis of critical barriers in implementing eco-design initiatives in the pulp and paper industry. Procedia Manuf 11:742–750. https://doi.org/10.1016/j.promfg.2017.07.175
25. Schmidt DM, Schenkl SA, Mörtl M (2014) Matrix-based decision-making for compatible systems in product planning concerning technologies for the reduction of CO_2-emissions. In: Marle F, Jankovic M, Maurer M, Schmidt DM, Lindemann U (eds) Risk and change management in complex systems. Carl Hanser Verlag GmbH & Co. KG, München, pp 107–116. https://doi.org/10.3139/9781569904923.011
26. Eppinger SD, Browning TR (2012) Design structure matrix methods and applications. MIT Press, Cambridge, UK
27. VDI 5207 (2020) Energy-flexible factory—fundamentals. Düsseldorf: VDI Publishing Group
28. Graßl M (2015) Bewertung der Energieflexibilität in der Produktion. PhD-Thesis TUM München: Utz

Applied Sensor Technologies

Abbas Dehghani-Sanij and Uriel Martinez-Hernandez

Abstract A key element of any mechatronics system is in its interaction with the environment within which it operates, and as such sensors and the processing of sensor data play a major role in the operation of such systems. Indeed, in many mechatronics applications from manufacturing to assistive technologies, and increasingly within EcoMechatronics, the role of the embedded sensors is key not only to the operation of an individual device, but also as a source of information impacting upon the wider environment within which that device is operating. In this chapter, the nature of sensing and sensor technology is considered in relation to mechatronic, and particularly EcoMechatronic, applications along with the means by which the resulting data may be analysed and interpreted, illustrated by examples drawn from wearable robotic technologies in particular.

Introduction

Human interaction with environment has already shown an adverse impact which is clearly manifested in global warming and the recent events around the world. These events, although are natural, such as unprecedented rain falls and flooding, and high temperature and mass forest fires as a consequence, are referred to as man-made events. This is because we as human beings have not been careful on how to look after the environment that we are living in and have caused catastrophic changes in the atmosphere. Great efforts are now needed to reverse any adverse impacts from our interactions with the environment.

A. Dehghani-Sanij (✉)
University of Leeds, Leeds, UK
e-mail: a.dehghani@leeds.ac.uk

U. Martinez-Hernandez
University of Bath, Bath, UK

P. Hehenberger et al. (eds.), *EcoMechatronics*,
https://doi.org/10.1007/978-3-031-07555-1_6

One of the ways to control the impact on the environment is to consider manufacturing. It is stated that "*about 80% of the environmental impact of a machine is determined by its design*" [1], this suggests that a proper design approach can considerably reduce this impact. That is why EcoMechatronics can play an important role.

A mechatronic system in principle includes sensing, actuation and control where an integrated approach will bring mechanics, electronics and computing together to fulfil the design of an intelligent system. If the systems under consideration are energy efficient and are designed to last longer and the end user comfort has been taken into account throughout the design and development process then such systems will have much reduced environmental impact.

Considering manufacturing as referred to earlier, an EcoMechatronics approach will ensure not only more efficient systems but also less impact on the environment and more sustainable outcomes. For example, noise pollution can harm those who work in manufacturing environments. Less noise will certainly improve the comfort of the workers and reduce harmful impacts.

This idea can be even further expanded to cover not just manufacturing but also a large number of mechatronics products being used by different end users. On the one hand, in the manufacturing of such products, EcoMechatronics can be applied but, on the other hand, the design of the products themselves could be influenced by this approach to enhance product efficiency, sustainability and end users' comfort.

In any mechatronic system, sensors have a profound role to play. They are crucial both in terms of sensing within the system itself and also in relation to external bodies and the environment. Therefore, the development of sensor technologies has greatly advanced mechatronic systems and will make great impacts on advancements of EcoMechatronic systems across various applications.

The purpose of this chapter is to consider some sensor technologies in general and to put these into the context of a typical application area and demonstrate how these technologies have already influenced the systems under consideration, and how they will do so in future both in such applications and in other similar applications alike.

One typical area that has benefitted hugely from sensor technologies is in relation to those mechatronic devices which are used as assistive technologies to help with the mobility of individuals with mobility impairments. Some typical sensors will be reviewed first and then typical examples in the form of case studies will be presented. Furthermore, ideas will be exposed for potential future of sensor technologies to enhance this particular area of application and therefore influence impact on environment and sustainability. The aspect of analysing data captured by sensors and how best to use them across a range of applications will also be taken into consideration in the context of data management.

Mobility monitoring and data management has a number of applications in, for example, health care, device development and direct feedback to end users. Accurate gait analysis in real time and real environments is essential both for patient management and well-being and for design, development and improvement of intelligent customized assistive devices. It is well known that the conventional methods to assess human gait are expensive, complex and limited to laboratory environments. Systematic reviews have identified the limitations and drawbacks of current sensor technologies. A major challenge here is capturing accurate, validated, verifiable and relevant mobility data from patients in real time and in real environments by using appropriate advanced sensor technologies and then to properly manage and interpret the data. Future research has a range of challenges, two of which are of considerable importance: (1) identifying and developing sensing methods that are accurate and reliable and can be used in an integrated real-time, real-environment patient mobility monitoring system and (2) appropriate analytical methods that can be used to deal with large amount of mobility data and analyse them in real time. The focus of this chapter is to address these two aspects by considering the latest developments in these areas and also looking into a few relevant case studies.

Sensor Technologies

There are a wide range of sensors which are used in mechatronic systems. However, in this section, the focus will be on applied sensor technologies for assistive devices which are to be used in contact and harmony with the human body. These devices can be divided into two broad categories: (i) those which are used for people with weakened body segments (such as in the older population) and they could be referred to as wearable devices (exoskeletons, orthotics) and (ii) those which are expected to replace a lost segment, e.g. in amputees (prosthetics).

These mechatronic devices require sensors within themselves along with other elements for their proper functionality. For example, to ensure an accurate knee angle position is achieved, a sensor should capture knee position at an appropriate frequency and feedback this to the controller in the system. The controller then compares this with the required knee angle position and sends the control signal to the actuator. However, in addition to these sensors, other sensors are also needed to ensure a proper interaction between the devices and human body to allow harmony and assistance to be applied to impaired mobility of end users.

Electromyography (EMG) Sensors

One of the major challenges for mechatronic/robotic assistive devices to work in harmony with human body is to capture user intent. This can be achieved either by using invasive or non-invasive methods. Due to problems associated with invasive methods (e.g. infections), it is very common to use non-invasive techniques. User intent can be captured from the brain using arrays of electrodes embedded in a cap (brain-computer interface, electroencephalography (EEG)). Research in this area is ongoing by a number of researchers around the world. In our case studies, we have not referred to this type of sensor technology.

The major sensor technologies that are most commonly used for intent recognition is electromyography (EMG). This type of sensor can be very useful for real-time, real-environment mobility data capture as well.

The movement in the body is based on contraction of muscles where they receive the relevant signals from the brain. Two types of EMG sensors may be used. A needle-based EMG will capture the electrical signal directly from within the muscle. On the other hand, surface EMG (sEMG) electrodes are placed on the surface of the skin on the belly of muscle to capture the electrical signal. The electrical signals captured by sEMGs require signal conditioning to remove noise and amplify the signal.

As the signals are usually weak, the positioning of the sEMG electrodes is very critical. Latest development of sEMG sensor arrays has played an important role in solving this problem to a great extent. However, the number of channels used and associated signal conditioning require more computer processing and time.

sEMG are used in many application areas such as prosthetic devices and exoskeletons for their control and also for human gait analysis. Various machine learning techniques are used for such analytical purposes.

It is also stated that "*an EMG measures muscle activities such as the voluntary or involuntary contractions. It can disclose muscle dysfunction, nerve dysfunction, and transmission problems between the nerve and muscle all causing gait impairments ...and... the surface EMG (sEMG) testing offers an assessment of various gait-related features like changes in muscle properties paresis and muscle stiffness and tension*" [2].

Inertial Measurement Units (IMUs)

An inertial measurement unit (IMU) combines three different sensors in one unit: accelerometer, gyroscope and magnetometer. It is a small and low-cost sensor which is used in many applications and is particularly useful for mobility data capture.

IMU sensor measures the linear acceleration, angular velocity, orientation and gravitational force. Some of these sensors have onboard processing and signal conditioning units and provide ready data such as absolute orientation with the update rate of 100 Hz. It is the most popular sensor for gait analysis application due to its various advantages including internal sensor fusion of sensor data [2].

Force Sensors

Another category of sensors which are commonly used in mobility data collection and gait analysis includes force or pressure sensors. These are particularly useful in cases where interaction with ground is required to be included in the gait analysis. Examples of such cases are determination of ground reaction force (GRF) or establishment of

Initial Contact (IC) or Terminal Contact (TC) in gait analysis. Arrays of such sensors may also be used to investigate foot pressure pattern. Force sensors are divided into four main categories: piezoelectric, resistive, capacitive and optical force sensors. These sensors can be implemented in the shoes and soles, for example [2].

Vision Sensors

In principle, laboratory vision systems for gait analysis are the gold standard and other systems are compared against these. They use large number of cameras in a laboratory setting which capture motions of subject by interacting with markers that are placed on different positions of the subject's body. Suitable software packages are then used to extract the required motion data from the captured image. The main problem with such systems is that they are only useful for laboratory scenarios and as such do not provide gait data for real environments. That is why wearable sensors are used for real-environment mobility data capture.

Wearable cameras are also available, and they can provide a lot of useful data as compared to other wearable sensors. For example, by using such sensors, contextual information about the environment can be acquired. A camera system can be worn on the user waist and as such it can provide continuous information from the environment in which the user is moving [3].

Analytical Methods

One of the challenges to have wearable systems capable of assisting humans safely and reliably, and indeed of any complex mechatronic system, is the development of methods for data processing, analysis and understanding of the data in real time. Several methods from heuristic approaches based on rules to expert systems and autonomous learning have been investigated during the last decades. A description of these computational methods used for monitoring, recognition and control is presented in the following sections.

Heuristic Methods

Heuristics-based methods employ a list of predefined rules and conditions obtained from the observation of the event to be addressed. A variety of heuristic methods have been used for recognition of motion, control and assistance in respect of humans. One of these heuristic approaches employed a threshold-crossing method to recognize the angle of the lower limbs to decide when to apply the required assistance [4]. This method was used with the robot suit HAL for the delivery of support during walking and sit-to-stand activities [5]. Rule-based methods also use features extracted from sensor data, e.g. mean value, standard deviation, zero-crossing, as input for the decision-making process in assistive robotics. This heuristic approach has been used with a single feature threshold-crossing algorithm for recognition of sit and stand activities, achieving a range of accuracies from 70 to 90% [6, 7]. Such heuristic methods have also been implemented in applications of bioinspired robots for interaction with the environment, mechatronic tools for disassembly and recycling, management of energy consumption and energy harvesting for sustainable systems [8–10].

Heuristic methods are relatively simple approaches that can perform well under controlled environments with invariant conditions. However, their nature of relying on a predefined and fixed set of rules make these methods highly susceptible to failure when even slight changes in sensor measurements or environment conditions are present in the system. This is related to the fixed set of rules that do not permit the system to adapt to unexpected slight changes. For this reason, it is preferable to use heuristic approaches together with other data processing methods. For example, activity recognition has been investigated using probabilistic methods, genetic algorithms and artificial neural networks with heuristic approaches to map specific features from sensor data and monitoring [11, 12].

Machine Learning Methods

Machine Learning (ML) offers a promising approach capable of perceiving and making decisions for the development of safe and reliable assistive robots. The purpose of ML methods is to find features and relations between datasets from one or multiple sensors that can be used to create rules learned from data instead of manually defining the set of fixed rules as in heuristic methods. This capability of learning directly from data allows the assistive device to adapt to slight changes observed in the data or slight changes in the environment.

One of the traditional methods used for activity recognition is Fuzzy Logic (FL), which has been employed with different sensing modalities and control of robot platforms [13]. FL and vision sensing have been able to detect sit-to-stand and transition activities with accuracies ranging from 69% to 94.6% [14]. A multilayer fuzzy logic architecture has also shown the potential of this approach for recognition of walking activities in level-ground walking and ramps using data from wearable inertial measurement units [15]. Principal Component Analysis (PCA) together with Support Vector Machines (SVM) has offered an approach for recognition of transitions from sit-to-stand activities with an accuracy of 92.94%. This method was able to achieve such high accuracy using a windowing approach and data from a large number of inertial measurement units, force sensors and potentiometers [16]. SVMs are usually employed for binary classification; however, some works have used this approach together with Decision Trees for multi-class recognition using visual data

[17]. This multi-class approach was able to improve the recognition accuracy of sit to stand and transition to 94.6%.

Bayesian methods and sequential analysis have achieved accurate recognition of sit-to-stand activities using a single wearable accelerometer [18]. The combination of traditional machine learning methods such as Hidden Markov Models (HMM) and artificial neural networks has ensured reliable prediction of sequential gait stages [19]. Another popular method for monitoring and detection of human activity is the Kalman filter. This approach has been able to successfully detect human fall events using arrays of wearable sensors and typically achieving detection accuracies ranging from 95% to 99.4% [20, 21]. Kalman filters have been combined with HMMs allowing assistive devices to track the human body accurately and classify with a mean accuracy of 93% multiple activities of daily living in real-time and well-controlled environments [22]. Sitting, standing and walking activities have also been detected accurately by the extraction of features and processing with Gaussian Mixture Models (GMM) [23].

Machine learning methods have been also used in other fields that require sustainable mechatronic systems. Examples of these applications include mechatronic systems for sustainable agriculture, adaptive energy management for hybrid electric vehicles, low-carbon systems for smart cities and collaborative farm robots [24–26].

Deep Learning Methods

Deep Learning methods are primarily focused on Neural Network (NN)-based architectures and have demonstrated their ability for accurate activity recognition with complex datasets where other machine learning methods fail [27].

Deep Learning (DL) techniques based on Convolution Neural Networks (CNN) have gained popularity for the identification and classification of human activities. CNNs have been successfully trained with 3D data sequences from vision sensors and inertial measurement units and able to detect a large number of activities achieving high accuracy between 98% and 99.78% [28]. Ensemble of CNNs has also been investigated for recognition of activities of daily living using wearable sensors such as vision, inertial units and electromyography sensors. This approach showed an improved activity recognition task with a mean accuracy of 99.68% [29]. Another recent method that has gained the attention from researchers is Recurrent Neural Networks (RNNs), which are useful for sequential analysis of sensor data. RNNs combined with CNN methods have shown the potential of deep learning methods for recognition of a large number of activities of daily living employing wearable sensors [30, 31]. Other examples of RNN-based methods are Long Short-Term Memory (LSTM) methods and DeepConvLSTM architectures [32, 33] which are becoming widely used for analysis and processing of sequential data from wearable sensors.

Deep Learning, together with a variety of sensing modalities, has also demonstrated its potential in a wider range of applications such as mechatronic systems for robot control, manufacturing processes in sustainable Industry 4.0, smart management processes, smart cyber-physical production systems, management of plants and plant diseases, and human–robot interaction and collaboration in assembly processes [34–39].

Case Studies

The sensing technologies and methods for data analysis shown in the previous sections have been used widely in assistive, rehabilitation and healthcare mechatronic/robotic systems. Examples of these applications are described in the following sections.

Wearable Assistive Robots for Locomotion Activities

There exist several wearable robots built with rigid and semi-rigid materials and structures that are commonly known as exoskeletons and they are employed for assistance in locomotion activities. These exoskeletons integrate multiple sensing technologies, and different methods for data processing, recognition and control. Figure 1 shows examples of lower limb exoskeletons for human assistance. The ReWalk robotic system with a semi-rigid structure can assist the knee and hip of adults with partial and complete mobility impairments [40], detecting and enhancing the user's walking action. They can assist humans to keep balance while walking but also to accomplish daily activities including sitting, climbing stairs and kneeling [41–46]. These lower limb devices can be configured to respond quickly to the user movement intention using data from electromyographic (EMG), inertial measurement unit (IMU) and torque sensors. Unfortunately, these devices tend to be bulky, heavy (lower limb robots weighing between 15 and 25 kg) and expensive.

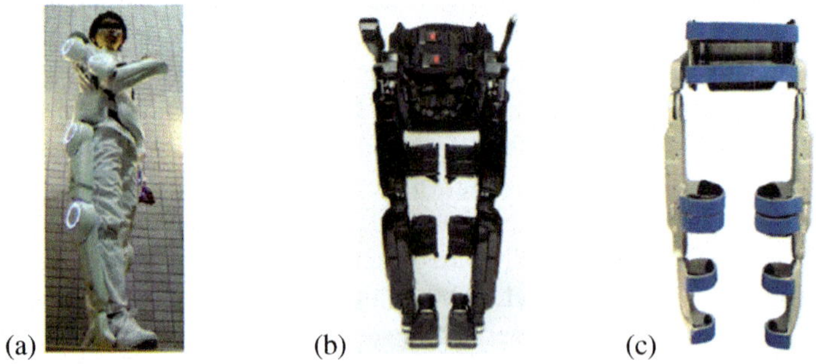

Fig. 1 Wearable assistive robots with rigid and semi-rigid structure: **a** HAL [41], **b** REX [45], **c** Vanderbilt exoskeleton [46]

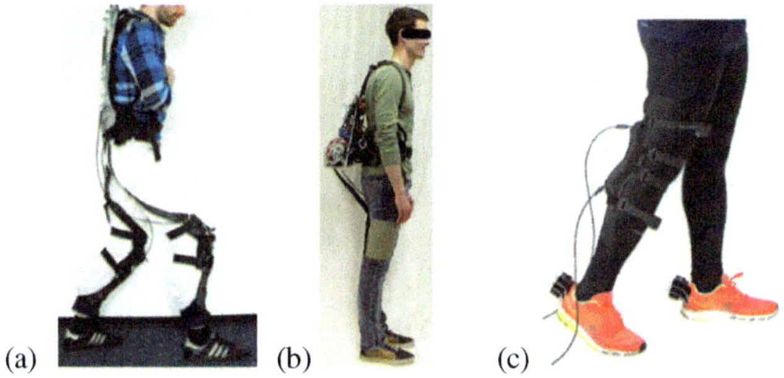

Fig. 2 Lower limb wearable assistive robots with soft materials and structure: **a** soft inflatable knee exosuit [47], **b** soft exosuit for hip assistance [49], **c** multi-articular hip and knee exosuit [51]

Another type of robots that use a variety of sensing technologies, recognition and control methods for locomotion assistance is the wearable soft robot. These robotics platforms are mainly built with soft materials making these robots lightweight and portable. Figure 2 shows examples of lower limb wearable soft robots.

Pneumatic artificial muscles, Boden cables, textiles and shape memory alloys are the main material technologies that have been employed in a variety of wearable soft robots. Soft wearable knee and ankle robots have been used to assist contraction/extension leg movements and also with foot movements in dorsiflexion, plantarflexion, inversion and eversion orientations [47, 48]. Exosuits using textiles and Boden cables are some of the most advanced and lightweight devices for assistance to the hip, leg and ankle–foot while walking on flat surfaces [49–51].

Recognition of Walking Activities and Assistance

Recognition of walking activities, composed of gait phases, periods and events, is a key aspect needed to provide assistance appropriately with robotic systems. This section presents a description of approaches for the recognition of locomotion activities using different sensor technology, recognition methods and robotic platforms.

Detection of gait events has been investigated for assistance of lower limb amputees. This work employs heuristics methods and foot switches placed underneath the foot of participants [52]. The proposed method is capable of recognizing Foot-Flat-Start (FFS) and Heel-Contact (HC) events while walking on flat surfaces. Figure 3 illustrates the experimental setup for data collection, data processing and control. Overall, the detection accuracy was found to be 100% for Initial Contact, FFS, and Toe-Off, and 98.3% for HC.

The use of arrays of inertial measurement units together with Dynamic Bayesian Networks has been employed for the recognition of walking on flat surfaces and ascending and descending ramps [53]. This work has also been employed with

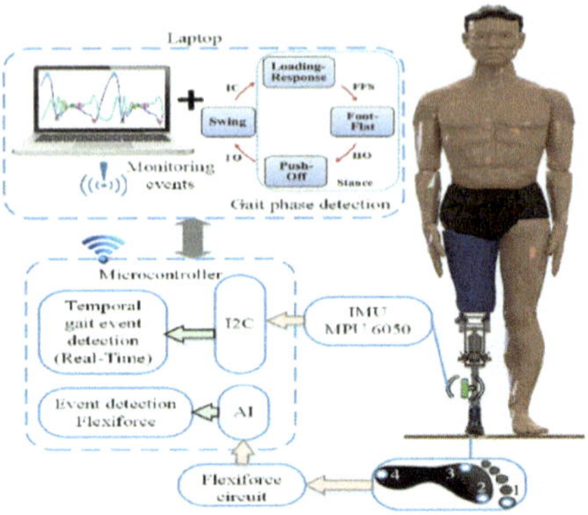

Fig. 3 Experimental setup for data collection, monitoring, data processing, control and recognition of gait events [52]

modules for combining multiple sources of information and prediction processes, which allows the system to speed up and improve the recognition accuracy. Figure 4 shows an example of the activities performed by participants and the flowchart of the recognition method.

An architecture for recognition of gait phases has been developed using an array of classifiers, feature extraction modules and finite-state machines using data from multichannel electromyography signals [54]. This approach is not only capable of accurately recognizing gait phases from lower limb amputees but it also adapts to

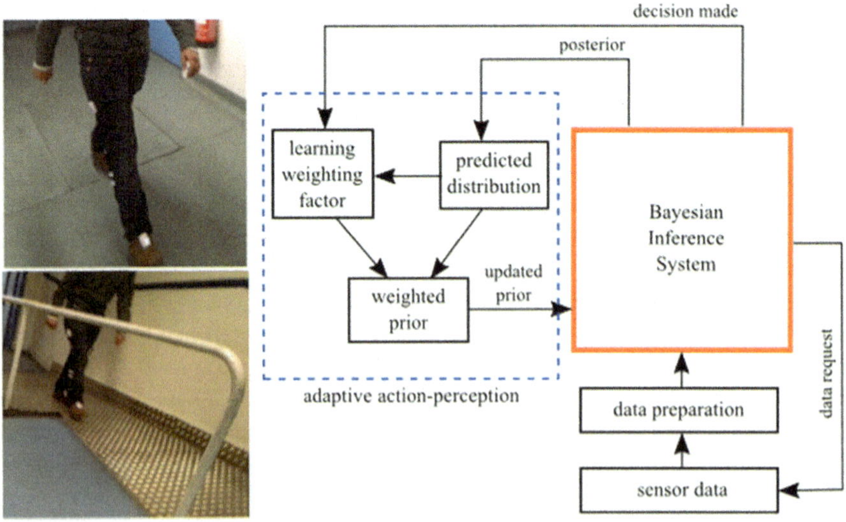

Fig. 4 Walking activities (level-ground walking and ramps) and dynamic Bayesian network for learning and prediction processes [53]

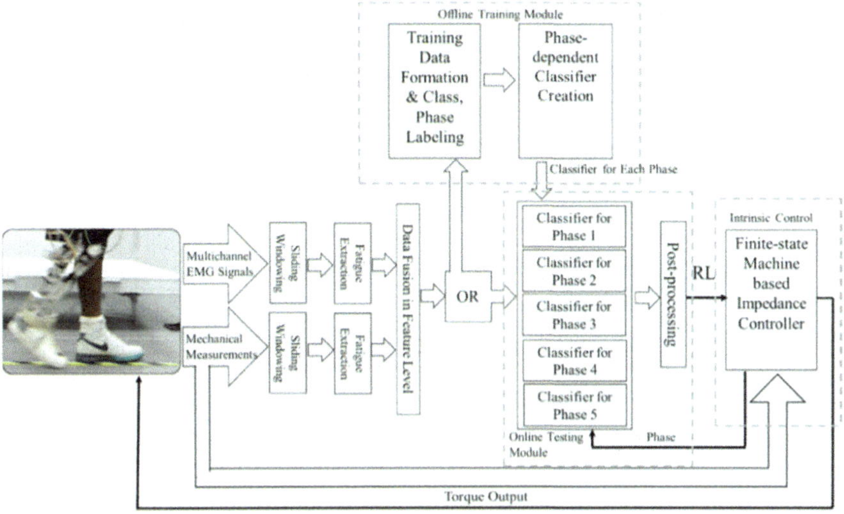

Fig. 5 The locomotion recognition system used for lower limb prosthesis control [54]

different terrains. Figure 5 shows the modules and their interconnections for the recognition and control processes. This work showed that using an array of classifiers, the assistive system can recognize all the gait phases during the walking cycle in real time and with a mean accuracy of 92.70%.

Future Directions in Research and Sensor Technologies

Research in different areas ranging from medical, industrial, construction, agricultural and environmental will be very much dependent on acquiring more accurate data. The impact of more accurate data will be evident in safer and more sustainable environments, as well as more productive and more efficient outcomes. For this to manifest itself, the huge amount of data that will become available as a result of advancements in sensor technologies will require methods for analysis and making sense of them to allow decision-making and accurate control.

Sensor technologies are advancing very rapidly in a number of areas. The trend clearly shows that this technology has been and is moving more and more towards miniaturization. By reducing the physical size of sensors and yet improving their accuracy, they are made even more efficient as they use less power. In close parallel to this trend is the increase in their capabilities, moving in the direction of developing more intelligent and smart sensors. Such sensors will have more autonomy in that they not only sense a physical characteristic but also provide some initial processing to the captured data.

Another clear direction is fusion of sensor data. There are many applications that may require a range of different type of data to be collected to allow a proper decision-making. For example, in some processes, temperature, humidity and pressure may be needed and the captured data should be processed by fusion of all three categories of data. In this context, the advancement is in two directions, on the one hand, the sensor technologies are moving towards integration of different type of sensors, and, on the other hand, methods such as machine learning are being developed to process data.

As a result of miniaturization, smart sensors and sensor fusion, it will become more and more feasible to use many more sensors and therefore collect huge amounts of data. This then necessitates technologies for transferring data in real time and facilities to allow this quantity of data to be stored and processed. Therefore, technologies for more efficient wireless communication, the cloud and big data will be required.

Conclusions

The purpose of this chapter was to briefly address two research challenges which are relevant to many applications and indeed to the area of EcoMechatronics.: (1) identifying and developing sensing methods that are accurate and reliable and (2) appropriate analytical methods that can be used to deal with large amount of data and analyse them in real time.

A few exemplar sensor technologies such as inertial measurement units, force sensors and vision sensors were briefly mentioned as they were directly related to the case studies to be followed in the chapter. The application of such sensors is very wide and has been used in many areas.

With respect to processing the captured data from sensors, a few analytical methods were also covered. These included heuristic, machine learning and deep learning methods.

These methods have been used in many applications such as the medical area of assistive technologies where some case studies were presented here. They have also been implemented in bioinspired robots for interaction with the environment, mechatronic tools for disassembly and recycling, management of energy consumption and energy harvesting for sustainable systems, mechatronic systems for sustainable agriculture, adaptive energy management for hybrid electric vehicles, low-carbon systems for smart cities and collaborative farm robots, mechatronic systems for robot control, manufacturing processes in sustainable Industry 4.0, smart management processes, smart cyber-physical production systems, management of plants and plant diseases, and human–robot interaction and collaboration in assembly processes.

The future trend in sensor technologies will be on miniaturization, smart sensors and fusion of sensing both in terms of developing sensors with multi-sensing functionality and on board processing and also development of methods to deal with huge data processing. This trend will result in safer and more sustainable environments and improve productivity and reduce costs.

Acknowledgements The authors would like to thank the postdoctoral and PhD researchers who were involved in the case studies referred to in this chapter. They have been referenced in the references.

References

1. "Eco-design of Energy-Related Products". European commission's directorate-general for energy. (Wikipedia, accessed Oct. 2021)
2. Abdul Saboor et al (2020) Latest research trends in gait analysis using wearable sensors and machine learning: a systematic review. IEEE Access Spe Sect Body Area Netw 8
3. Vallabh P (2018) Reza Malekian. Fall detection monitoring systems: a comprehensive review. J Ambient Intell Human Comput 9:1809–1833
4. Tsukahara A, Kawanishi R, Hasegawa Y, Sankai Y (2010) Sit-to-stand and stand–to-sit transfer support for complete paraplegic patients with robot suit hal. Adv Robot 24(11):1615–1638
5. Suzuki K, Mito G, Kawamoto H, Hasegawa Y, Sankai Y (2007) Intention-based walking support for paraplegia patients with robot suit hal. Adv Robot 21(12):1441–1469
6. Capela NA, Lemaire ED, Baddour N (2015) Improving classification of sit, stand, and lie in a smartphone human activity recognition system. In: Medical measurements and applications (MeMeA), 2015 IEEE international symposium on, IEEE, pp 473–478
7. Haché G, Lemaire ED, Baddour N (2011) Wearable mobility monitoring using a multimedia smartphone platform. IEEE Trans Instrum Meas 60(9):3153–3161
8. Luneckas M et al (2021) Hexapod robot gait switching for energy consumption and cost of transport management using heuristic algorithms. Appl Sci 11(3):1339
9. Favi C et al (2019) A design for disassembly tool oriented to mechatronic product de-manufacturing and recycling. Adv Eng Inf 39:62–79
10. Noll M-U, Lentz L, von Wagner U (2020) On the improved modeling of the magnetoelastic force in a vibrational energy harvesting system. J Vib Eng Technol 8(2):285–295
11. Feuz, Dillon K, Cook DJ (2014) Heterogeneous transfer learning for activity recognition using heuristic search techniques Int J Pervas Comput Commun
12. Chen L, Hoey J, Nugent CD, Cook DJ, Yu Z (2012) Sensor-based activity recognition. IEEE Trans Syst Man Cybern Part C (Appl Rev) 42(6):790–808
13. Kiguchi K, Tanaka T, Fukuda T (2004) Neuro-fuzzy control of a robotic exoskeleton with emg signals. IEEE Trans Fuzzy Syst 12(4):481–490
14. Banerjee T, Keller JM, Skubic M, Abbott C (2010) Sit-to-stand detection using fuzzy clustering techniques. In: Fuzzy systems (FUZZ), 2010 IEEE international conference on. IEEE, pp 1–8
15. Rubio-Solis A, Panoutsos G, Beltran-Perez C, Martinez-Hernandez U (2020) A multilayer interval type-2 fuzzy extreme learning machine for the recognition of walking activities and gait events using wearable sensors. Neurocomputing 389:42–55
16. Doulah A, Shen X, Sazonov E (2016) A method for early detection of the initiation of sit-to-stand posture transitions. Physiol Measur 37(4):515
17. Qian H, Mao Y, Xiang W, Wang Z (2010) Recognition of human activities using svm multi-class classifier. Pattern Recogn Lett 31(2):100–111
18. Martinez-Hernandez U, Dehghani-Sanij AA (2019) Probabilistic identification of sit-to-stand and stand-to-sit with a wearable sensor. Pattern Recognit Lett 118:32–41
19. Zhao H, Wang Z, Qiu S, Wang J, Xu F, Wang Z, Shen Y (2019) Adaptive gait detection based on foot-mounted inertial sensors and multi-sensor fusion. Inf Fusion 52:157–166
20. Anania G, Tognetti A, Carbonaro N, Tesconi M, Cutolo F, Zupone G, De Rossi D (2008) Development of a novel algorithm for human fall detection using wearable sensors. In: Sensors, 2008 IEEE. IEEE, pp 1336–1339

21. He J, Bai S, Wang X (2017) An unobtrusive fall detection and alerting system based on Kalman filter and Bayes network classifier. Sensors 17(6):1393
22. Wu JK, Dong L, Xiao W (2007) Real-time physical activity classification and tracking using wearable sensors. In: Information, communications & signal processing, 2007 6th international conference on, IEEE, 2007, pp 1–6
23. Piyathilaka L, Kodagoda S (2013) Gaussian mixture based HMM for human daily activity recognition using 3D skeleton features. In: 2013 IEEE 8th conference on industrial electronics and applications (ICIEA), pp 567–572. IEEE
24. Grau JB et al (2009) Sustainable agriculture using an intelligent mechatronic system. 2009 35th annual conference of IEEE industrial electronics. IEEE
25. Li G, Görges D (2018) Ecological adaptive cruise control and energy management strategy for hybrid electric vehicles based on heuristic dynamic programming. IEEE Trans Intell Transp Syst 20(9):3526–3535
26. Guo J, Xu T (2021) Intelligent low-carbon emission design concept for industrial cities driven by artificial intelligence. 2021 5th international conference on intelligent computing and control systems (ICICCS). IEEE
27. Hatcher WG, Yu W (2018) A survey of deep learning: platforms, applications and emerging research trends. IEEE Access 6:24411–24432
28. Nunez JC, Cabido R, Pantrigo JJ, Montemayor AS, Velez JF (2018) Convolutional neural networks and long short-term memory for skeleton-based human activity and hand gesture recognition. Pattern Recogn 76:80–94
29. Ijjina EP, Mohan CK (2016) Hybrid deep neural network model for human action recognition. Appl Soft Comput 46:936–952
30. Ignatov A (2018) Real-time human activity recognition from accelerometer data using convolutional neural networks. Appl Soft Comput 62:915–922
31. Ordóñez, Javier F, Roggen D (2016) Deep convolutional and lstm recurrent neural networks for multimodal wearable activity recognition. Sensors 16(1):115
32. Hammerla NY, Halloran S, Ploetz T (2016) Deep, convolutional, and recurrent models for human activity recognition using wearables. In: Proceedings of the IJCAI international joint conference on artificial intelligence, New York, NY, USA, 9–15 July 2016, vol 2016, pp 1533–1540
33. Ordóñez FJ, Roggen D (2016) Deep convolutional and LSTM recurrent neural networks for multimodal wearable activity recognition. Sensors 16:115
34. Xing K et al (2018) Hand gesture recognition based on deep learning method. 2018 IEEE third international conference on data science in cyberspace (DSC). IEEE
35. Al GA, Estrela P, Martinez-Hernandez U (2020) Towards an intuitive human-robot interaction based on hand gesture recognition and proximity sensors. 2020 IEEE international conference on multisensor fusion and integration for intelligent systems (MFI). IEEE
36. Afolalu, Sunday A et al (2021) Enviable roles of manufacturing processes in sustainable fourth industrial revolution–a case study of mechatronics. Mater Today Proc 44:2895–2901
37. Ampatzidis Y, De Bellis L, Luvisi A (2017) iPathology: robotic applications and management of plants and plant diseases. Sustainability 9(6):1010
38. Andronie M et al (2021) Artificial intelligence-based decision-making algorithms, internet of things sensing networks, and deep learning-assisted smart process management in cyber-physical production systems. Electronics 10(20):2497
39. Male J, Martinez-Hernandez U (2021) Collaborative architecture for human-robot assembly tasks using multimodal sensors. 2021 20th international conference on advanced robotics (ICAR). IEEE
40. Chen B, Ma H, Qin LY, Gao F, Chan KM, Law SW, Qin L, Liao WH (2016) Recent developments and challenges of lower extremity exoskeletons. J Orthop Transl 5:26–37
41. Herr H (2009) Exoskeletons and orthoses: classification, design challenges and future directions. J Neuroeng Rehabil 6:1–9
42. Esquenazi A, Talaty M, Packel A, Saulino M (2012) The ReWalk powered exoskeleton to restore ambulatory function to individuals with thoracic-level motor-complete spinal cord injury. Am J Phys Med Rehabil 91:911–921

43. Barbareschi G, Richards R, Thornton M, Carlson T, Holloway C (2015) Statically vs. dynamically balanced gait: Analysis of a robotic exoskeleton compared with a human. In: Proceedings of the 2015 37th annual international conference of the IEEE engineering in medicine and biology society (EMBC), Milan, Italy, 25–29 August 2015, pp 6728–6731
44. Sankai Y (2010) HAL: Hybrid assistive limb based on cybernics. In: Robotics research; Springer: Berlin/Heidelberg, Germany, pp 25–34
45. Birch N, Graham J, Priestley T, Heywood C, Sakel M, Gall A, Nunn A, Signal N (2017) Results of the first interim analysis of the RAPPER II trial in patients with spinal cord injury: ambulation and functional exercise programs in the REX powered walking aid. J Neuroeng Rehabil 14:1–10
46. Yan T, Cempini M, Oddo CM, Vitiello N (2015) Review of assistive strategies in powered lower-limb orthoses and exoskeletons. Robot Auton Syst 64:120–136
47. Sridar S, Qiao Z, Muthukrishnan N, Zhang W, Polygerinos P (2018) A soft-inflatable exosuit for knee rehabilitation: assisting swing phase during walking. Front. Robot. AI 5:44
48. Park YL, Chen BR, Young D, Stirling L, Wood RJ, Goldfield EC, Nagpal R (2014) Design and control of a bio-inspired soft wearable robotic device for ankle-foot rehabilitation. Bioinspir Biomim 9:016007
49. Asbeck AT, Schmidt K, Walsh CJ (2015) Soft exosuit for hip assistance. Robot Auton Syst 73:102–110
50. Asbeck AT, De Rossi SM, Holt KG, Walsh CJ (2015) A biologically inspired soft exosuit for walking assistance. Int J Robot Res 34:744–762
51. Khomami AM, Najafi F (2021) A survey on soft lower limb cable-driven wearable robots without rigid links and joints. Robot Auton Syst 144:103846
52. Maqbool HF, Husman MAB, Awad MI, Abouhossein A, Iqbal N, Tahir M, Dehghani-Sanij AA (2018) Heuristic real-time detection of temporal gait events for lower limb amputees. IEEE Sens J 19(8):3138–3148
53. Martinez-Hernandez U, Dehghani-Sanij AA (2018) Adaptive Bayesian inference system for recognition of walking activities and prediction of gait events using wearable sensors. Neural Netw 102:107–119
54. Liu M, Zhang F, Huang HH (2017) An adaptive classification strategy for reliable locomotion mode recognition. Sensors 17(9):

MBSE for Mechatronic Systems Design with Human, Energetic, Cyber, and Physical Aspects

Faïda Mhenni and Jean-Yves Choley

Abstract Technological advances are completely reshaping the world we live in by allowing the design of increasingly sophisticated, fast, accurate systems that serve and assist human beings in different domains. These systems increasingly rely on information technology to process and exchange data to provide enhanced services. Additionally, energy and material resources are getting scarce. To take into account all these aspects during the design, we can then consider that today's systems are characterized by four dimensions: human, energetic, cyber, and physical. On the other hand, machines and equipment are invading our daily life more than ever. The production and use of these products damage the planet in two ways: firstly, by greedily drawing on the earth's limited resources (fossil fuels, rare materials, forests, etc.), and secondly by emitting harmful and polluting emissions. Faced with this imminent danger, awareness is growing and regulations are being enforced to control the way we design and produce our systems. With this increasing complexity, current design approaches are hitting their limits and require adaptation. In this perspective, we propose a conceptual framework to structure current systems. This framework will serve as the basis for a new design approach, of which we briefly present an outline with a focus on sustainability aspects. An insulin pump system is used as a case study to illustrate the proposed work.

Introduction

Mechatronics is a key technology with relevant innovative collaborative design methodologies, for today's newly designed systems. It allows the coupling of the physical and material world with cyberspace, mainly in cyber-physical systems (CPS) and IoT. Interestingly, such complex systems increasingly interact with human beings (operators, users, etc.), consume and/or produce or harvest energy of all kinds.

In this new context characterized by four main facets, namely cyber, physical, human, and energetic, designing such systems has to deal with the environment,

F. Mhenni (✉) · J.-Y. Choley
ISAE-Supméca, Saint-Ouen, France
e-mail: faida.mhenni@isae-Supmeca.fr

P. Hehenberger et al. (eds.), *EcoMechatronics*,
https://doi.org/10.1007/978-3-031-07555-1_7

climate change, and sustainability concerns, and also has to take into account social and corporate responsibility. It means that designers have to cope with energy and physical resources considerations and human-friendly constraints while taking advantage of the ability to be connected to cyberspace with relevant data, information, and knowledge storage, as well as remote computing resources, including complex AI ones.

Regarding their intrinsic complexity, it is obvious that designing such systems relying on mechatronics to be efficiently interfaced with those four facets (or spaces/pillars) has to summon model-based systems engineering (MBSE) for better integration of different domains (mechanics, electronics, computing, etc.) and physics (thermal, vibration, electromagnetism, etc.). MBSE methodologies shall be then enhanced and enriched with specific system views in order to cope with environmental, human, and life-centered concerns.

With regard to the environmental aspects, eco-design is an approach based on the assessment and the improvement of performances, measured with environmental performance indicators (EPI). An exhaustive list of EPI is provided in [1] and may help to define the relevant specific indicators related to mechatronic systems. In [2], a literature review of eco-design methods for mechatronic systems is proposed. Authors highlight the need for reduced models in the development phase for the calculation of the carbon footprint which moreover may take into account uncertainties to be reduced at each stage of the design process. Regarding the production processes, the availability of detailed models is also considered to be important.

Mechatronic system end-of-life (EoL) is a crucial problem to be managed for mechatronic systems because they closely integrate mechanical parts and electronic devices. Thus, the ease of disassembling such systems has to be rationalized. In [3], the authors proposed a method and a design tool for disassemblability and recyclability that may be adapted for mechatronic systems. However, as highlighted in [4], some new design methods and tools have to be developed in order to make circular economy (CE) an unavoidable concern for mechatronic systems designers. In [5], Bougain et al. proposed a methodology to integrate the environmental impacts of a mechatronic product during its design. This is obtained through a parametric diagram in which the cumulative energy demand is computed for four life cycle phases: extraction, manufacturing, use, and end-of-life.

In this work, we rather propose a conceptual framework [6] to better structure newly designed systems in this constraining context. The different EPIs and other eco-design criteria can then be integrated into this framework. Based on an MBSE methodology adapted from [7], this framework is briefly presented with an application on an insulin pump system to illustrate the proposed work.

This chapter is organized as follows. Firstly, a new conceptual framework for mechatronic systems is presented. Secondly, an MBSE approach is proposed in order to support EcoMechatronic design. Then, an application to an insulin pump is presented before giving the conclusion and some future works.

New Conceptual Framework for Mechatronic Systems

Systems today have reached an exceptional level in terms of functionality offered to users. Indeed, thanks to technological progress, modern man can be assisted in all areas, such as transportation, medicine, and production machines, to cite only a few. In an effort to improve the quality of life, the development of information technology has made it possible to have near-real-time information in several areas. Passengers can have instantaneous real-time information about traffic on the streets and about flights and other public transportation. Doctors can remotely perform surgeries and get follow-up information on their patients. As a matter of fact, we can see that the cyber dimension is now omnipresent through monitoring and data processing to improve the management of existing physical systems and provide users with more functionality. Several research works focus on this transition from mechatronic systems to cyber-physical systems [8].

In [9], the authors consider that mechatronic systems have their place at the interface of cyber and physical spaces enabling these distinctive worlds to communicate. However, as a consequence of these advances in modern systems and their tight interaction with humans, the level of complexity and interconnection of modern systems has considerably increased.

Additionally, the increasing number of systems in use, often energy-intensive, that constantly collect and process the required data, on the one hand, and the scarcity of natural resources, on the other hand, make the energy and material aspects of primary importance during systems design.

Based on these statements, one may consider that today's world relies on four main interconnected dimensions that must be thoroughly dealt with during the design process:

- **The human dimension**. Indeed, systems are mainly intended to respond to a need or provide service to humans. Humans may also be part of or closely interacting with the systems, thus strongly impacting or contributing to their behavior. On the flip side, the modern world relies on technical systems for all kinds of tasks and domains, which may cause several disturbances and constraints on humans such as the increasing pollution (air, sounds, etc.), the induced environmental changes, the scarcity of some resources, etc.
- **The physical dimension**. The physical dimension enables the system to act on and sense all kinds of materials and objects be they static or dynamic, in a 3D physical environment. The system itself is a 3D physical object made of materials.
- **The informational (cyber) dimension**. Modern systems increasingly rely on data either collected internally from the different sensors or from external sources within the system environment such as other interacting systems or different operators and users. Being able to deal with such information allows systems to optimize and adapt their behavior taking into account the different parameters. However, this comes at the price of exposure to cyberattacks and bugs.

- **The energetic dimension**. The energetic concerns are of paramount importance in today's systems design. Indeed, systems need the energy to perform the required functions but high energy consumption has both economic and ecological implications. Designs must optimize the energy consumption and whenever possible try to recover lost energy.

A complex system is thus immersed in a four dimensions space, and mechatronic components (i.e. sensors, actuators, embedded system software and hardware, HMI, etc.) act mainly as interfaces between the four corresponding spaces, as shown in Fig. 1.

Then, in order to better take into account the context of the new systems, we propose a set of postulates and associated hypotheses as follows already presented in [6].

These postulates and hypotheses will provide a framework that enables a better consideration of the different facets during the design process.

- **System Core Postulate** (Abbreviated as CoPo): A system is articulated around four complementary and synergistic domains, namely the physical (P), cyber/information (C), energy (E), and human (H) spaces. As a hub providing an interface between these domains, mechatronics enables synergies, based on

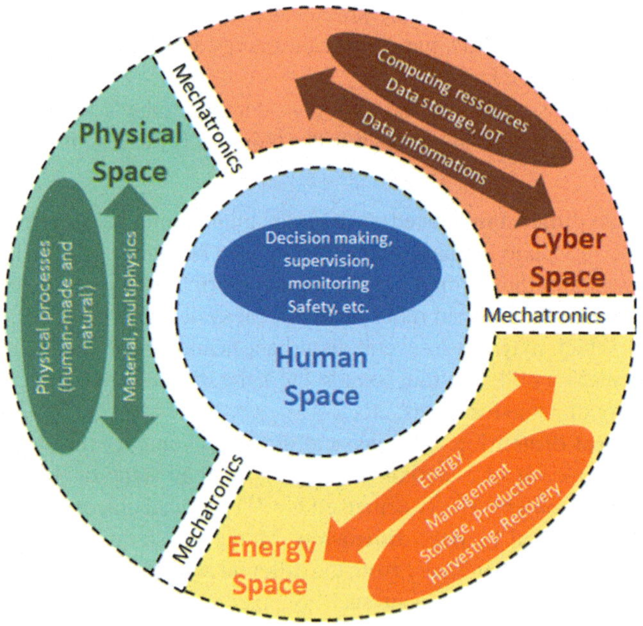

Fig. 1 Human, energy, cyber, and physical domains with mechatronics in new systems

a backbone built with sensors, actuators, real-time embedded systems (hardware and software), human system interface (HSI) devices, etc. as shown in Fig. 1;

- **Safety Critical System Hypothesis** (Abbreviated as SCHy): Systems, as described in the basic system postulate, have (exhibit) a sophisticated structure and complex behavior. They are therefore potentially critical in terms of safety and, more generally, subject to dependability issues (reliability, availability, maintainability, safety, and security);
- **System Rationale Postulate** (Abbreviated as SRPo): The rationale of a system is to respond to a need, by achieving a mission. This need is expressed by a set of requirements, most of which constrain the parameters and variables of the system. The system mission is formalized by operational scenarios describing the expected behavior, in a specific context, for each phase of the system life cycle;
- **Dynamic Requirements and Context Hypothesis** (Abbreviated as DRHy): Requirements continue to change throughout the life cycle of the system, especially in an uncertain and evolving context (technology, standards, etc.);
- **System Structure and Behavior Postulate** (Abbreviated as SBPo): The mission of a given system can be decomposed into a set of interacting functions, each of which being achieved by a relevant component. This results in the structure of the system (topology) and its overall behavior emerging from the interconnections and interactions of the components. This emerging behavior of the system must meet the expected behavior, Consequently, there is a duality between the structural view and a behavioral view;
- **Unintended Behavior Hypothesis** (Abbreviated as UBHy): Interconnections between physical components can give rise to unintended behaviors, generated, for example, by multi-physical and geometric interference and interactions, as well as by improper or inconsistent interfacing.

MBSE as a Support to EcoMechatronics

Mechatronic systems are inherently complex and are increasingly so in the new context with increased functionality and more dimensions to take into account. Hence, their design requires a global system approach to tackle the whole aspects, which is why there is a need for a systems engineering and mainly an MBSE approach.

Our proposal relies on an MBSE methodology presented in [7] and consists of two complementary phases (a black-box and a white-box phase). In this methodology, a system architecture is developed progressively, based on the requirements, and the system model is accordingly built in SysML. However, we intend to further develop this methodology by integrating the postulates and hypotheses in order to take into account the different MBSE concerns and also by structuring it according to the four facets: human, energy, cyber, and physics.

From a model-based systems engineering point of view, human, energy, and resources considerations can be taken into account both:

- During the "black box" analysis phase in order to elicit, capture and classify requirements regarding human, energy, cyber, and physical aspects;
- During the "white box" analysis phase, in order to quote functional and components architectures regarding some parameters related to energy and resources, to reduce the number of components with fewer connections and interconnection components (but probably with some recyclability concerns).

To do so, the idea is then to type the elements (functions or physical components depending on the view) and links (representing the exchanged flows) in the functional and component views in the system model. The proposed types refer to the main dimension of the element or link (H, E, C, and P). Based on this typing, for each type, provide designers with criteria, guidelines, recommendations, etc. to efficiently consider the constraints relative to these facets. It is obvious that the criteria and guidelines will be numerous and vary a lot according to the application area. Establishing a generic list can only be accomplished for a specific field and the list has to be updated continuously in order to take into account the technological advances as well as the knowledge acquired during different projects. In this work, we propose to provide a list of example guidelines with a focus on the eco aspects applied to a critical system: the insulin pump.

Illustration with a Case Study

The case study considered to illustrate this work is an insulin pump, also named artificial pancreas. It is a device intended to regulate the glucose level in diabetes patients' blood. To this aim, the device shall continuously sense the glucose rate in the patient blood, compute the required amount of insulin and then, when needed, and infuse the right insulin dose into the patient body at the right time.

To this end, the device is composed of a "glucose sensor" to sense the glucose level, a "touchscreen" to communicate with the patient, a "transceiver" to receive and emit wireless data, a "controller" to manage the whole system and compute the required insulin dose, a "battery" to store electrical energy, and a "pump" to drive insulin from a cartridge to an "infusion cannula" fitted on the patient's skin.

A Layered System Model for the Insulin Pump

The system structure is described in a SysML internal block diagram (IBD) given in Fig. 2.

In this model, one can notice that the flows between components are typed regarding each domain: human (presented in black), energy (presented in green), cyber (presented in blue), and physical (presented in red). This classification can have many benefits. It can for instance help in extracting different viewpoints (or layers) of the system according to one or more facet(s) in order to ease analyses performed by relevant domain designers. As an example, the different layers of the insulin pump system are given in Fig. 3.

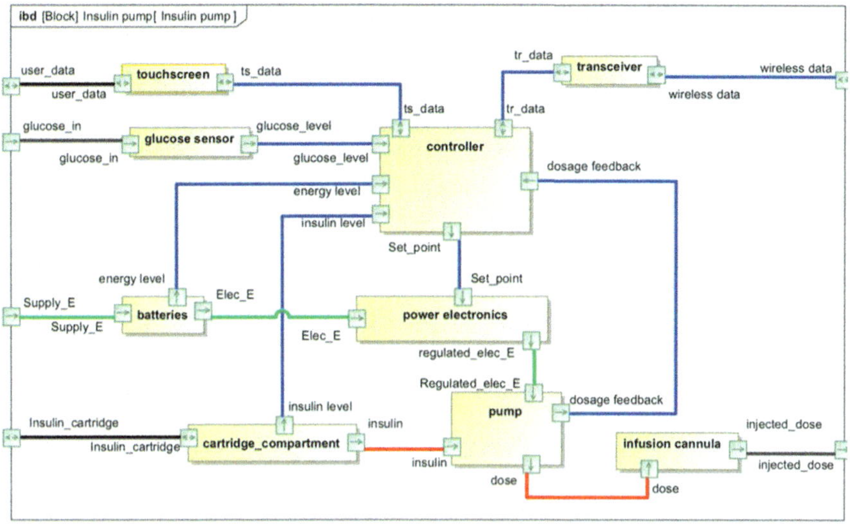

Fig. 2 Insulin pump IBD diagram with flows classification

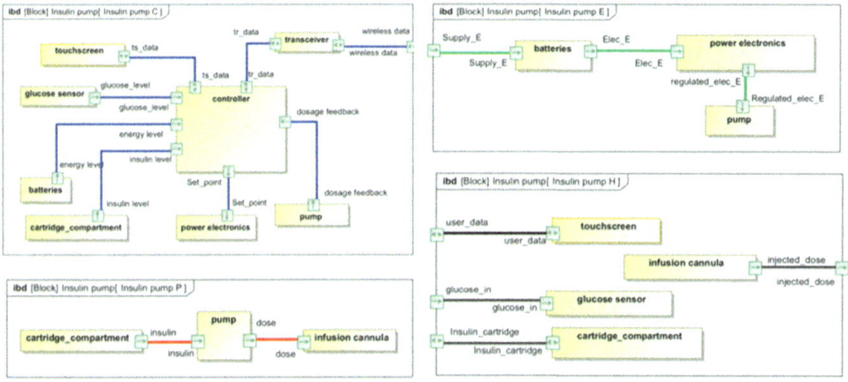

Fig. 3 Model layering regarding the four facets (human, energy, cyber, and physical)

A classification of components would be more intricate since most of the components have more than one facet. For instance, many components need energy to operate (even if sometimes it may be negligible). All components (except software ones if represented separately from their hardware supports) are physical and thus subject to physical constraints and concerns. Physical concerns may contribute to reducing the environmental impact through an adequate material choice, and friction reduction to reduce energy consumption and dust. Cyber and human facets on the other hand usually concern fewer components. Nevertheless, we can attribute different labels to components according to the different types of flows that cross them.

Specification Matrix for EcoMechatronics Design

The system core postulate (CoPo) allows us to build an implication mapping of all other postulates and hypotheses regarding each of the four proposed system facets for a given system. Thus, as shown in Table 1, it is possible to specify a set of complementary constraints, engineering proposals, guidelines, standards, templates, etc. Each cell of this table has to be customized for a given industrial or engineering domain (energy, transportation, medicine, etc.) and can help capitalize specifications.

In the MBSE design process, during a black-box analysis, a white-box analysis, and a potential safety analysis, all the information registered in this table have to be taken into account in order to better define the requirements, the system context, the functional and dysfunctional behavior of the system, and the functional and components-based system architectures. As an illustration, Table 2 proposes an instantiation of this EcoMechatronics specification matrix applied for the insulin pump.

We can notice that information in columns "MBSA/Safety critical hypothesis" and "MBSE/Unintended behavior hypothesis" may overlap, be complementary or not. Indeed, an unintended behavior may not be safety-critical. In the same way, safety considerations may not be related to unintended behavior.

Conclusions, Future Work

In this chapter, we have proposed a new conceptual framework taking into account the new context for mechatronics with associated constraints such as eco-design, low energy and resources consumption, and human friendly. Some postulates and associated hypotheses are also proposed, to structure the design activities while taking

Table 1 Specifications matrix for EcoMechatronics design

Mapping of constraints, engineering proposals, guidelines, standards, templates, etc. for EcoMechatronic design of a given system					
System views according to	MBSA analysis	Black box (MBSE) analysis		White box (MBSE) analysis	
CoPo system core postulate	SCHy safety critical hypothesis	SRPo system rational (reqs, etc.) postulate	DRHy dynamic requirements and context hypothesis	SBPo structure and behavior postulate	UBHy unintended behavior hypothesis
Human	H-SCHy	H-SRPo	H-DRHy	H-SBPo	H-UBHy
Energy	E-SCHy	E-SRPo	E-DRHy	E-SBPo	E-UBHy
Cyber	C-SCHy	C-SRPo	C-DRHy	C-SBPo	C-UBHy
Physical	P-SCHy	P-SRPo	P-DRHy	P-SBPo	P-UBHy

Table 2 Specification matrix for the insulin pump

System views

System	MBSA (Rams)	MBSE (Black box)		MBSE (White box)	
P1: system core post	H1: safety critical hypothesis	P2: system rational postulate	H2: dynamic reqs and context hypothesis	P3: structure & behavior post	H3: unintended behavior hyp
Human	– User may damage system – System may harm user	– Autonomous process – Ergonomic requirements – Medical relevance – Economy relevance	– New medical practices' ability – Context (home, office, leisure) – Cleaning context	– Touchscreen IHM – May include a mobile phone	– Wrong dosage – No monitoring – No communication
Energy	– Prohibit dangerous energy source – Prohibit dangerous energy usage (consumption, transformation)	– High efficiency required: low consumption, autonomy – Use green energy – Harvest energy	– Day and night mode – Warm/cold conditions – Waterproof – Modularity to handle new kind of energy technology	– PV, micro fuel cell – Sudation chemical energy – Mechanical energy (micro flywheel, spring)	– Short circuit – Faulty connection – Overconsumption
Cyber	– Internal flaws (failure, bug) – Security (cyber attack) – EMC	– Ability to be connected (network, notebook, telecom)	– Upgradable for new IT and telecom (xG) – Software modularity	– May include a mobile application – Dev. language – EMC protection	– Faulty connection – Cyber attack – Bug/failure – EMC pollution
Physical	– Maintainability – Reliability – Availability – Vibrations and thermal conditions	– Trade-off between space/mass/duration/price optimization versus recyclability	– Upgradable for new technologies/material – Components modularity	– Materials, production (RAMS, pollution, toxicity, cost) – Multiphysics to be taken into account	– Leak or clogged – Overheating – Toxic emission – Material or component defect

into account the four complementary domains (human, energy, cyber, and physical) on which today's systems rely with mechatronics acting as interfaces between these domains.

We also proposed some guidelines to enhance an existing MBSE methodology based on the layering of the system model and a specification matrix for EcoMechatronics, with respect to the four domains. This matrix could contain constraints (such as integrating specific components or technologies or standard specifications), design guidelines, and other specifications as well as different EPIs, and other eco-design criteria relative to the system to be developed. The multilayer modeling allows verifying that all the specifications have been taken into account.

A further step could be the implementation of the model layering according to the four above-mentioned aspects allowing the designer to extract views with selected aspects. A database of criteria relative to each domain could also be established and updated all along the projects to support the designer in considering all the appropriate constraints.

References

1. Heslouin C, Perrot-Bernardet V, Cornier A, Perry N (2017) A user oriented framework to support environmental performance indicators selection. Procedia CIRP 61:709–714. https://doi.org/10.1016/j.procir.2016.11.211
2. Merschak S, Hehenberger P (2019) Ecodesign methods for mechatronic systems: a literature review and classification. In: 2019 20th international conference on research and education in mechatronics (REM), Wels, Austria, mai 2019, pp 1–8. https://doi.org/10.1109/REM.2019.8744105
3. Favi C, Marconi M, Germani M, Mandolini M (2019) A design for disassembly tool oriented to mechatronic product de-manufacturing and recycling. Adv Eng Inform 39:62–79, janv. 2019. https://doi.org/10.1016/j.aei.2018.11.008
4. Chouinard U, Pigosso DCA, McAloone TC, Baron L, Achiche S (2019) Potential of circular economy implementation in the mechatronics industry: an exploratory research. J Clean Prod 239:118014, déc. 2019. https://doi.org/10.1016/j.jclepro.2019.118014
5. Bougain S, Gerhard D (2017) Integrating environmental impacts with SysML in MBSE methods. Procedia CIRP 61:715–720. https://doi.org/10.1016/j.procir.2016.11.196
6. Mhenni F, Choley J-Y (2021) Mechatronics in a new context: a structuring conceptual framework proposal for MBSE. In: 2021 18th international multi-conference on systems, signals & devices (SSD), Monastir, Tunisia, mars 2021, pp 1459–1464. https://doi.org/10.1109/SSD52085.2021.9429445
7. Mhenni F, Choley JY, Penas O, Plateaux R, Hammadi M (2014) A SysML-based methodology for mechatronic systems architectural design. Adv Eng Inform 28(3):218–231. https://doi.org/10.1016/j.aei.2014.03.006
8. Penas O, Plateaux R, Patalano S, Hammadi M (2017) Multi-scale approach from mechatronic to cyber-physical systems for the design of manufacturing systems. Comput Ind 86:52–69, avr. 2017. https://doi.org/10.1016/j.compind.2016.12.001
9. Choley J-Y, Mhenni F, Nguyen N, Baklouti A (2016) Topology-based safety analysis for safety critical CPS. Procedia Comput Sci 95:32–39. https://doi.org/10.1016/j.procs.2016.09.290

Concurrent Multi-domain Modelling and Simulation for Energy-Efficient Mechatronic Systems

Stanislao Patalano and Ferdinando Vitolo

Abstract Nowadays, mechatronic systems are used in every field and all around us, going through micro, small, medium and large power systems. Due to their widespread use, mechatronic system design plays an important role in energy sustainability. Mechatronic systems are generally complex systems from a modelling point of view due to the presence of several domains and a set of related interfaces for energy exchanges. Such a context needs a multi-domain and integrated design to correctly evaluate mechatronic system performances. Modelling and simulation tasks are generally adopted to evaluate the behaviour of mechatronic systems in order to meet system requirements. Unfortunately, energy saving is not always considered among the set of primary requirements, while it is tackled only after reaching the expected system behaviour. The present chapter tackles the multi-domain modelling and simulation of mechatronic systems according to V-model steps and using the RFLP approach driven by sustainability requirements. The key idea is to perform a concurrent evaluation of energy and domain performances. It summarises a case study, by using Matlab/Simscape environment, to illustrate the used approach. Moreover, it introduces the design of mechatronic systems pulled by environmental sustainability as a new paradigm in opposition to the design pushed by system performances.

Introduction

The research and development of new technical solutions for mechatronic systems provided a continuous improvement for such systems; we assisted over time, for example, to different integration levels in mechatronic systems: (i) separated; (ii) joined; (iii) included; (iv) merged [1]. Such a technological evolution induced a wide diffusion of mechatronic systems, and such systems operate by managing micro,

S. Patalano · F. Vitolo (✉)
Fraunhofer J-Lab IDEAS, Department of Industrial Engineering, University of Naples Federico II, 80125 Naples, Italy
e-mail: ferdinando.vitolo@unina.it

S. Patalano
e-mail: patalano@unina.it

P. Hehenberger et al. (eds.), *EcoMechatronics*,
https://doi.org/10.1007/978-3-031-07555-1_8

small, medium and large power. Therefore, today, it is mandatory to address the energy efficiency of mechatronic systems, as they can significantly impact the carbon footprint of final consumer products.

In such a context, multi-domain object-oriented modelling is a significant and useful methodology for energy consumption assessment, but several challenges are still open. A significant survey evidenced the actual contribution in the literature to methods devoted to eco-design and the actual challenges related to eco-design issues [2]. Even if mature software environments are today available for numerical simulation [3], several deficiencies are still existing, and they are related to methodological issues. In particular, two challenges could be highlighted: (i) how to assure, focusing on the design and development phase, a sustainable mechatronic system; (ii) how to perform an effective concurrent multi-domain modelling, in the presence of opposite needs related to different domains involved (for example, electrical and mechanical domains).

Performance-Based Versus Sustainability-Based Design

Let's take a carrying system composed of a drive wheel engaged with a gearmotor actuated by a control unit that modulates the system acceleration and deceleration (Fig. 1).

Let's compare the requirements defined in a classical design process, i.e. a performance-based process and the requirements defined assuming energy consumption as primary requirements, i.e. a sustainability-based process. Table 1 summaries the requirements classified as follows:

- Basic requirements are the requirements specified to assure the basic performances of the system;
- Primary requirements are the requirements that have to be met with the highest priority; therefore, they mainly drive the design process.

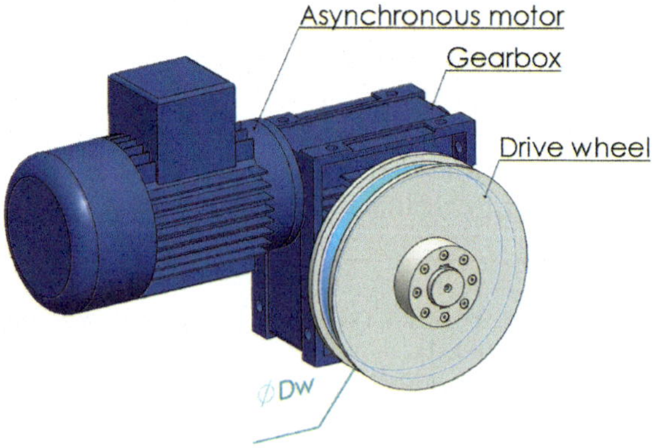

Fig. 1 Drive wheel system

Table 1 Different requirement lists for performance-based and sustainability-based designing process of a drive wheel system

	C1: System requirements pushed by performances	C2: System requirements pulled by sustainability
Basic requirements	System has to carry a mass m = 10000 kg	
	Stroke s = 20 m	
	Operating speed $0.15\,\text{m/s} \le v \le 0.30\,\text{m/s}$	
	Wheel Diameter $D_w = [0.2, 0.25, 0.30]\,\text{m}$	
Primary requirements	Acceleration time $t_{tr} = 5\,\text{s}$	Energy consumption $E \le 0.55\,\text{kWh}$

In the carrying system development phase, the requirements are different for the transitional state: within a performance-based process (see C1 in Table 1) only mechanical performances are defined and, in particular, the acceleration time is a primary constraint; within a system requirements definition pulled by sustainability (see C2 in Table 1) and, taking into account energy consumption, a different primary requirement is fixed, accordingly.

Let's see the carrying system. By assuming for both cases:

- a constant resistive force $F_r = 10000 * D_w[N]$;
- worm gearbox (speed ratio $1/\tau = [40, 50, 60, 70, 80, 90, 100]$ and efficiency $\eta = [0.80, 0.77, 0.73, 0.7, 0.67, 0.630.6]$) engaged with an three-phase asynchronous motor with 4 poles characterised by an angular speed $n_m = 1400\,\text{rpm}$ at nominal torque,

Let's define the admissible gearbox speed ratio. It can be computed as follows:

$$\frac{1}{\tau} = \frac{n_m * \frac{2\pi}{60} * D_w}{2 * v} = \begin{cases} 97.7 \rightarrow 100 & \text{per} \quad D_w = 0.20\,m, v = 0.15\,\text{m/s} \\ 49 \rightarrow 50 & \text{per} \quad D_w = 0.20\,m, v = 0.3\,\text{m/s} \\ 122 \rightarrow na & \text{per} \quad D_w = 0.25\,m, v = 0.15\,\text{m/s} \\ 61 \rightarrow 70 & \text{per} \quad D_w = 0.25\,m, v = 0.3\,\text{m/s} \\ 146 \rightarrow na & \text{per} \quad D_w = 0.30\,m, v = 0.15\,\text{m/s} \\ 73 \rightarrow 80 & \text{per} \quad D_w = 0.30\,m, v = 0.3\,\text{m/s} \end{cases} \tag{1}$$

Case C1—Neglecting energy aspects, it is plausible the event of covering the total stroke occurs within a lesser time. This theory brings about the highest speed $v = 0.3\,\text{m/s}$.

Let's take into account the inertia force as an added resistive force during acceleration. The inertia force can be computed as follows:

$$F_i = m * a = m * \frac{v}{t_{tr}} = 10000 * \frac{0.3}{5} = 600\,\text{N} \tag{2}$$

The system is characterised by a steady state (sr) and transient regime (tr) which require a different power to perform the task. The resistive torque at the motor shaft and the required motor power are computed as follows:

$$M_{tr} = (F_r + F_i) * \frac{D_w}{2} * \frac{tau}{\eta} = \begin{cases} \frac{(10000*0.20+600)*0.20}{2*50*0.77} = 6.75\,\text{Nm} \\ \frac{(10000*0.25+600)*0.25}{2*70*0.70} = 7.91\,\text{Nm} \\ \frac{(10000*0.30+600)*0.30}{2*80*0.67} = 10.07\,\text{Nm} \end{cases} \tag{3}$$

Selecting the configuration with the minimum resistive torque, the system could be characterised by a $D_w = 0.2m$ and a gearbox with $1/\tau = 50$ and $\eta = 0.77$.

$$M_{sr} = (F_r) * \frac{D_w}{2} * \frac{tau}{\eta} = \frac{(10000 * 0.2) * 0.20}{2 * 50 * 0.77} = 5.2\,\text{Nm} \tag{4}$$

$$P_{tr} \geq M_{tr} * n_m * \frac{2\pi}{60} = 6.75 * 1400 * \frac{2\pi}{60} = 0.99\,\text{kW} \tag{5}$$

$$P_{sr} \geq M_{tr} * n_m * \frac{2\pi}{60} = 5.2 * 1400 * \frac{2\pi}{60} = 0.76\,\text{kW} \tag{6}$$

During the total stroke, it can be defined into three phases: acceleration phase, steady-state phase and deceleration phase. Assuming the same slope for the acceleration and deceleration phases, the stroke time (t_s) and the power average (P) can be computed as follows:

$$x_{tr} = \frac{v}{2} * t_{tr} = \frac{0.3}{2} * 5 = 0.75\,\text{m} \tag{7}$$

$$t_{sr} = \frac{s - 2 * x_{tr}}{v} = \frac{20 - 2 * 0.75}{0.3} \approx 62\,\text{s} \tag{8}$$

$$t_s = t_{sr} + 2 * t_{tr} = 62 + 2 * 5 = 72\,\text{s} \tag{9}$$

$$P = \frac{P_{sr} * t_{sr} + 2 * P_{tr} * t_{tr}}{t_s} = \frac{0.76 * 62 + 2 * 0.99 * 5}{72} = 0.79\,\text{kW} \tag{10}$$

Case C2—Let's take into account a maximum motor power $P = 0.55\,\text{kW}$. In this condition, it is plausible the event of covering the total stroke occurs by using a lesser admissible speed. By Eq. (1), there is only one admissible wheel diameter $D_w = 0.2\,\text{m}$. Therefore, assuming the efficiency $\eta = 0.6$ for gearmotor verification, the speed ratio can be defined as follows:

$$\frac{1}{\tau} = \frac{(F_r + F_i) * \frac{D_w}{2}}{\frac{P*\eta}{n_m} * \frac{60}{2\pi}} = \frac{(10000 * 0.2) * \frac{0.2}{2}}{\frac{550*0.6}{1400} * \frac{60}{2\pi}} = 89 \rightarrow 90 \tag{11}$$

$$v = n_m * \frac{2\pi}{60} * \tau * \frac{D_w}{2} = 1400 * \frac{2\pi}{60} * \frac{1}{90} * \frac{0.2}{2} = 0.16\,\mathrm{m/s} \tag{12}$$

$$M_m = \frac{P}{n_m} * \frac{60}{2\pi} = \frac{550}{1400} * \frac{60}{2\pi} = 3.75\,\mathrm{Nm} \tag{13}$$

By Eqs. (2) and (3), the acceleration time t_{tr} can be computed that can be provided by

$$t_{tr} = \frac{m * v}{F_i + F_r} = \frac{m * v}{\frac{M_m * 2 * \eta}{D_w * \tau} - F_r} = \frac{10000 * 0.16}{\frac{3.75 * 2 * 0.63 * 90}{0.2} - 10000 * 0.2} = 12.7\,\mathrm{s} \tag{14}$$

$$F_i = m * a = m * \frac{v}{t_{tr}} = 10000 * \frac{0.16}{12.7} = 126\,\mathrm{N} \tag{15}$$

$$M_{tr} = (F_r + F_i) * \frac{D_w}{2} * \frac{tau}{\eta} = \frac{(10000 * 0.2 + 126) * 0.20}{2 * 90 * 0.63} = 3.75\,\mathrm{Nm} \tag{16}$$

$$M_{sr} = F_r * \frac{D_w}{2} * \frac{tau}{\eta} = \frac{(10000 * 0.2) * 0.20}{2 * 90 * 0.63} = 3.53\,\mathrm{Nm} \tag{17}$$

$$P_{tr} \geq M_{tr} * n_m * \frac{2\pi}{60} = 3.75 * 1400 * \frac{2\pi}{60} = 0.55\,\mathrm{kW} \tag{18}$$

$$P_{sr} \geq M_{sr} * n_m * \frac{2\pi}{60} = 3.53 * 1400 * \frac{2\pi}{60} = 0.52\,\mathrm{kW} \tag{19}$$

$$x_{tr} = \frac{v}{2} * t_{tr} = \frac{0.16}{2} * 12.7 = 1.02\,\mathrm{m} \tag{20}$$

$$t_{sr} = \frac{s - 2 * x_{tr}}{v} = \frac{20 - 2 * 1.02}{0.16} \approx 112.3\,\mathrm{s} \tag{21}$$

$$t_s = t_{sr} + 2 * t_{tr} = 112.3 + 2 * 12.7 = 137.7\,\mathrm{s} \tag{22}$$

$$P = \frac{P_{sr} * t_{sr} + 2 * P_{rt} * t_{tr}}{t_s} = \frac{0.52 * 112.3 + 2 * 0.55 * 12.7}{137.7} = 0.53\,\mathrm{kW} \tag{23}$$

Comparing Eqs. (10) and (23) figures out 0.26 kWh of energy saving. Assuming 20 years of operating time and 10 h of operating time per day, an energy saving of 18,980 kWh equal to 1329 € (for 0.07 €/kWh) and 4840 kg of saved CO_2 (0.255 kg CO_2/kWh) can be estimated.

The Design and Development of Sustainable Mechatronic Systems

The methodology to assure the design and development of sustainable mechatronic systems has been definitively accomplished and validated. Model Based Systems Engineering (MBSE) is considered the most effective methodology for mechatronic system development, verification and validation by adopting the V-cycle approach [4–8]. In particular, the VDI2206 guideline is a practice-oriented guideline for the systematic development of a mechatronic system, and it provides a functional modelling methodology based on the V-model. VDI2206 proposes the integration of different methods to create a system model able to capture and process: (i) information about the system purpose, and (ii) the functions of its components to fulfil the system purpose [9]. A significant contribution of VDI2206 is that mechatronic system design can be structured into four major phases, called "system design", "domain-specific design", "system integration" and "assurance of properties" (Fig. 2).

A declination of the V-model proposed for mechatronic systems is the RFLP approach. The descending branch of V-model is organised into 4 views: Requirement engineering view, Functional view, Logical view and Physical view. In particular, in the Logical view, multi-domain software environments and tools are used to model and carry out numerical analysis, such as Modelica, Matlab/Simulink and Bondgraphs.

Finally, an effective eco-design methodology for mechatronic systems cannot fail to consider the entire life cycle of the system. In particular, an eco-design methodology must consider the four main phases of the life cycle, as stated in [2]: (i) Development; (ii) Production; (iii) Usage, and (iv) End-of-Life (Fig. 3). Taking into account these four phases, it is possible to integrate the influences of Production, Usage and EoL phases in the Development phase; such integration also allows the ecological assessment of the mechatronic system. The integration of "successive" phases of the system life cycle is generally performed by collecting data related to

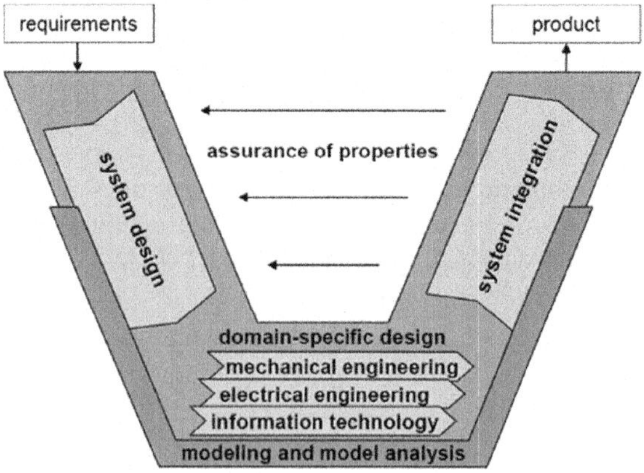

Fig. 2 V-cycle from VDI2206 [10]

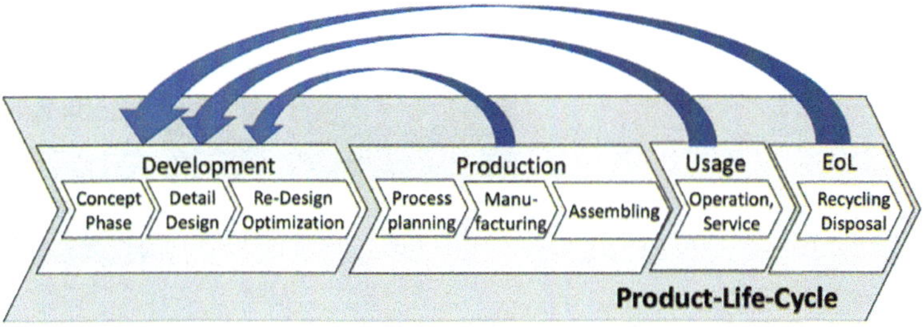

Fig. 3 Life cycle of a mechatronic system [2]

Development (CAD data, Material data), Production (Tool data, Production capacity and Logistic data), Usage (Use data, Operating conditions), and EoL (Recycling data, Disposal Data) [2].

The Concurrent Multi-domain Modelling and Simulation for Mechatronic System Development

The present approach for mechatronic system development tackles the mechanical and electrical domains considered by VDI2206 as mechatronic system performances are widely influenced by the development phase involving such domains.

The key concept of the approach is that the mechatronic system development is not "pushed by" system performances but is "pulled by" environmental sustainability, as depicted in Fig. 4.

Furthermore, in the presence of opposite needs related to mechanical and electrical domains, concurrent multi-domain modelling is used according to the VDI2206 V-cycle and the corresponding declination using RFLP views. In particular, concurrent modelling refers to the concurrent development of the model, related to the involved domains, by performing a zig-zagging among them in order to accomplish the same level of detail—for subsystems and component development—through all domains. In such a context, there does not exist a "domain-specific design" but only the concurrent design of the whole system. Simulation is then the way to assign operational input and verify system performances thus anticipating the integration of the usage phase with the development phase.

Case Study: Hangar Sliding Door System

Huge sliding doors are the usual equipment of aircraft hangars and shipyard plants. it is machinery characterised by redundant safety to avoid failures and collisions with itself, aircraft and/or human beings. Its general configuration is composed of

Fig. 4 Key concept of the approach for mechatronic system development

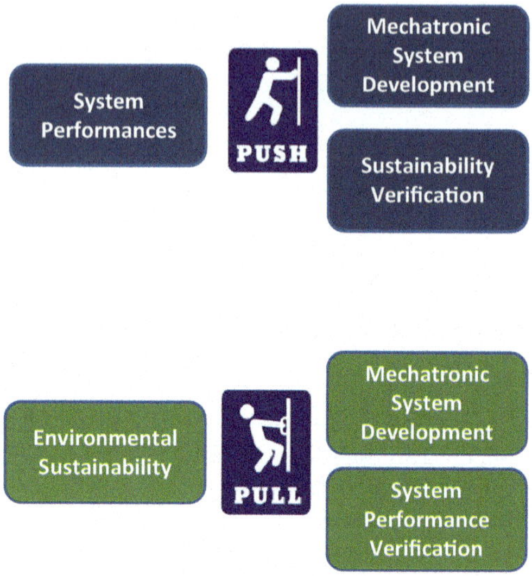

an embedded control unit and gearmotors which drive the side sliding of leaves guided by bottom and top rails that are either continuous or staggered (Fig. 5). Due to overall doorway size (up to width = 100 m height = 30 m), the hangar sliding door could require up to 30 kWh energy consumption. Therefore, based on operating energy consumption, the hangar sliding door can be classified as an energy-eater mechatronic system. This section compares the performance-based and sustainability-based design approaches applied to a hangar sliding door system by figuring out workflow and result differences.

The case study faces a hangar sliding door design with two continuous parallel rails; it closes with a doorway width of 33.5 m and height of 10 m. The automation level is dead-man which means that the operator has to press the button for sliding the door; otherwise, the door is fixed.

Requirement Definition

The first step of the RFPL approach consists of requirement definitions. Table 2 shows all the defined requirements for both performance-based and sustainability design approaches, and it highlights the basic and the primary requirements.

The requirement lists have been implemented within Simulink Requirements Environment. For each requirement, a hierarchy and a class have been assigned. Initially, all the requirements were assumed as functional requirements that should be implemented and verified; clearly, during construction, it was necessary to make changes distinguishing

- Functional—the requirements that could be implemented and/or verified;
- Container—the requirements that served to contain sub-requirements and that did not deserve to be implemented and verified;

Table 2 Different requirement lists for performance-based and sustainability-based designing process of an Hangar Sliding Door System

	HD1 Performance-based requirements	HD2 Sustainability-based requirements
Basic requirements	System has to carry a mass $m = 10000\,\text{kg}$	
	Leaves	
	The system must be composed of 4 leaves	
	The system has to have a sliding mass of 3500 kg	
	The system must be transportable	
	Drive wheel diameter = [200 230 260] mm	
	Leaf must have an embedded powertrain	
	Leaf must be actuated by a three-phase asynchronous motor	
	Power supply must be 400 versus 50 Hz	
	Power connection must slide with leaf	
	Safety requirements	
	Door must not overturn	
	The system must avoid collision with people and itself	
	The system must have a redundant collision detector	
	Maintenance	
	The system must guarantee access to electrical and mechanical parts	
	Material	
	The material must not be flammable	
	The visible material colour must be anti-reflective	
	The system must be resistant to corrosion	
	The system must be impact-resistant and shock-proof	
	The system must guarantee thermal insulation	
Primary requirements	The sliding speed must be 0.17 m/s (no exceed)	The sliding speed must be 0.17 m/s (no exceed)
	Leaf acceleration and deceleration time is 5 s	Overall energy consumption must not exceed 1 kWh

- Informational—the requirements that provide additional information and do not contribute to implementation and verification;
- Justifications—all those requirements that are verified outside Matlab such as requirements that have been developed in a CAD or that simply complied with European regulations.

Functional Architecture

The functional architecture of the hangar sliding door has been implemented as depicted in Fig. 6. The "*obstacle signal*", "*operator command*" and "*power source*" ports have been defined as inputs. The "*Receive command data signal*" block received

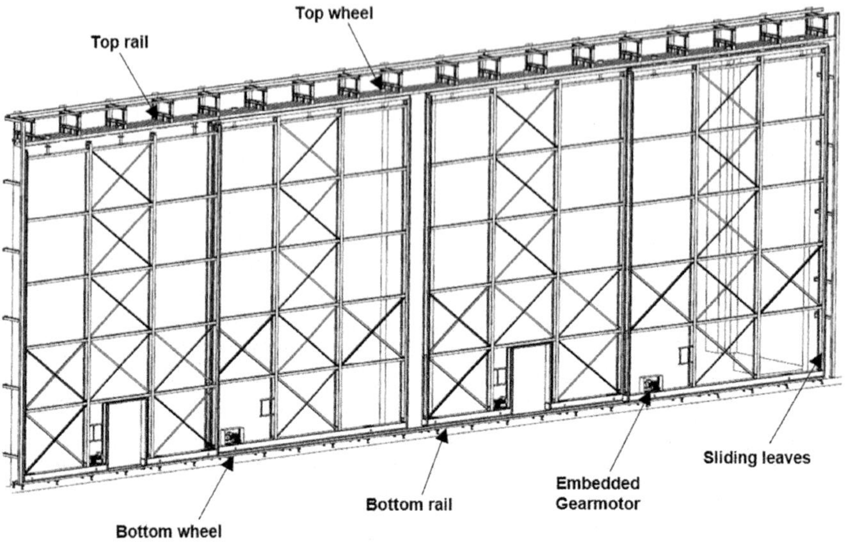

Fig. 5 Hangar sliding door system –2 + 2 continuous parallel sliding leaves

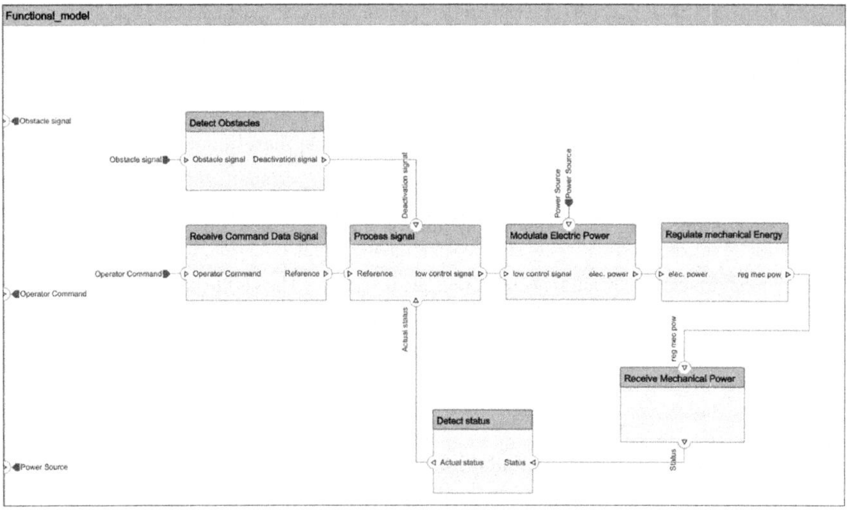

Fig. 6 Hangar sliding door system—door functional architecture

a command from the operator, and sent a reference signal to the "*Process signal*" which compares it with the signal coming from the "*Detect obstacles*" block which is responsible for detecting obstacles during operating states. The "*Process signal*" output reaches the "*Modulate electric power*" block that modulates electric power in order to supply "*Regulate mechanical energy*" block which regulates power for

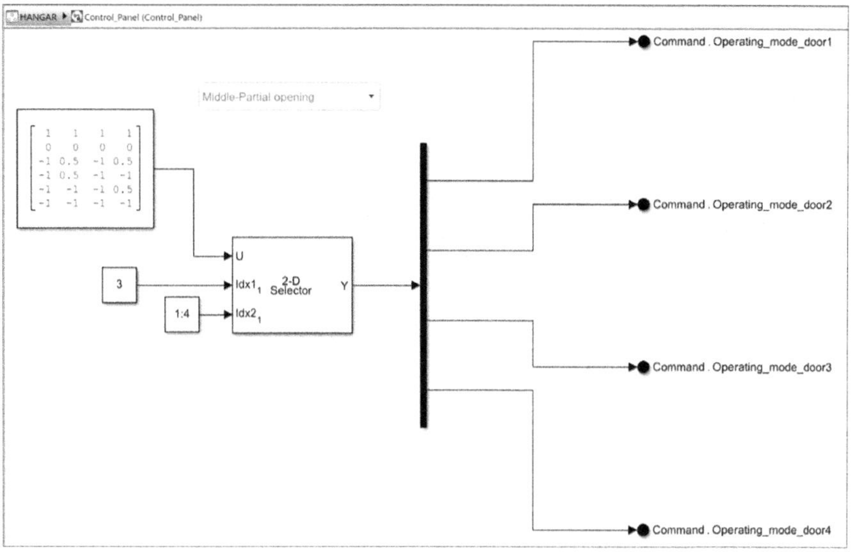

Fig. 7 Logical architecture of the hangar sliding door system—control panel block

mechanical parts of the system represented by the "*Receive mechanical power*" block.

Moreover, "Modulate electric power" block has got an external link through the power source port.

Logical Architecture

Logical architecture is composed of two main blocks:

- Control Panel (Fig. 7) that represents the operator interface;
- Hangar system (Fig. 8) that is the whole electromechanical system.

This last logical block has been decomposed into the logical elements necessary for the low-level description of the system.

The hangar system block is composed of

- an electromechanical block for each door;
- a controller block for each door;
- an obstacle block that represents both photocells (used to detect obstacles) and load cell (used to detect collisions);
- a multibody block that gives a 3D representation of the whole system.

Multi-domain Modelling

A digital model of the real system is created by using blocks belonging to the different domains involved within the system. The multi-domain physical model mainly characterises the performance-based design from the sustainability-based design; in fact,

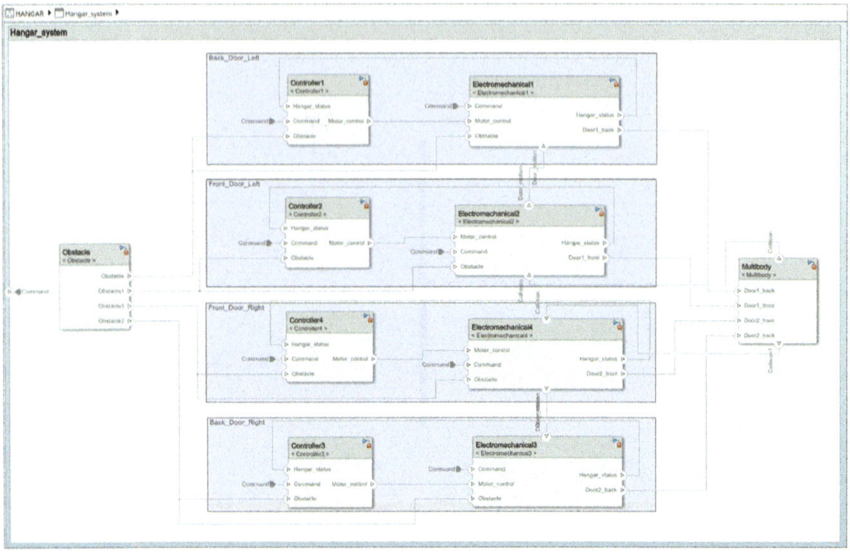

Fig. 8 Logical architecture of the hangar sliding door system—hangar system block

the two respective models are only separated by the control logic for pursuing the speed target; both models present the same architecture as depicted in Fig. 9. The controller block of the performance-based model modulates the leaf acceleration by using an inverter programmed in order to obtain an acceleration slope of 5 s. The controller block of the sustainability-based model modulates the supply of energy in order to not overcome the specified limit for energy consumption.

The electromechanical model of a single leaf is shown in Fig. 8. The model is composed of a three-phase voltage source of 400 V (VRMS) and a frequency of 50 Hz. The source is connected to a circuit breaker; its goal is to turn off the power whenever a collision is detected by the collision sensor. An electric motor is driven by an inverter composed of a three-phase rectifier, an intermediate layer and an inverter. The intermediate layer is used to stabilise the rectified current, to avoid peak of current and to separate the source from the system. The inverter is commanded through the signal generated by the control block. That control signal can assume three values:

- value 1 is associated with the motion of the door in the same direction as the reference;
- value -1 is associated with the motion of the door in the opposite direction of the reference;
- value 0 is associated with the quiet condition of the door.

A three-phase asynchronous motor is represented by an ad-hoc subsystem block characterised by a heat power port. The source of heat power represents the active power dissipated according to the power factor and the efficiency of the machine.

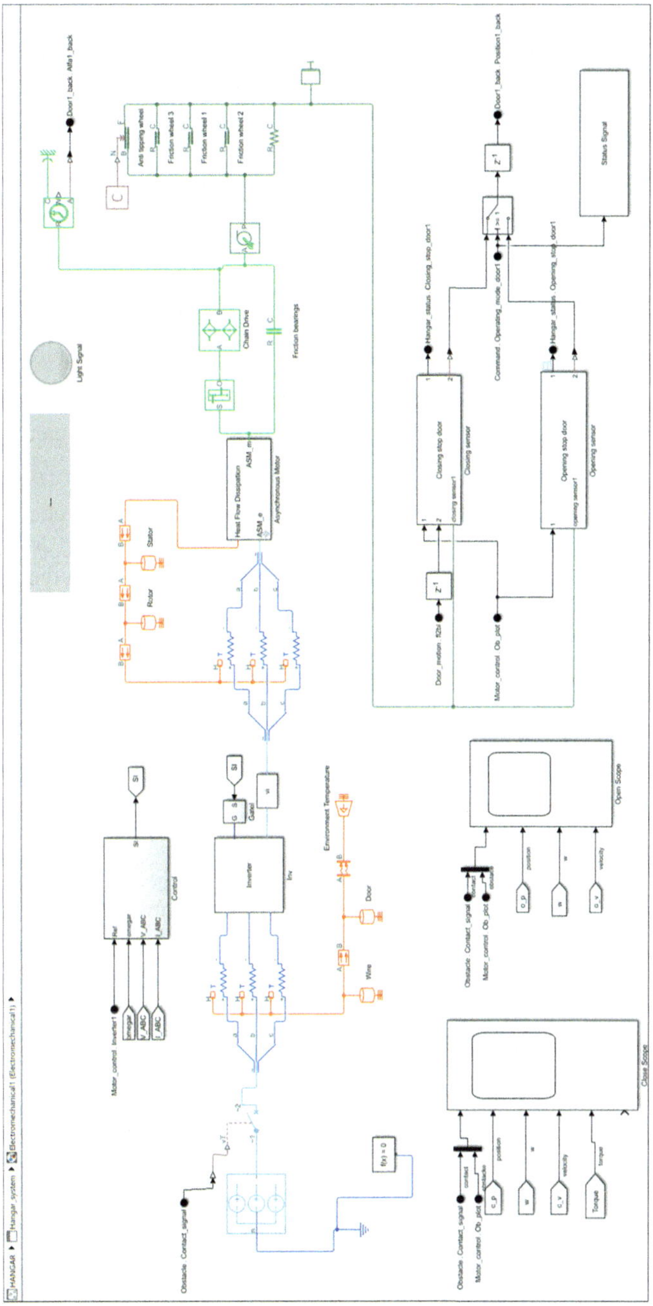

Fig. 9 Hangar sliding door system—electromechanical system of a single leaf modelled in Simscape environment

Motor power reaches the chain transmission and the rotational-translational conversion of the motion. The mechanical behaviour of the door is generated by using several parallel blocks: a block models the mass, a block models the stiffness, three blocks model the friction between the motion wheel and the rail track, and a block models the friction between the antitipping wheel and the upper track due to the wind pressure. Each door has been equipped with two limit switches: one for closing and the other for opening. The closing limit switch is modelled by a position sensor. According to the position, the limit switch will provide a Boolean signal equal to 1 (24 V signal) if the door is moving, or a signal equal to 0 (0 V signal) if the door reaches the limit position. In case of an obstacle, the reverse of the motion is provided to guarantee the re-opening of the door. In case of collision between leaves, the turn off of the system is provided.

The controller block has the goal to calculate the low power signal to command the inverter. According to the operating mode, the command signal can assume three different values:

- 1 (+24 V)—the operating mode is total closing. The limit switch signal is used as the second input and signal 1 will be sent to the inverter if the door is far from the limit switches; a signal 0 is sent if the door is in touch with the limit switch. In case of an obstacle, the signal to the inverter is set to −1 to have the reverse of the motion;
- 0 (0 V) the door has to keep the quiet condition, therefore, a 0 signal is sent to the inverter;
- −1 (−24 V)—the leaf must be opened. So, the inverter will be sent a −1 signal, when the door is between the limit switches, otherwise, a 0 signal will be sent to the inverter. In case of an obstacle, a 0 signal will be sent to the inverter to shut down the power supply.

The obstacle block has the goal to model both obstacles (detected by the photocells) and the collisions (detected by the load cells). The obstacle block is composed of trigger blocks commanded by random signals.

Model Simulation

The developed model of the hangar sliding door has been simulated by using "System Composer". The reference systems to compute the displacement of each leaf are depicted in Fig. 10.

Each leaf has been assumed as a point-mass (red dot), which slides during operating mode; the maximum clearance available is 33.8 m but two side pockets have been provided to accommodate the doors when they are fully open.

Figures 11 and 12 show the simulation results of the performance-based model (speed performance set on 0.1 m/s) for whole stroke and obstacle detection, respectively. Diagrams show that during operating mode without obstacle (Fig. 10), the obstacle signal is null, and the leaf slides for 16.9 m (starting from 25.35 up to 42.25 m) by reaching a maximum speed of about 0.1 m/s. Instead, in the case of obstacle detection (Fig. 11) the leaf stops its sliding when the obstacle signal changes from 0 to −1.

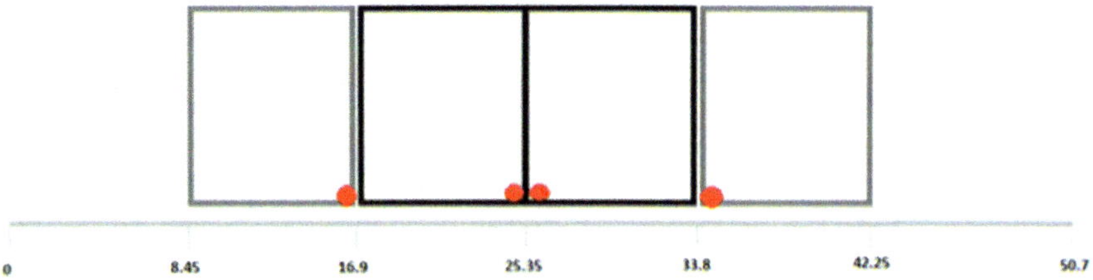

Fig. 10 Reference systems for hangar leaves

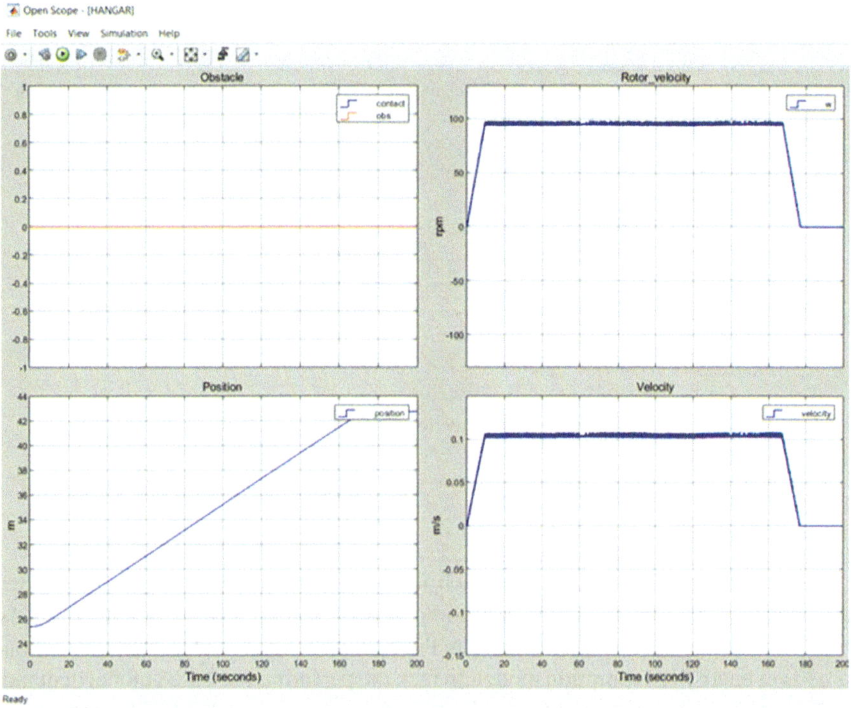

Fig. 11 Simulation results without obstacles

Figure 13 compares the speed and power supply during leaf sliding for performance-based and sustainability-based models set according to the requirement lists (Table 2). Diagrams show results without noises (low-pass filtered).

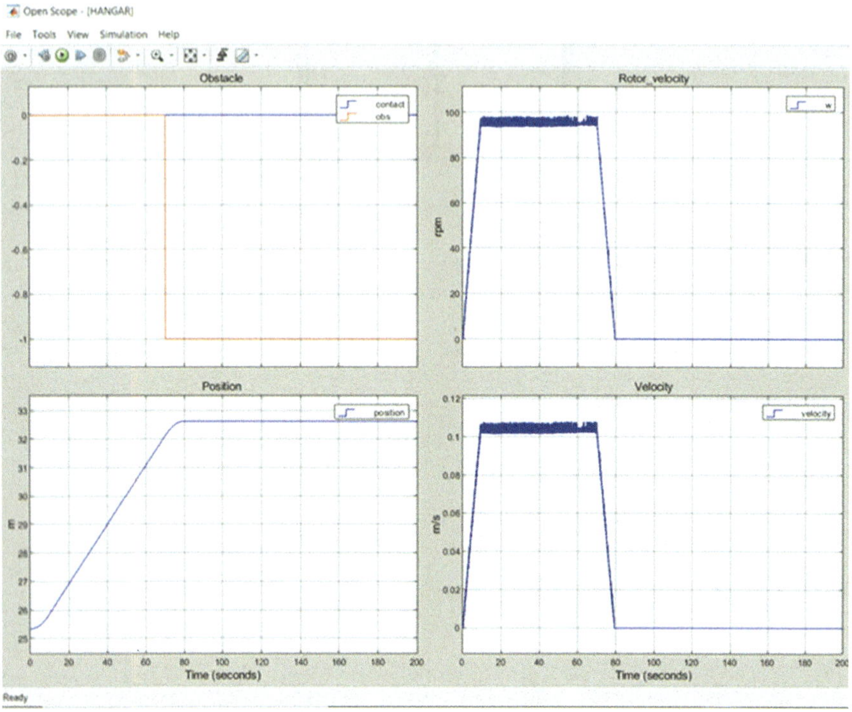

Fig. 12 Simulation results with obstacle detection

Discussion

As expected, different design approach lead to different results. Figure 13 shows that the leaves reach the reference sliding speed for both models. Although the performance-based model is faster than the sustainability-based model, the last one is a lower energy consumption model. In fact, the performance-based model demands 0.313 kW/h while the sustainability-based model demands 0.203 kW/h. It means an energy saving of 0.11 kW/h.

Assuming 20 years of operating time and 10 h of operating time per day, an energy saving of 8030 kWh equal to 562 € (for 0.07 €/kWh) and 2048 kg of saved CO_2 (0.255 kg CO_2/kWh) can be estimated.

Conclusions

Mechatronic systems are today widely used and therefore it is mandatory to introduce new paradigms and new approaches to both design and development phases in order to assure system environmental sustainability. Even if a consistent eco-design could

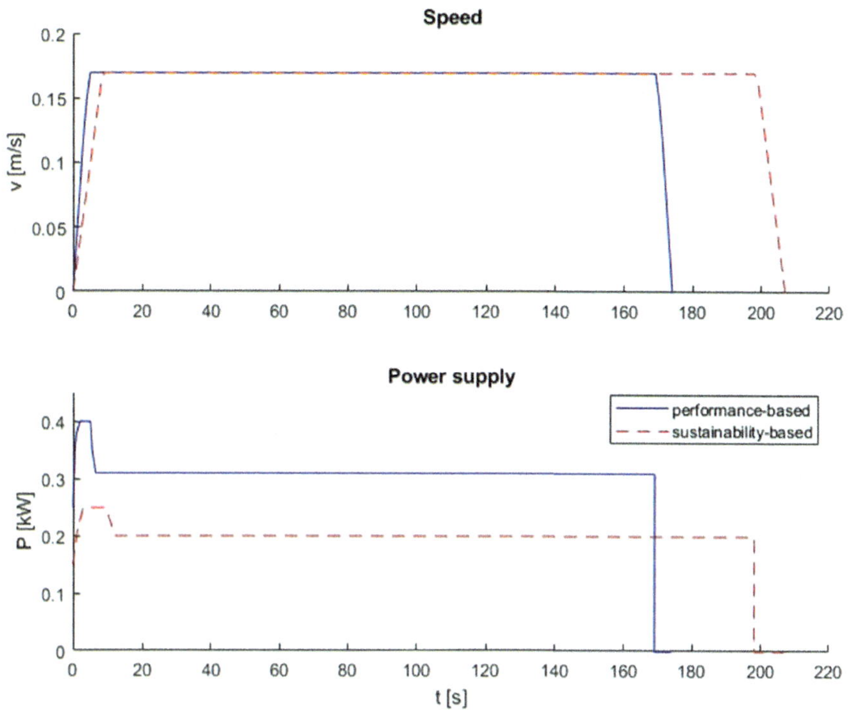

Fig. 13 Comparison between performance-based and sustainability-based simulation results

be fully accomplished only by tackling the whole system life cycle, it is possible to tend to an increasingly enhanced sustainability by adopting, the key concept that the mechatronic system development is not "pushed by" system performances but is "pulled by" environmental sustainability. The present work has demonstrated that mechatronic system design "pulled by" environmental sustainability assures energy saving as well as a reduction of CO_2 production with huge benefits thus enabling green and healthy design.

Further efforts must be made to perform a design "pulled by" environmental sustainability during the early phases of mechatronic system development as the high uncertainty for several design parameters could create a further limitation for the development of behavioural models and simulations.

Acknowledgements This study was developed with the economic support of European Commission under the remit of the H2020 project "ENERgy-efficient manufacturing system MANagement—ENERMAN", DT-FOF-09-2020, n. 958478.

References

1. Zheng C, Bricogne M, Le Duigou J, Eynard B (2014) Survey on mechatronic engineering: a focus on design methods and product models. Adv Eng Inform 28:241–257
2. Merschak S, Hehenberger P (2019) Ecodesign methods for mechatronic systems: a literature review and classification. In: 20th international conference on research and education in mechatronics (REM), pp 1–8. https://doi.org/10.1109/REM.2019.8744105
3. Ranjana S (2021) Simulation and analysis of modern power systems, Mc Graw-Hill
4. Akundi A, Lopez V (2021) A review on application of model based systems engineering to manufacturing and production engineering systems. Procedia Comput Sci 185:101–108
5. Kleiner S, Kramer C (2013) Model based design with systems engineering based on RFLP Using V6. In: Proceedings of the 23rd CIRP design conference. https://doi.org/10.1007/978-3-642-30817
6. Mhenni F, Choley J-Y, Penas O, Plateaux R, Hammadi M (2014) A SysML-based methodology for mechatronic systems architectural design. Adv Eng Inform 28:218–231
7. Hammadi M, Iebba M, Camargo Rosa F, Veneri O, Capasso C, Patalano S, Copertino G (2015) System approach to the pre-design of electric propulsion systems for road vehicles. In: Proceedings of IEEE international symposium on systems engineering, ISSE 2015, Rome, Italy, pp 229–236. ISBN: 978-147991920-8.https://doi.org/10.1109/SysEng.2015.7302762
8. Patalano S, Vitolo F, Lanzotti, A (2015) Object-oriented model validation for single bowden power window system development. In: Proceedings of IEEE international symposium on systems engineering, ISSE 2015, Rome, Italy, pp 150–154. https://doi.org/10.1109/SysEng.2015.7302748
9. VDI2206, DesignHandbook2206 (2003) Entwicklungsmethodik fur mechatronische Systeme/Design methodology for mechatronic systems. VDI Publishing Group, Dusseldorf
10. Bathelt J, Jonsson A, Bacs C, Dierssen S, Meier M (2005) Applying the new VDI design guideline 2206 on mechatronic systems controlled by a PLC. In: International conference on engineering design (ICED 05), Melbourne, Australia, pp 1–14
11. Capasso C, Hammadi M, Patalano S, Renaud R, Veneri O (2015) RFLP approach in the designing of power-trains for road electric vehicles. In: Sousse MH et al (eds) Multiphysics modelling and simulation for systems design and monitoring: proceedings of the multiphysics modelling and simulation for systems design conference, MMSSD 2014. Springer International Publishing, pp 249–258. ISBN 978-3-319-14531-0

Artificial Intelligence, Ethics and Privacy

Roger White and David Bradley

Abstract Mechatronics-based technologies are associated with the collection of large volumes of data associated with both systems operation and the users of those systems. This data has the potential to form a major resource in respect of applications ranging from healthcare to energy systems, transport and manufacturing. However, the analysis and interpretation of the data, particularly when associated with the application of Artificial Intelligence (AI), presents many challenges in relation to individual privacy, as well as wider ethical issues associated both with the individual and with the ability of such systems to autonomously make decisions which can impact on the wellbeing of both individuals and larger groups. The chapter, therefore, sets out not to provide specific answers or solutions to the many and various issues associated with the deployment of AI and Big Data analytics, but to help systems designers in identifying those questions that need to be asked. For this reason, the chapter includes a wide-ranging, but by no means all-inclusive or fully comprehensive, bibliography as well as more conventional references as a means of directing and aiding systems designers in identifying problem-specific solutions.

Introduction

The growth in the numbers and capability of devices able to capture and record a wide range of data, including personal data, while making it available for analysis is creating significant issues and raising significant questions in respect of system ethics and individual privacy, the needs of which need to be balanced against the potential benefits that could potentially be achieved through appropriate data interpretation and analysis in areas such as health.

R. White (✉)
RC2 Inc., Arden, ON, Canada
e-mail: rjw_rc2@icloud.com

D. Bradley
Abertay University, Dundee, UK

P. Hehenberger et al. (eds.), *EcoMechatronics*,
https://doi.org/10.1007/978-3-031-07555-1_9

Consider for instance the operation of a device such as the *AirTag* introduced by *Apple* in 2021. In essence, this enables a user to determine the location, typically within metres, of an object fitted with such a tag anywhere in the world, so (hopefully) no more lost suitcases at airports! However, the tag relies for its operation not on access to its owner's *iPhone* or *iPad*, which may be some significant distance away,[1] but on the ability to communicate via Bluetooth with other *Apple* devices in its vicinity which then act to relay the information to the owner's system [1]. While arguing that privacy is maintained as the *AirTag* itself does not contain user data, and that its data can only be accessed by the owner with the intermediary devices, and their owners, remaining anonymous, a number of vulnerabilities have been suggested, for instance, the owner of a device used by the tag as a remote node potentially being faced with roaming charges [2]. Other privacy issues exist with home-based devices and technologies such as video doorbells which have been used to generate profiles of behaviour, and hence establish if a house is occupied or not [3, 4].

While there is now considerable, and increasing, attention, being given to issues such as privacy in respect of applications of Artificial Intelligence (AI), for instance through the introduction of the General Data Protection Regulations (GDPR) in Europe [5], and more recently in the form of a white paper on the application of AI [6], this interest has only developed relatively recently. It is in this context that along with the usual list of references, the chapter also includes an extensive bibliography. The chapter itself comprises four main elements; a discussion on AI and related issues is followed by sections looking at ethics and ethical issues and privacy after which a number of exemplars are provided before the conclusions.

Artificial Intelligence

Despite the hype about the use of *Machine Learning* (ML) as a route to *Artificial Intelligence (AI)* in recent years, with the usual claims that "*we are just 5 years away from commercialising self driving cars, home robot maids,* etc.", there is a growing awareness that the current technology has a long way to go to meet the standards required for *Human Intelligence* (HI). Of potentially even more concern, governments, corporations and financial institutions are deploying machine learning in ways that have the potential to manipulate people into altering their behaviour to better fit the models being deployed to make many administrative processes more "*efficient*". In a quest for cost savings in the name of "*fairness*", "*diversity*" and "*preventing discrimination*", algorithms have been deployed by governments for identifying potential criminals, by corporate human resources (HR) in assessing potential employees, by universities in selecting potential students and more. As this practice has expanded, levels of technological and business innovation have

[1] Including on the other side of the world!

in general declined in advanced economies, workplaces have become less diverse in terms of socio-economic background and race, and the life chances of young offenders permanently blighted.

On the plus side, as a result of political turmoil as well as the acceleration of digital working through the Covid-19 pandemic, many issues surrounding the use of data-intensive modelling are now starting to be realised by the political establishment and wider public. Aside from obvious privacy concerns about the collection of large volumes of data from individuals by a small cohort of technology corporations whose raison d'être is to package and sell access to their databases for the training of algorithms, there is a growing awareness that algorithms based on historical data are only as good as the human assumptions that are built-in during their creation.

There are many areas where machine learning provides powerful insights that can assist researchers to identify patterns in complex data sets that would take decades for a human to uncover, if ever. This can dramatically enhance areas such as drug discovery, vaccine development, x-ray imaging and elsewhere where many interrelated mechanisms are in operation that make traditional analysis difficult. However, it is also becoming clear that making real-world decisions based on correlations between human behavioural factors using machine learning is fraught with ethical and practical issues that prevent any equivalence with human intelligence in the manner suggested by many AI cheerleaders, something which has on occasion led to the overselling of capability. For instance, the capabilities of the Watson Health system developed by IBM for cancer diagnosis and management were exaggerated by marketing while the data collection and training processes resulted in potentially biased outcomes [7].

At its core, the problem comes down to being able to state with confidence that correlation is demonstrating causation, i.e. when something has been shown to happen in a particular way historically, it does so as a result of an underlying invariant causal mechanism. Data reduction in bio-medical research is successful because by and large this condition is met, and a limited set of assumptions need to be applied to the analysis as performed. Once Human Intelligence starts to interact with Artificial Intelligence, causation becomes much more difficult to identify from historical correlations, and outputs tend to become a function of the assumptions applied by human programmers when developing algorithms.

How these assumptions are applied will, to a large extent, determine how close an AI prediction gets to a correct real-world prediction. So, in essence, AI can only be as good as the HI that develops it, leading to a range of perverse outcomes from reducing workforce diversity and increasing social inequity to reducing innovation.

Global supply chain issues are a classic example of the shortcomings of the AI-driven algorithmic resource planning that underpins global "*Just-in-Time*" delivery. While the global economy was acting in a historically predictable, steady-state fashion, then systems designed to ensure efficient use of supply chains were able to predict future trends with a high level of confidence. However, built in to this highly tuned model of operation were the seeds of its own destruction. Once the Covid-19 pandemic created the need for countrywide lockdowns in advanced economies, the predictive capability of these automated systems broke down as there was no longer relevant historic data on which to base decisions. As a result, the initial drop in economic activity was over predicted, while the equally unexpected upswing in online commerce was under predicted. With no slack in such a finely tuned system, chaos has resulted. The main lessons to draw from recent experiences are:

1. Machine learning is an effective method of extracting useful information from large complex sets of data requiring significant mathematical analysis, but is generally poor at predicting behaviour when unpredictable (human) factors change the existing paradigm.
2. Human intelligence is poor at using pure logic to extract knowledge from large, complex data sets, but good when applying empathic skills to determine solutions to problems not easily accessible to pure logic.

To improve the ability of AI as a predictive tool, without some completely novel approach to the computational processes that underpin it, will require a massive expansion of data capture capabilities as well as structural changes in how the algorithms that use the data collected are developed and deployed. This will represent either a major shift of power from democratic governments to technology corporations that collect the data, or a takeover by governments of the data collected on the citizens within their borders—with all the potential implications first laid out by George Orwell in *1984*. An additional practical outcome of the growth in data capture would be the unsustainable increase in the need for power generation with all the climate implications this would entail. A final, more philosophical question to answer with regard to AI is:

> If the software could be developed that matched human capabilities in areas such as empathy such that it could be said to be truly intelligent, would it actually perform any better than the human brain?

While the discussions on the use of automation in general, and AI in particular, have tended to focus on a binary good/bad argument, with technology companies pushing the efficiency benefits of AI, while social scientists stress the damage to mental health and social equity of replacing humans with machines, there is an alternative. Increasingly, as the focus has shifted towards the ethics of AI and its social impact, the discussion has moved to Government level regulation to mitigate the privacy and ethical concerns of the wider public. In a global digital society, however, this is always going to be an impossible task as it leads to a growth in perverse scenarios created by the imperfections in regulatory processes coupled with fragmented, nationalistic politics that reduce enforcement possibilities.

A more profitable approach may, therefore, be to recognise the benefits of automation and human thought and create systems that augment human performance limitations through the use of robotics and machine learning. In this way, human intellect can be freed up to expand on knowledge derived from a synergistic mixture of machine learning to pull out obscure results from complex data sets that can then drive true innovation.

Ethics and Ethical Issues

The question of the ethics of systems incorporating or based on the use of AI is one that is becoming increasingly significant. Of particular concern are questions of transparency of operation where the AI can evolve, as a result of its own internal learning processes, to function and take actions in ways that were not foreseen by either the programme designers or users. An illustration of this was seen during the Go challenge match which took place in 2016 between the AI programme *AlphaGo* and the 9th dan Go master Lee Sedol of South Korea when the 19th stone by *AlphaGo* in game 2 was variously referred to by commentators at the time as "*creative*" and "*unique*"[2] and "*a rare and intriguing shoulder hit*",[3] such was its uniqueness.

Other ethical issues associated with the operation of AI revolve around questions of discrimination and fairness, and in particular, issues where the AI operates in such a way as to display an operational bias against different groups in society. Referred to as "*Algorithmic Bias*" or "*Algorithmic Discrimination*", this can often arise unintentionally as a result of the training sets being used or the way the programme is deployed in the real world [8–10]. Other areas of ethical concern relate to the impact of the operation of the AI on user privacy, and the ability to potentially aggregate individual items of anonymised data such as to be able to identify an individual.

As a consequence, consideration is increasingly being given to the introduction of standards, rules and regulations, for instance, in the manner expressed in Europe in 2020 in the form of the White Paper *On Artificial Intelligence—A European approach to excellence and trust* [11] which sets out to try to establish the basis for such regulations. There are also unresolved issues as to the responsibilities associated with the operation of the AI, is it for instance the programmer or the person who devised the training set, or indeed, is it the AI itself that has evolved as a result of its inbuilt learning strategies to develop new modes of operation? [12, 13]. In this context, the World Economic Forum has suggested an approach to the management of AI based on the concepts of [14]:

Active Inclusion—the requirement that applications should be based on a diversity of inputs that reflect the nature of individual populations and societies.

Fairness—That the nature of what is fair and right in the context of the application is prioritised in the design, development and implementation of that application.

Right to Understanding—Where the operation of the AI has the potential to impact upon individual rights, then this must be disclosed and an explanation provided, which is understandable to all users, as to the decision-making processes involved.

Access to Redress—All involved in the design, development and implementation of an AI system are responsible for identifying any potential negative impacts and must provide appropriate avenues for redress.

---

[2] Michael Redmond, 9th dan.

[3] An Young-gil, 8th dan.

The challenge for any systems designer is, therefore, to be able to incorporate the appropriate and relevant AI structures into system operation while ensuring that the appropriate and relevant safeguards are in place. In relation to EcoMechatronics, this means taking the relevant environmental issues into consideration from the onset of the design process.

Privacy and Privacy Issues

The question of personal privacy is one that has increasingly come to the fore with the ability of a wide range of mechatronic-based systems to gather information about an individual, their behaviour and movements. Even when the use of such information may be considered beneficial either in terms of the individual, for instance, through the recording of health-related data to support that individual in relation to, say, their ability to live independently or, as during the Covid-19 pandemic, for the greater public good as part of a track-and-trace system, there are legitimate concerns over its collection and use.

Even when measures such as data anonymisation are deployed in relation to data being gathered, this needs to be considered in relation to the use of such data in combination with other data from other sources to ensure that individuals cannot be identified as a result of the combination of information derived from multiple data sources. This has, in turn, led to the development of *Privacy by Design* [15–17] thinking in which the concepts and structures of privacy are embedded at all levels within the design process [18–20]. The implementation of mechatronic components and devices within both Cyber-Physical Systems and the Internet of Things is generally associated with all aspects of data collection, often involving personal or user data with the potential to create larger data sets through aggregation of data from and across multiple users. This aggregation of data carries with it implications for the privacy and security of both individual and aggregated users [21–23]. Table 1 then illustrates how the concepts of privacy have been involved in relation to deployed technologies in the period from 1945 to date.

This then suggests a Privacy by Design strategy for mechatronics along the following lines:

1. Establish the nature of the data being collected and whether it can be linked to an individual, and hence identify the associated privacy issues.
2. Evaluate privacy implications associated with the authorisation, authentication, validation and verification of the data and its onward transmission.
3. Evaluate the privacy implications associated with the authorisation, authentication, validation and verification of data on receipt.
4. Establish the protocols required to allow a user to exercise control over the authorisation, authentication, validation and verification of data for both onward transmission and receipt.
5. Embed these user protocols within the core of the mechatronic system.

Table 1 Privacy eras {after [24, 25]}

Period	Characteristics
1945–1960 Privacy Baseline	• Limited information technology developments • High public trust in government and business sector • A general comfort with the information collection
1961–1979 First Era of Privacy Development	• Rise of information privacy as an explicit social, political, and legal issue • Early recognition of potential dark sides of the new technologies • Formulation of the Fair Information Practices (FIP) Framework • Government regulatory mechanisms such as the Privacy Act of 1974
1980–1989 Second Era of Privacy Development	• Rise of computers and network systems • Database capabilities • Legislation designed to channel the new technologies into FIP, including the Privacy Protection Act of 1984 • National data protection laws for both the private and public sectors
1990–2010 Third Era of Privacy Development	• Growth of the Internet • Reported privacy concerns increase
2010 to date Current Privacy Era	• Social media • Mobile devices • Online activity and eCommerce • Cyber crime • Trading and transfer of personal data

6. Establish the system protocols required for the authorisation, authentication, validation and verification of data for both onward transmission and on receipt.
7. Embed these system protocols within the core of the mechatronic system.

These considerations then need to be integrated with the established methods and procedures associated with the design of mechatronic systems.

The Role of Big Data

The term Big Data is generally applied to the analysis and interpretation of large and complex data sets for which conventional data processing methods and techniques are inadequate [26–28]. Such sets are in many instances, as for instance in the case of health related data, structured around an individual's personal data and can be added to, often without the knowledge of the individual, by devices such as those used to measure and record exercise levels. The analysis of such large data sets has resulted in the evolution of methods such as predictive analytics, knowledge discovery and data mining as a means of extracting information, and hence knowledge, from such data. However, the ability to extract such knowledge also carries with it privacy implications for those individuals whose data is incorporated into the overall data set.

In recognition of this potential conflict between the privacy of the individual and the use of Big Data methods to extract both information and knowledge, in the US, the President's Council of Advisors on Science and Technology (PCAST), 2014, produced a report dealing with issues of individual privacy in relation to Big Data [29]. This concluded that:

- Encryption is not a perfect solution for securing data, it is, however, a valuable component of a comprehensive privacy strategy.
- Third parties would create privacy profiles for consumers who would then select their profile such that data holders would be required to differentiate data use between users based on the user's adopted privacy profile.
- Anonymisation and de-identification have limited relevance as linked data points tend to take on other identifiable attributes.
- Deletion and non-retention policies are not effective means of protecting individual privacy.

It went on to recommend that:

- Concentration should be on data use rather than collection and analysis.
- Policies and regulations should be expressed in terms of intended outcomes and not technological solutions.
- Research into privacy technologies should be strengthened.
- There is a need for more education and training in the area of privacy protection.
- There is a need for policies that stimulate the introduction of practical privacy protection policies.

When taken together with the analysis of the links between the Internet of Things and Big Data, a number of issues can be identified, including:

- It is no longer adequate to rely on hard methods to protect privacy.
- It is the responsibility of everyone involved in the data chain to manage and ensure privacy.
- In protecting privacy, the use to which the data is to be put is more significant than its collection and analysis.
- Education and training have important roles to play in increasing awareness of privacy issues and solutions at all levels from design to implementation.

In the UK, the Commercial and Markets Authority in 2015 issued a report on a survey into the ways in which consumer data was being collected and used [30]. The conclusions reached were that:

- Consumers should know when and how their data is being collected and used and be able to decide whether and how to participate.
- They should have access to information as to how their data is collected, stored and used.

- Firms should compete on all issues that matter to consumers, including the provision of clear and useable controls to enable consumers to manage data sharing.
- Consumers and firms should share the benefits of using consumer data. Consumers may get a new or better service or lower prices because firms become more efficient, or even trade their data for a direct financial reward. Firms may gain more sales or market share or become more profitable.
- The regulation of the collection and use of data should ensure the protection of essential rights such as privacy. The market can help achieve this goal where regulations encourage competition and choice, allowing a '*race to the top*' by firms to offer consumers better service.
- Non-compliance with regulation should be tackled proportionately and effectively, so that firms and consumers can feel confident that the rules are being applied fairly.

Within the general context of Big Data, mechatronic systems are most usually associated with the collection and onward transmission of data which can then be integrated with similar data from other users. The mechatronic system is, therefore, functioning as a data source for the wider networks, which has implications for both component and system operation and for design in order to ensure the privacy of the user(s). Considering the need to ensure individual privacy, there is also a requirement to take into account the fact that in areas such as healthcare [31, 32], the ability to analyse related data across large numbers of individuals brings with it the possibility of generating new knowledge of potential benefit to all users. Such conflicts of interest between the need to preserve individual privacy and anonymity and the greater benefit to the community as a whole present significant challenge at the system level, and hence to system designers.

Exemplars

Algorithmic Discrimination

Algorithmic discrimination refers to the actions of the machine learning algorithms that form a part of an AI system in providing results that effectively discriminate against sectors of the population to whom the algorithm is applied [8]. For instance, the choice of training data will significantly influence the behaviour of the AI system in operation, but the available training data sets may not represent a broad spectrum of the population, resulting in a potential bias against those members of the community not represented in the training data. Similarly, local regulatory considerations, as for instance whether certain specific types of data specific may be collected and used, may influence outcomes. Even where adequate training data is available, there still remains significant potential for discrimination. Possible sources of error include:

Model design—The fact that an AI construct proved effective in one context does not mean that it will operate effectively in another, perhaps superficially similar, environment. Thus, a model used for predicting, say, loan eligibility in one country may prove discriminatory if deployed in a country with different financial models.

Unintended bias—The users of AI models are likely to have an idea of what constitutes a successful outcome and may adjust the model so that the results conform with their perception of success, reflecting what are often unexpressed and unintended prejudices. For instance, a model which uses geographical data on crimes to allocate police resources may result in discrimination against specific social groupings depending on how the algorithm is implemented in respect of the type of crimes being considered. As for instance, between high volume, low level crime in respect of low volume, high impact crime.

Lack of oversight—Once an AI construct appears to be functional, human supervision may be withdrawn with the result that the AI begins to take prejudicial actions which would be easily detected by a '*human in the loop*' model to provide confirmation of a decision. Thus, a diagnostic programme may wrongly identify individuals with specific conditions who would otherwise have automatically been removed from the diagnostic pathway by a clinician.

Unpredictability—It needs to be clear to the users of AI how decisions are arrived at. However, as an AI evolves through its own internal processes, the ability to predict, and indeed understand, the nature of the response in specific circumstances may well decrease.

Inappropriate training data—This arises across all aspects of human activity where the data sets collected to be used for training are themselves not truly representative (facial recognition, education, employment, commercial credit, and so forth), leading to enhanced levels of discrimination rather than less.

Intentional discrimination—It is possible for the designers of AI systems to build in intentional bias, as opposed to unintended bias arising out of an inadequate data set or unforeseen and unanticipated modes of use. This suggests a need for regulation to ensure user redress.

In illustration of the above, consider the following. Police forces in the US attempted to use machine learning to model and predict criminal behaviour as part of a drive towards preventative policing. Data sets from state criminal records were used to train software used by parole boards when deciding on the prisoner release, and in some cases by regional forces looking to plan deployment levels based on predictions of likely future "*criminality*". However, by assuming that past criminal behaviour is a predictor of future conduct, the ability of an offender to rehabilitate is reduced by the perpetuation of the belief that "*once a thief, always a thief*". More insidiously, because the decisions made are based on historical criminal records in a justice system in which the individuals included are disproportionately poor and from ethnic minorities, the training set itself will hard wire in these discriminatory factors, targeting communities simply because they are overrepresented in the data.

One potential workaround for the limitations of this AI-centric approach to predicting human behaviour, though perhaps questionable ethically, is for governments, financial institutions and corporations to attempt to "*train*" populations to behave in ways that allow machine learning to better predict outcomes. A credit score is an obvious example of this approach, where financial institutions "*educate*" customers on how to behave to enhance their score, and hence enhance their ability to borrow money.

An alternative, and perhaps more ethically acceptable approach, is to use machine learning algorithms to augment human intelligence through the rapid analysis of data patterns that will flag obvious issues relating to the decision in hand, then pass this information on as part of an HI assessment of the overall picture based on data, experience and intuition with automated post-decision safety checks used to check for discrimination as part an audit process.

These, and related, concerns led in 2017 to the setting out of what is referred to as the *Asilomar AI Principles* [33, 34] which try to establish a series of questions and protocols to be associated with the development and implementation of AI systems to minimise potential impacts.

Business Models

There has been a growing belief in recent years that increased efficiency and reduced discrimination can be achieved by replacing "*subjective*" HI management processes with "*objective*" AI systems across most areas of business practice from HR, through production/stock control to financial planning/credit control and marketing. The belief is that by removing high-cost human employees from routine decision-making, discrimination and bias can be avoided and labour costs reduced.

Thus, the use of algorithms to credit score loan applications, allowing banks to reduce costs through the elimination of experienced human lending managers while pretending it is a more "*objective*" process, is likely to lead to less bad debts or discrimination, but also acts to reduce the numbers of good, entrepreneurial ideas coming to market. Predicting from past history how successful a given start-up will be in a specific geographic and technology market will almost invariably reduce innovation. Any algorithm needs data to work on, while a start-up based on a novel idea has, by definition, none available. As a result, any genuinely novel idea becomes a high-risk outlier in machine learning terms, leading to (ridiculous) requests from lenders that a start-up looking for funding provides 3 years of financial history to base their decision on. Real innovation levels have been in decline as a result since the 1970s, while levels of bad debt have increased rather than reduced. More importantly, the squeezing of entrepreneurial activity that has resulted has lead to increased concentrations of corporate power amongst the few, largely well-connected companies that were able to get funding, dramatically boosting inequality and reducing the overall efficiency of free-market capitalism.

That is not to say that all business AI has a negative impact. Just-in-Time manufacture on a global scale has been implemented using AI-driven predictive modelling to control costs of production. Business efficiency can be enhanced in this way where AI is an invaluable tool in analysing production flows to reduce the effect of capital held in unused stock. This is particularly true in markets where product lifetimes are short such as cell phones.

As with all systems that place too much reliance on the outputs of algorithms using historical data to predict the future, the short-medium term cost savings derived from eliminating "*waste*" from the supply chain can be overwhelmed by the cost of the system breaking down when an unexpected (i.e. not included in the training data) event occurs. While there are many interrelated issues behind the global supply chain failures associated with the Covid-19 pandemic, one thing is clear; corporate decisions to assume that a free flow of goods would continue unfettered by World affairs making it unnecessary to build in sufficient surplus stockpiles in regional centres, has not been helpful. The AI behind a lot of the logistics issues has been excellent at predicting steady-state trade flows, but has no model on which to base contingency planning since it has been programmed to optimise "*business efficiency*". Combining AI pattern analysis with experienced logistics, HI could have enabled slack to be built into the system based on human intuition and experiences.

Corporate marketing has also become highly dependent on algorithmic surveillance capitalism practises developed by Google et al., in the belief that through advertisement micro-targeting, sales conversion rates will be enhanced as the material should be being viewed by just the individuals it has been designed to influence. With the increased focus on the principles behind surveillance capitalism seen in recent years, research groups are starting to examine whether this is true as an underlying truth. The contra-argument is that while you reduce the cost of reaching people already interested in buying your product, there is no direct evidence to suggest that they become more likely to buy as a result, since they would have received the same messaging from traditional mass marketing. No strong increases in sales conversion rates have been apparent with the shift to micro-targeting, though research is at a very early stage. Conversely, opportunities to reach people who do not yet realise they need a company's products or services are actually reduced, potentially constraining market growth in the medium-long term as individual needs and wants change over time.

Recruitment of "*the best people*" based on algorithmic analysis of resumes should also be easier, boosting corporate innovation levels, reducing churn of high-value employees, and reducing/eliminating discrimination. One of the key issues around the use of machine learning to screen university candidates or employment resumes centres on a key assumption—that the metrics Governments have decided to use as measures of academic achievement are an accurate guide to future performance. To get to a "*good*" university, specific examination grades need to be obtained in qualifying examinations as these are the main metric used to forecast likely degree marks. However, people develop academically at very different rates and are stimulated to perform by differing learning environments. Going back to the 1970 and 1980s' in the UK, there was a lot of anecdotal evidence that the inverse may actually be true with students with lower A-level grades at entry ending up with the highest degree grades on graduation. This was postulated to be a result of both a change in the learning environment between school and university and a jolt to underperforming students to work harder to avoid further low grades.

Studies have been undertaken to try and identify the truth of these arguments, the results of which seem to suggest that this is no statistically valid correlation between 18+ and degree level grades [35]. Thus, to base university application decisions on automated screening will simply leave out late developers, unconventional thinkers, or poorer students unable to afford private tuition from ever coming in front of a human decision maker, leading to increasing homogeneity in student intake and increased inequality as those with the means to pay for professional support gain an advantage. Over-reliance on degree certificates in employment screening simply amplifies this problem—how would Einstein reached the position he did in a modern automated World!

Recently, a body of research has been accumulating to suggest that human replacement in business management methods is far more complex than media commentators and business leaders would have people believe. Algorithms are written by humans and trained using data collected by and from humans. As a result, when used to predict outcomes in recruitment by initial screening of resumes, there is a tendency for the pool of successful candidates to become increasingly monolithic in their skills, history and social attributes. Worst still, recruitment advisors have developed techniques for predicting what should go in a resume to stand a better chance of passing the automated screening phase, leaving potentially far more suitable candidates in the reject column. Human interview panels are increasingly asked to select candidates that are clones of each other and well suited in theory to maintain the status quo in an organisation.

Successful team development and innovation in business require a mix of generalist and specialist skills, with the former providing lateral thinking to solve problems in novel ways. Unfortunately, in any model of human behaviour, such individuals are outliers in any statistical analysis of behavioural attributes related to employment skills and are often filtered out of automated selection models based on historical training sets and optimised correlation factors that form the basis of machine learning algorithms.

As a result of all these factors, the use of AI to increase diversity in the search for equality and innovation in a corporate workplace can actually work in opposition to these goals since increasingly the "*mavericks*" needed in any team to provide essential dynamism in group thinking do not make it to a human interview panel where the managers present may have seen through experience and intuition that a person had the right "*spark*" to be successful despite not meeting the perceived attribute scores that have been decided upon through machine learning.

Health Care

Home-based healthcare solutions using personal robotic systems coupled with AI have been promoted in recent years as a cost-effective solution to the increasing problem of supporting an ageing population. Significant investment has gone into both Smart Home and wearable technologies coupled to Cloud-based AI to monitor the health of an individual in their own home, and hence reduce the workload of community medical staff while delaying the need for long-term hospitalisation and/or care and support.

Until recently, the focus has been on ways to reduce the direct costs of supporting an ageing population through a reduction in direct contact between individuals and overstretched health professionals, relying instead on remote telemetry and algorithmic diagnostics. While this approach has many attractions economically, and perceived benefits relating to maintaining health and independent living for the elderly in particular, there are also concerns as to the long-term effects. This came to a head in the Covid-19 pandemic, where it has been realised that the break in the direct link between doctor and patient can lead to a range of mental health issues as well as an increase in excess deaths in the home due to delays in treatment of diseases such as cancers where lateral (human) diagnostic skills at an early stage can be crucial in patient long term survival.

Reliance on technology heavy approaches also creates a number of ethical and privacy issues that need to be addressed. As well as obvious concerns regarding data security and personal privacy in a medical environment, a number of key questions will need to be answered if the goal of enhancing the quality of life over an extending lifespan, while at the same time, containing costs to a manageable level is to be achieved. These include:

1. Who "*owns*" the data collected from an individual, and how do you ensure that professionals requiring access can do so without breaching privacy?
2. The development of robust diagnostic models for specific conditions will require population-level monitoring to avoid biassing data through selecting groups that are atypical, including monitoring those without any signs of illness. A degree of compulsion may well be necessary to achieve such a level playing field, as an ability to opt-out of screening will skew training data towards the "*worried well*" and those already known to have, or be prone to developing, the condition of diagnostic interest. This will require stringent regulation and policing, which could lead to increased anxiety-driven mental health issues, social unrest, for instance, is seen within anti-vaxxer movements, and significant additional government spending to administer.
3. Algorithmic approaches to diagnosis, while better in some instances at picking up early stage conditions, also suffer from being over-specific, i.e. they diagnose based on the strongest correlation between symptoms, bio-markers and genetics to provide a diagnosis based on a "*best-fit*" model without considering that the individual, or their lifestyle, may not follow the historical trends used in training.

4. For such systems to work correctly, remote rural communities will need to have the same level of internet access as large conurbations. Maximum cost benefits for using remote, AI-driven healthcare will be obtained through deployment in these communities, where different medical issues will also exist that need to be built into the predictive diagnostics. Even in advanced economies such as North America, up to 30% of the population remain either unconnected, or using dial-up speeds that make deployment impossible without large-scale infrastructure upgrades.
5. Social inequity is also an issue that needs to be considered, as it will be predominantly those in lower socio-economic groups where participation in AI-based healthcare will be lowest—as is the case more generally with healthcare models that require an element of patient funding. AI approaches will need to overcome the inherent bias that this creates both in existing medical data used to train systems, and in the ongoing evolution of such algorithmic support systems. Without this, social inequity will be enhanced as decision-making becomes increasingly biassed by data from the economically comfortable capable of participating.
6. Expansion of the infrastructure to accommodate both rural communities and the exponential growth that will be needed in Cloud data storage will also result in a potentially unsustainable increase in the carbon footprint of the modern internet. Already, it has been demonstrated that the current global data network has an equivalent footprint to that of air transport, largely as a result of the need for data warehousing [36]. How the expansion of the internet infrastructure to accommodate healthcare using current correlation-based AI is a crucial element of any expanded rollout in the future.

Alternative approaches do exist that may provide answers to these questions while maintaining the essential face-to-face link between patients and health professionals. Using decentralised robot "*pets*" (or even household appliances in certain circumstances) in the homes of people living alone to monitor and map everyday activity, maintain contact with friends and family and provide key medical data in the event of an emergency while collecting data on a continuous basis for storage on an isolated hub in the home. This hub would upload specified information on a daily basis to the individual's medical centre for incorporation into their medical records. With correct permissions in place, this data could then be shared with medical researchers.

The main benefit of such a system is that it would allow a doctor to be better informed of an individual's overall wellbeing on a semi-continuous basis, and allow him/her to schedule visits, etc. It would also allow for early signs of decline to be flagged, such as increasing apathy about moving around the house or going out that can lead to growing incapacity, mental decline through loneliness and accelerated need for in-home care to be provided to someone living alone. Social care budgets can be stretched a lot further by delaying the need for in-home carer support for even a relatively short period of time.

Military Systems

The possible creation of a future generation of what military systems which incorporate the possibility of acting autonomously in deploying lethal force, referred to as Lethal Autonomous Weapons Systems (LAWS), is both a matter of concern and debate and led in 2015 to the publication at the International Joint Conference on Artificial Intelligence of an open letter calling for a ban on such systems [37]. Current military systems are restricted to incorporating a *"human-in-the-loop"* who must authorise the use of any weaponry. There are, however, exceptions for certain types of what are referred to as *'defensive'* systems, such as the *Iron Dome* missile and air defence system developed in Israel [38], acting in response to perceived threats, the classification of which is generally determined by embedded AI [39].

Drone Technologies

Drones are perhaps the most often cited example of a technology that could be adapted through onboard AI to act autonomously when deploying advanced weaponry. A range of drone types is deployed for both civilian (mapping, surveying, aerial photography, cinematography) and military (surveillance, reconnaissance, target strikes) purposes [40]. At the moment, weapons-armed drones such as the MQ-9 *Reaper* [41] have a human operator to make command decisions but are potentially capable of operating autonomously [42]. Other drone types, and in particular, nano drones present their own particular problems in relation to autonomous operation, something graphically illustrated by the *Slaughterbots* video released by the Future of Life Institute [43, 44].

Vehicle Systems

Modern vehicles are increasingly configured around a range of smart components and systems, including Cyber-Physical Systems, which can be integrated within an Internet of Vehicles (IoV). This will involve not only communication between individual vehicles, but also links into Cloud-based systems for traffic management and route planning [45–47]. Consider the apparently simple question of finding a parking space. Studies suggest that considerable time, and hence cost is involved in drivers locating parking spaces, reaching in 2017 an average annual search time for a parking space of 67 h for a driver in London at an annual cost of £1,104 (~£1,200 in 2020). In the same year in New York, it was reported as 107 h at an annual cost

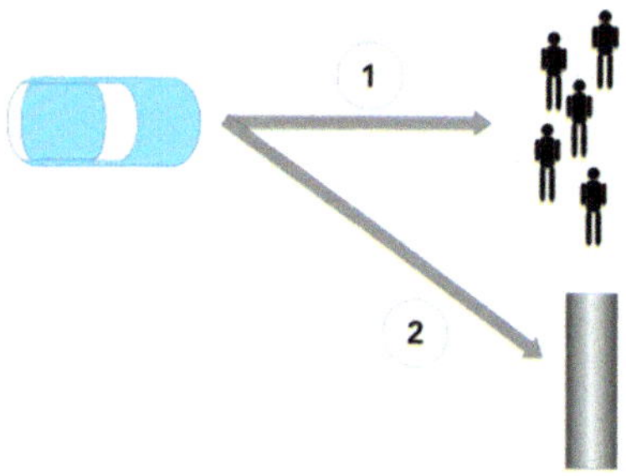

Fig. 1 The basic trolley problem

of \$2,243 (~\$2,370 in 2020) [48, 49]. By allowing vehicles to negotiate both with each other and with parking locations, this time penalty and cost could be effectively eliminated, resulting in significant savings in CO_2 emissions[4] and more effective and efficient use of space [50–52].

An ability for individual vehicles to autonomously negotiate with other vehicles, and indeed other road users, also has the potential to significantly impact other areas of traffic management. A major issue in towns is that of traffic flow, which is often of a stop–start nature. While traffic flows can be improved by linking the operation of traffic lights to traffic density, using the information provided by the vehicles themselves, an alternative, which has been shown to be potentially effective in smoothing out flow, is to remove traffic lights completely and rely on the negotiation between vehicles to optimise traffic flow at junctions [53].

Finally, by linking vehicles to their digital twin, constant monitoring and check on performance can be obtained, enabling a much more rapid response to fault conditions. It is, therefore, possible to imagine a vehicle, in cooperation with its digital twin which will, in turn, be drawing on information from all cars of that type in operation, detecting an incipient fault when it will check the user's diary and book itself in for a service. At the garage, instead of having to hold a wide range of components and sub-assemblies, with all the implications for resource extraction, manufacturing and distribution that implies, the requisite parts will be produced at the garage on-demand using additive manufacturing methods.

Autonomous Vehicles

However, perhaps the most contentious area of vehicle technology is that of autonomous operation, and in particular how such a vehicle should respond to emergency situations. Often cited as the '*Trolley Problem*' [54–58] and expressed in the form of a decision to be made in the form shown in Fig. 1 where the vehicle is presented with two options either (1) continue straight ahead and run into the group

[4] Assuming that the average distance travelled in London while looking for a parking location is 5 km, this translates to around 0.6 kg of CO_2 per vehicle based on data from www.eea.europa.eu/ims/co2-performance-of-new-passenger (accessed 8 December 2021).

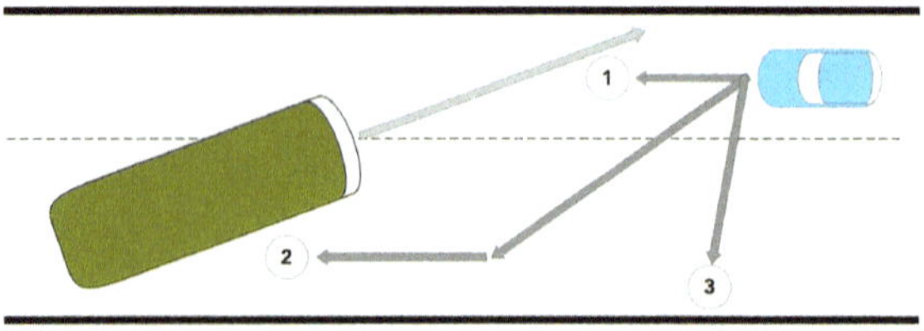

Fig. 2 A more complex scenario

of pedestrians, saving the occupants of the vehicle but killing the pedestrians or (2) swerving to hit the barrier, killing the occupants of the car but saving the pedestrians.

A more complicated scenario is that of Fig. 2, where an autonomous car is presented with the case where a bus suddenly swerves out of its lane and heads towards the crash barrier. Under option (1), the car continues towards the bus risking casualties on board both the car and the bus at differing levels of severity. This is sometimes referred to as the Humanist or Egalitarian option as it spreads the outcomes across all participants. Under option (2), the Protectionist option, the car takes all necessary action to safeguard its passengers, even though this may well increase the likelihood of severe casualties on the bus as it hits the crash barrier a the side of the road. Option (3), the Sacrifice option, would result in the probability of severe casualties on both the car and the bus with no benefit to either.

The situation is further complicated by individual preferences regarding the outcomes of any incidents, something which is demonstrated by the results of the Moral Machine Experiment carried out by MIT [59] in which individuals were asked to express their preferred outcomes in relation to a number of different scenarios, resulting in issues such as preferences for sparing pedestrians over passengers, the young over the elderly or the more over the fewer. The results also illustrated that views and opinions differed significantly from country to country!

Conclusions

Artificial Intelligence has in recent years been moving from the laboratory into real-world environments where the simplicity of computer logic meets the complexity of human evolutionary experience. With the Covid-19 pandemic forcing the pace on remote working and learning, coupled with a growing awareness of the political implications of AI, there is a need to consider the detailed costs and benefits of deploying deep learning algorithms in specific applications. Thus:

- Before deploying an AI construct to model and control any given application, can it be stated with confidence that correlation is demonstrating causation. Namely that when something has been shown to happen in a particular way, it does so as a result of an underlying invariant causal mechanism. Such relationships are found in many biological and mechanical systems, but can the same be said for human behaviour?

- AI constructs are developed by humans using a set of assumptions as to how outcomes should be determined, and human selected data sets are used to train the algorithms. Computers will then provide the "*optimum*" solution to any issue based on mathematical logic, which may not be ethical in human terms. What regulations need to be put in place globally to ensure that AI works to promote equality and reduce discrimination rather than exacerbate current tensions?
- The main difference between today and earlier attempts at commercialising AI is the ability to collect vast arrays of data on people through the promotion of online activities such as social media and shopping. Who owns this data, and how permission is gained to use it, is becoming a major issue as the influence of surveillance capitalism business models, as well as the sustainability of mega data-centres with a carbon footprint equivalent to the airline industry, have come under the political spotlight in recent years.
- How do AI systems trained on such data ensure that an individual's privacy is not compromised through ill-informed consent provisions in service contracts and thus prevent customers turning into a mineable raw material?
- If regulations allowing opt outs from data collection are introduced, how can bias in training data be avoided when privacy concerns are not consistent across the range of human behaviour that goes into the creation of big data training sets?
- An AI construct is only as good as the training data used in its construction, leading to problems with unconscious bias and discriminatory outcomes in the models based on a given model. Examples of such bias based on unrepresentative data sets have resulted in historical discrimination against minority groups that feed into and influences the training models used and which has led to increases in discrimination in, amongst others, policing, human resource management and education.
- There has been a tendency in recent years, encouraged by the broader political imperative of reducing the role of government by relaxing regulations on the corporate business sector to encourage economic growth, for technologists to assume that populations can be manipulated into behaviour patterns that allow the limitations of correlation-based predictions to be overlooked, i.e. change human behaviour to better match the idealised world predicted by any given AI model. This assumption has been shown to be inadequate in the face of unexpected events outside the scope of collected training data sets, such as the effects of the current pandemic on global political and economical systems, and climate change. Deep learning works best when history is predictable but struggles due to the limitations inherent in the assumptions used in the face of sudden lateral shifts in events, as can be seen in part with the current global supply chain issues. How can future systems be designed to allow for automated modelling systems to ensure sufficient "*inefficiency*" is included to allow for the unexpected?

AI, once again, is at a crossroads. Should it continue in a largely technocratic way, absorbing individual personality into a corporate body controlled by a small elite, banned as a threat to humanity, or developed as an assistive tool that marries HI with AI to enhance productivity and innovation while promoting equality and working to eliminate discrimination.

References

1. www.apple.com/uk/airtag/. Last Accessed 18 May 2021
2. https://theconversation.com/remember-apple-airtags-and-find-my-app-only-work-because-of-a-vast-largely-covert-tracking-network-160781. Last Accessed 16 Dec 2021
3. Jia Y, Xing L, Mao Y, Zhao D, Wang X, Zhao S, Zhang Y (2020) Burglars' IoT paradise: understanding and mitigating security risks of general messaging protocols on IoT clouds. In: IEEE symposiyum security &privacy, pp 465–481
4. Chalhoub G, Kraemer MJ, Nthala N, Flechais I (2021) "It did not give me an option to decline": a longitudinal analysis of the user experience of security and privacy in smart home products. In: Proceedings of 2021 CHI conference on human factors in computing systems, pp 1–16
5. Watt S, Milne C, Bradley D, Russell D, Hehenberger P, Azorin-Lopez J (2016) Privacy matters–issues within mechatronics. IFAC-PapersOnLine 49(21):423–430
6. https://ec.europa.eu/info/sites/default/files/commission-white-paper-artificial-intelligence-feb2020_en.pdf. Last Accessed 2 June 2022
7. Kahan R (2021) IBM bids goodbye to Watson health, its interactive AI system. www.calcalistech.com/ctech/articles/0,7340,L-3895071,00.html. Last Accessed 3 Jan 2022
8. O'Neill C (2016) Weapons of math destruction: how big data increases inequality and threatens democracy. Penguin Random House
9. Žliobaitė I (2017) Measuring discrimination in algorithmic decision making. Data Min Knowl Disc 31(4):1060–1089
10. Kleinberg J, Ludwig J, Mullainathan S, Sunstein CR (2018) Discrimination in the age of algorithms. J Leg Anal 10:113–174
11. European Commission (2020) On artificial intelligence-a European approach to excellence and trust, COM(2020) 65 final. https://ec.europa.eu/info/sites/default/files/commission-white-paper-artificial-intelligence-feb2020_en.pdf. Last Accessed 2 June 2022
12. Hagendorff T (2020) The ethics of AI ethics: an evaluation of guidelines. Mind Mach 30(1):99–120
13. Coeckelbergh M (2020) AI ethics. MIT Press
14. World Economic Forum (2020) How to prevent discriminatory outcomes in machine learning. https://www3.weforum.org/docs/WEF_40065_White_Paper_How_to_Prevent_Discriminatory_Outcomes_in_Machine_Learning.pdf Accessed 2 June 2021
15. Cavoukian A (2009) Privacy by design: the 7 foundational principles. Information and Privacy Commissioner of Ontario, Canada
16. Cavoukian A (2012) Privacy by design and the emerging personal data ecosystem. www.ipc.on.ca/wp-content/uploads/Resources/pbd-pde.pdf. Last Accessed 24 Aug 2021
17. Mulligan DK, King J (2011) Bridging the gap between privacy and design. U Pa J Const Law 14(4):989–1034
18. Cavoukian A, Jonas J (2012) Privacy by design in the age of big data. Information and Privacy Commissioner of Ontario, Canada
19. Perera C, McCormick C, Bandara AK, Price BA, Nuseibeh B (2016) Privacy-by-design framework for assessing internet of things applications and platforms. In: Proceedings of 6th international conference on the Internet of Things, pp 83–92
20. Bednar K, Spiekermann S, Langheinrich M (2019) Engineering privacy by design: are engineers ready to live up to the challenge? Inf Soc 35(3):122–142
21. Bu F, Wang N, Jiang B, Liang H (2020) "Privacy by Design" implementation: information system engineers' perspective. Intl J Inf Manag 53:102124
22. Porter ME, Heppelmann JE (2014) How smart, connected products are transforming competition. Harv Bus Rev 92(11):64–88
23. Sadeghi AR, Wachsmann C, Waidner M (2015) Security and privacy challenges in industrial Internet of Things. In: Proceedings of 52nd annual design automation conference, pp 54–59
24. Smith HJ, Dinev T, Xu H (2011) Information privacy research: an interdisciplinary review. MIS Q 35(4):989–1016

25. Westin AF (2003) Social and political dimensions of privacy. J Soc Issues 59(2):431–453
26. Hashem IAT, Yaqoob I, Anuar NB, Mokhtar S, Gani A, Khan SU (2015) The rise of "big data" on cloud computing: review and open research issues. Systems 47:98–115
27. Herschel R, Miori VM (2017) Ethics & big data. Technol Soc 49:31–36
28. Wang J, Yang Y, Wang T, Sherratt RS, Zhang J (2020) Big data service architecture: a survey. J Internet Technol 21(2):393–405
29. PCAST (2014) Big data and privacy: a technological perspective. www.whitehouse.gov/sites/default/files/microsites/ostp/PCAST/pcast_big_data_and_privacy_-_may_2014.pdf. Last Accessed 27 Jan 2016
30. UK Competition & Markets Authority (2015) The commercial use of consumer data. www.gov.uk/government/uploads/system/uploads/attachment_data/file/435817/The_commercial_use_of_consumer_data.pdf. Last Accessed 19 Sept 2016
31. Dash S, Shakyawar SK, Sharma M, Kaushik S (2019) Big data in healthcare: management, analysis and future prospects. J Big Data 6(1):1–25
32. Price WN, Cohen IG (2019) Privacy in the age of medical big data. Nat Med 25(1):37–43
33. https://futureoflife.org/2017/08/11/ai-principles/. Last Accessed 2 June 2022
34. https://futureoflife.org/bai-2017/ Last Accessed 2 June 2022
35. Sear K (1983) The correlation between a level grades and degree results in England and Wales. High Educ 12(5):609–619
36. Obringer R, Rachunok B, Maia-Silva D, Arbabzadeh M, Nateghi R, Madani K (2021) The overlooked environmental footprint of increasing Internet use. Resour Conserv Recycl 167:105389
37. https://futureoflife.org/2016/02/09/open-letter-autonomous-weapons-ai-robotics/. Last Accessed 2 June 2021
38. Block J (2017) A laws of war review of contemporary land-based missile defence system 'Iron Dome'. Sci Milla: S Afr J Mil Stud 45(2):105–128
39. Carayannis E, Wilson RL, Iftimie IA, Iftimie MC (2020) The ethics of autonomy and lethality. In: Proceedings of 18th international conference ethical & social impacts of ICT, ETHICOMP 2020, pp 280–283
40. Mátyás P, Máté N (2019) Brief history of UAV development. Repüléstudományi Közlemények 31(1):155–166
41. www.af.mil/About-Us/Fact-Sheets/Display/Article/104470/mq-9-reaper/. Last Accessed 29 Nov 2021
42. Wilson RL (2019) Requirements for making the MQ-9 fully autonomous: an anticipatory ethical analysis. In: European conference impact of artificial intelligence and robotics, ECIAIR 2019, pp 369–376
43. www.youtube.com/watch?v=6Ipkq-BASaM. Last Accessed 29 Nov 2021
44. https://futureoflife.org/2021/02/25/stuart-russell-and-zachary-kallenborn-on-drone-swarms-and-the-riskiest-aspects-of-lethal-autonomous-weapons/. Last Accessed 2 June 2022
45. Contreras-Castillo J, Zeadally S, Guerrero-Ibañez JA (2017) Internet of vehicles: architecture, protocols, and security. IEEE Internet of Things J 5(5):3701–3709
46. Gerla M, Lee EK, Pau EG, Lee U (2014) Internet of vehicles: from intelligent grid to autonomous cars and vehicular clouds. IEEE World Forum Internet of Things 241–246
47. Xu W, Zhou H, Cheng N, Lyu F, Shi W, Chen J, Shen X (2017) Internet of vehicles in big data era. IEEE/CAA J Autom Sin 5(1):19–35
48. https://inrix.com/press-releases/parking-pain-uk/. Last Accessed 1 Mar 2021
49. https://inrix.com/press-releases/parking-pain-us/. Last Accessed 1 Mar 2021
50. Azshwanth D, Koshy MT, Balachander MT (2019) Automated car parking system. J Phys Conf Ser 1362(1):012059
51. Aliedani A, Loke SW (2019) Cooperative car parking using vehicle-to-vehicle communication: an agent-based analysis. Comput Environ Urban Syst 77:101256
52. Zheng Y, Rajasegarar S, Leckie C (2015) Parking availability prediction for sensor-enabled car parks in smart cities. In: IEEE 10th international conference intelligent sensors, sensor networks and information processing, pp 1–6

53. World Economic Forum (2020) Guidelines for city mobility. http://www3.weforum.org/docs/WEF_Guidelines_for_City_Mobility_2020.pdf Last Accessed 2 June2022
54. Smith CH (2018) The Tesla problem: trolleys, proximity, and duty, ephemeris, the undergraduate. J Philos 18(1):22–32
55. Keeling G (2020) Why trolley problems matter for the ethics of automated vehicles. Sci Eng Ethics 26(1):293–307
56. Goodall NJ (2014) Machine ethics and automated vehicles. Springer, In Road vehicle automation, pp 93–102
57. Lawlor R (2021) The ethics of automated vehicles: why self-driving cars should not swerve in dilemma cases. Res Publica 1–24
58. De Freitas J, Censi A, Smith BW, Di Lillo L, Anthony SE, Frazzoli E (2021) From driverless dilemmas to more practical commonsense tests for automated vehicles. Proc Natl Acad Sci 118(11):9
59. Awad E, Dsouza S, Kim R, Schulz J, Henrich J, Shariff A, Bonnefon JF, Rahwan I (2018) The moral machine experiment. Nature 563(7729):59–64

Bibliography

Amoore L (2020) Cloud ethics: Algorithms and the attributes of ourselves and others. Duke University Press. ISBN-13: 978-1-4780-0831-6

Anderson M (ed) (2018) Machine ethics. Cambridge University Press. ISBN: 9781108461757

Aoun J (2017) Robot-proof: higher education in the age of artificial intelligence. MIT Press, ISBN: 9780262037280

Barrat J (2015) Our final invention: artificial intelligence and the end of the human era. Macmillan. ISBN: 9781250058782

Bartneck C, Lütge C, Wagner A, Welsh S (2020) An introduction to ethics in robotics and AI. Springer. ISBN: 9783030511098

Bergstrom CT, West JD (2020) Calling bullshit–the art of scepticism in a data-driven world. Allen Lane. ISBN: 9780241327234

Broussard M (2019) Artificial unintelligence: how computers misunderstand the world. MIT Press. ISBN-13: 978-0-262-53,701-8

Burns L (2018) Autonomy: the quest to build the driverless car-and how it will reshape our world. William Collins. ISBN: 978-0008302061

Carr N (2016) The glass cage–who needs humans anyway? Vintage. ISBN: 9780099597452

Chamayou G (2015) Drone theory. Penguin. ISBN: 978-0241970348

Christian B (2020) The alignment problem; machine learning and human values. WW Norton. ISBN-13: 978-0–393-63582-9

Coeckelbergh M (2020) AI ethics. MIT Press. ISBN: 9780262538190

Crawford K (2021) Atlas of AI: power, politics, and the planetary costs of artificial intelligence. Yale University Press. ISBN-13: 978-0-300-20957-0

Daub A (2020) What tech calls thinking: an inquiry into the intellectual bedrock of Silicon Valley. FSGO/Logic. ISBN-13: 978-0-374-53864-4

Davies A (2021) Driven: the race to create the autonomous car. Simon & Schuster. ISBN: 978-1501199431

Denardis L (2020) The internet in everything: freedom and security in a world with no off switch. Yale University Press. ISBN: 978-0300233070

Du Sautoy M (2019) The creativity code: art & innovation in the age of AI. Belknap Harvard. ISBN-13: 978-0-674-98813-2

Epstein D (2019) Range: why generalists triumph in a specialized world. Riverhead Books. ISBN-13: 978-0-735-21448-4

Eubanks V (2019) Automating inequality: how high-tech tools profile, police, and punish the poor. Picador. ISBN-13: 978-1-250-21578-9
Ford M (2015) The rise of the robots. Oneworld. ISBN: 9781780747491
Fosch-Villaronga E (2020) Robots, healthcare, and the law: regulating automation in personal healthcare. Routledge. ISBN-13: 978-0-367-07661-0
Galliott J, Macintosh D, Ohlin JD (eds) (2020) Lethal autonomous weapons: re-examining the law and ethics of robotic warfare. Oxford University Press. ISBN: 978-0197546048
Gawdat M (2021) Scary smart: the future of artificial intelligence and how you can save our world. Bluebird Books. ISBN: 978-1529077186
Greenguard S (2021) The Internet of Things. MIT Press. ISBN: 978-0262542623
Haque E (2020) The ultimate modern guide to artificial intelligence: including machine learning, deep learning, IoT, data science, robotics, the future of jobs, required upskilling and intelligent industries, independent. ISBN: 9798691930768
Hare S (2022) Technology is not neutral: a short guide to technology ethics. London Publishing Partnership. ISBN: 978-1907994975
Helfand DJ (2016) A survival guide to the misinformation age. Columbia University Press. ISBN: 9780231168724
Hwang T (2020) Subprime attention crisis; advertising and the time bomb at the heart of the internet. FSGO/Logic. ISBN-13: 978-0-374-53865-1
Kane E, Kane M (2021) Future modes of travel 2020's & 2030's, independent. ISBN: 979-8591491758
Kearns M, Roth A (2020) The ethical algorithm. Oxford University Press. ISBN-139780190948207
Kissinger HA, Schmidt E, Huttenlocher D (2021) The age of AI: and our human future. Little, Brown & Company. ISBN-13: 978-0-316-27380-0
Kreps SE, Kaag JJ (2014) Drone warfare. Polity Press. ISBN: 978-0745680989
Lanier J (2018) Ten arguments for deleting your social media accounts right now. Bodley Head. ISBN: 9781847925398
Lepore J (2020) If then: how one data company invented the future. Liveright. ISBN-13: 978-1-631-49610-3
Liao SL (ed) (2020) Ethics of artificial intelligence. Oxford University Press. ISBN: 978-0-19-090504-0
Lin P, Jenkins R, Abney K (eds) (2019) Robot ethics 2.0: from autonomous cars to artificial intelligence. Oxford University Press. ISBN: 9780197503584
Marcus G, Davis E (2019) Rebooting AI: building artificial intelligence we can trust. Pantheon Books. ISBN-13: 978-1-524-74825-8
McNamee R (2019) Zucked: waking up to the facebook catastrophe. Harper Collins. ISBN: 9780008318994
Mitchell M (2019) Artificial intelligence: a guide for thinking humans. Pelican. ISBN: 9780241404829
Morgan SL, Winship C (2015) Counterfactuals and causal inference: methods and principles for social research, 2nd edn. Cambridge University Press. ISBN-13: 978-1-107-69416-3
O'Neill C (2016) Weapons of math destruction: how big data increases inequality and threatens democracy. Penguin Random House. ISBN: 9780553418811
Pak R, De Visser EJ, Rovira E (eds) (2020) Living with robots: emerging issues on the psychological and social implications of robotics. Academic Press. ISBN-13: 978-0-12-815367-3
Pasquale F (2020) New laws of robotics–defending human expertise in the age of AI. Harvard University Press. ISBN: 9780674975224
Pasquale F (2015) The black box society: the secret algorithms that control money and information. Harvard University Press. ISBN-13: 978-0-674-97084-7
Pearl J, Mackenzie D (2018) The book of why: the new science of cause and effect. Basic Books. ISBN-13: 978-0-465-09760-9
Ratti C, Claudel M (2016) The city of tomorrow: sensors, networks, hackers, and the future of urban life. ISBN: 9780300204803

Rirchie S (2020) Science fictions: exposing fraud, bias, negligence and hype in science. Bodley Head. ISBN: 9781847925657

Rouhiainen L (2018) Artificial intelligence: 101 things you must know today about our future. CreateSpace. ISBN: 9781982048808

Russell S (2019) Human compatible; artificial intelligence and the problem of control. Viking. ISBN-13: 978-0-525-55861-3

Singer PW (2009) Wired for war: the robotics revolution and conflict in the 21st century. Penguin. ISBN: 9781594201981

Tegmark M (2018) Life 3.0: being human in the age of artificial intelligence. Allen Lane. ISBN: 9780241237199

Wallach W (2010) Moral machines: teaching robots right from wrong. Oxford University Press. ISBN: 9780199737970

West DM (2019) The future of work: robots, AI, and automation. Brookings Institution Press. ISBN: 978-0815737865

Wooldridge M (2020) The road to conscious machines: the story of AI. Pelican. ISBN: 9780241396742

Yates K (2019) The math of life and death: 7 mathematical principles that shape our lives. Scribner. ISBN-13: 978-1-982-11187-8

Zuboff S (2019) The age of surveillance capitalism: the fight for a human future at the new frontier of power. PublicAffairs. ISBN-13: 978-1-610-39569-4

Applications

Mechatronic Applications in Rail Systems and Technologies

Christopher Ward, Roger Goodall, Tim Harrison, and Will Midgley

Abstract Rail systems are the longest established of the land-based mass transit forms. With a fixed infrastructure, the possibility of being fully electrified, huge capacity potential and low 'energy loss' running dynamics, they make a compelling case for being at the core of any 'net-zero' future transport system. These systems though are constrained by these same aspects that provide the benefits (fixed routes, formations and signalling systems), constraining capacity and limiting flexibility. One enduring concept to unlock these benefits is the use of mechatronics in the vehicle and the infrastructure. But why have these concepts not been more widely applied before and how can they make a system wide difference?

Introduction

Rail systems make a compelling case to be at the core of any future low-carbon transport system. 'Steel on steel' rolling contact is exceptionally efficient in terms of energy loss [1], the capacity potential of a mainline railway line is comparable to several multi-lane roads [2], and electrification making the use of renewable sources is relatively straightforward [3]. They also offer true city centre to city centre travel that cannot be matched by any other mode. This is remarkable for a system that was conceived and first implemented in the early 1800s [4].

However, the Victorian origins that provided so many benefits are also the cause of many of the constraints in the implementation of contemporary systems globally. Signalling systems that control the flow of rolling stock have been developed for very good reasons, primarily driven by safety. As such the capacity of a railway is not constrained by physical space and driver behaviour, as on a highway, but by the signalling logic (and to a lesser extent the physical capacity of the volume that trains pass through, the 'gauging'). Taking the UK as an example, four aspect signalling is very common (red, single yellow, double yellow and green) [5]. Signals control

C. Ward (✉) · R. Goodall · T. Harrison · W. Midgley
Wolfson School of Mechanical, Electrical and Manufacturing Engineering, Loughborough University, Loughborough L11 3TU, UK
e-mail: c.p.ward@lboro.ac.uk

P. Hehenberger et al. (eds.), *EcoMechatronics*,
https://doi.org/10.1007/978-3-031-07555-1_10

the entry to 'blocks' of track, for a 200 km/h line these are typically 1.1 km long (governed by braking distance appropriate to the speed). To run at full speed the driver must see a green aspect, meaning three blocks in front are clear. This roughly equates to a minimum of 3.3 km of track being unoccupied between trains. This is very conservative in its outlook and means tracks are virtually empty, even at peak times. Re-routing of trains is also very difficult and has ripple effects that can be felt across an entire network. The formation (and therefore capacity) of trains is often fixed and cannot be changed in the short term. In common with road transportation, railways are a very expensive endeavour with a large amount of 'embedded carbon' required to create the fixed infrastructure [6]. Attempts to improve this situation are usually classified as conservative, incremental, and reactionary to current issues.

With this background in mind, railways around the world are undergoing huge changes with ambitious targets to reduce energy use and primarily improve the customer experience. Incremental progress is not going to see these aims fulfilled, and it is the conjecture here that disruptive, fundamental changes are required.

Industry Vision

In the UK, the industry has created a series of visions for improvement. The 2012 Rail Technical Strategy (RTS) [7] created a 30-year vision, key aspects of this being:

1. Running trains closer together
2. Minimal disruption to train services
3. Efficient passenger flow through stations and trains
4. More value from data
5. Optimal energy use
6. More space on trains
7. Services timed to the second
8. Intelligent trains
9. Personalised customer experience
10. Flexible freight
11. Low-cost railway solutions
12. Accelerated research, development, and technology deployment

This was updated in 2021 [8] and picks up many of the same themes. However, the vision now concentrates on five main areas that are less focussed on technical specifics:

1. Easy to use for all
2. Low emissions
3. Optimised train operations
4. Reliable and easy to maintain
5. Data driven

This is backed by a wider European outlook with the Shift2Rail [9] initiative under EC Horizon 2020, with the follow-up work streams currently being decided in the Horizon Europe framework [10].

It is the authors' opinion that, welcome as these initiatives are, they again only provide an incremental view of how the railways should develop and that the traditional path of incremental improvements is not going to deliver the embedded vision of any of the strategies. This is where mechatronics comes in as an enabler to see the visions into fruition.

Embedding Mechatronics in Transport

Many of the key industrial competitors to rail have systematically embedded mechatronics systems design over many decades. Mechatronics is not seen as an addition to 'cure ills', but as a fundamental design paradigm. The following are transport system examples over a similar time-period (post-war to the modern day).

Military Aerospace

Military aerospace was perhaps the first industry to realise the true potential of mechatronics in transportation systems. Figure 1 is an example of two frontline fighter jets employed by the Royal Air Force (RAF). The first is the de Havilland Vampire that was first in service in the late 1940s. This was conventional by most metrics with the design (except for the twin boom tail) being of mixed wood and metal construction, swept leading edge wings, stable configuration, and manual controls. The second of these is the most modern fighter in the arsenal, the F-35B Lightning, that entered service in 2018. A very different aircraft it encapsulates an advanced composite construction, relaxed stability, full digital flight controls, short take off and vertical landing capability, and stealth. This can be seen as the current pinnacle

Fig. 1 de Havilland Vampire and Lockheed Martin F-35B (Chris Ward)

Fig. 2 Mercedes Benz 300SL and S-Class (Phil D and Martin Katler on Unsplash)

of a deployed mechatronic flight control system that began with the development of the methodology in the middle of the twentieth century.

In the 1960s there was a significant lobby against digital flight controls that were said to be dangerous, too slow to react, un-maintainable, and would never be accepted by pilots [11]. The exact opposite of these concerns is why mechatronics is so embedded in everything military aerospace undertakes in the modern era.

Automotive Systems

This trend is not limited to the aerospace sphere. In the automotive sector mechatronics is embedded in both the drivetrains and the chassis. Figure 2 shows a comparison for the development of chassis technology. The 1950s era Mercedes-Benz 300SL had a conventional suspension system with passive springs and dampers. The stability and safety of the vehicle was fundamentally a function of the operation of the driver to perceive and not go beyond the 'mechanical limit'. In the modern era the Mercedes-Benz S-class is always a technology showcase. The suspension system is now via air springs, with continuously variable damping, tilt control, and feedforward scanning. The stability of the vehicle is actively augmented and will prevent the vehicle (or recover it from) exceeding the limits of the tyres.

Along the lines of aerospace, embedding these concepts at the outset of design is now a fundamental part of automotive engineering culture as well as being enshrined in law. As of 2012 all new vehicles sold in the United States are required to be fitted with a minimum level of electronic stability control (ESC), with the EU following this lead in 2014 [12].

Rail

Rail has a history of innovation and embedding novel thinking into technology. In the UK much of the innovative work on understanding rail vehicle dynamics was led by British Rail Research, who were centred in Derby. An example of this

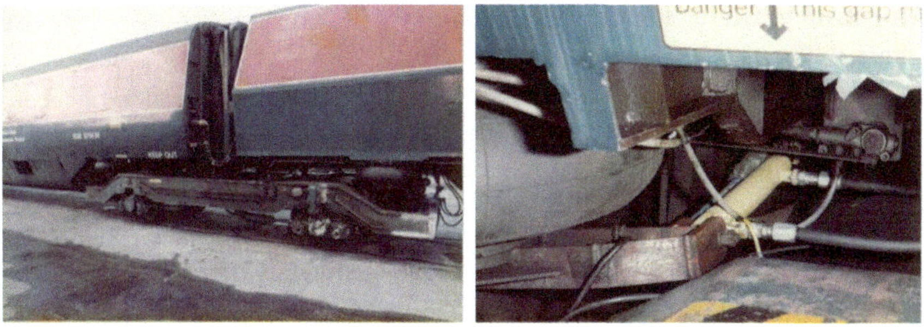

Fig. 3 Active lateral suspension experimental implementation

Fig. 4 Mark 1 and mark 5 UK coaching stock (Loughborough University and Chris Ward)

innovative thinking is seen in Fig. 3, the first example of an active lateral suspension control. Concepts such as active steering and guidance, low speed magnetically suspended vehicles, and signalling moving from mechanical relay based to solid state interlocking were also pioneered there [13].

However, if we look at the development of rail systems in service, many of the innovations are not embedded as a matter of course. Figure 4 shows two train formations over the same period as the aerospace and automotive examples. The vehicles on the left of the figure being hauled by a steam locomotive are British Rail mark 1 coaches, the guidance system for which is based on solid axle wheelsets, bogie systems, and two stages of passive suspension. The train on the right is the modern incarnation of a locomotive hauled train, a class 68 Diesel locomotive, with mark 5 coaches. The modern vehicle set is demonstrably better than its predecessor in terms of performance, maintainability, and customer experience, but the operational fundamentals remain the same. The guidance system is still solid axle wheelsets, bogies, and two stages of passive suspension.

As such the vehicle is designed to operate on a system that will struggle to fulfil the aims of the RTS and still operates in an environment constrained by a conservative signalling system.

The Rail Vehicle 'Opportunity Space'

There is a particular focus on the vehicle side of the implementation of mechatronics in this chapter. There is a very specific reason for this in terms of the vehicle representing the biggest use of 'operational carbon' and that the infrastructure is associated with 'embedded carbon' [14]. Mechatronics offers the possibility of simplifying infrastructure by transferring complexity to the vehicle (reducing embedded carbon), and at the same time minimising operational energy requirements (reducing operational carbon).

A rail vehicle has several fundamental top-level functions that are defined in European standard [15]. These can be summarised as the ability to carry out the following functions with the elements in **bold** highlighted as being the key targets for the use of vehicle mechatronics.

- Carry and protect passengers, train crew and load
- Provide appropriate conditions to passenger, train crew and load
- **Provide access and loading**
- Connect vehicles and/or consists
- **Provide energy**
- **Accelerate, maintain speed, brake and stop**
- Provide train communication, monitoring and control
- **Support and guide the train on the track**
- Integrate the vehicle into the complete railway system

Rail vehicles can be broken down also in terms of subsystems, with the items in bold the focus of the following sections. From the EC ModTrain project [16] these were defined as.

- Car body, fittings, interiors and lighting
- **Bogies and running gear**
- Power and propulsion system
- Auxiliary systems
- **Braking system**
- **On-board vehicle control**
- Passenger Information System
- Communication systems
- Cabling and Cabinets
- Door System
- HVAC
- **Tilt system**
- Couplers

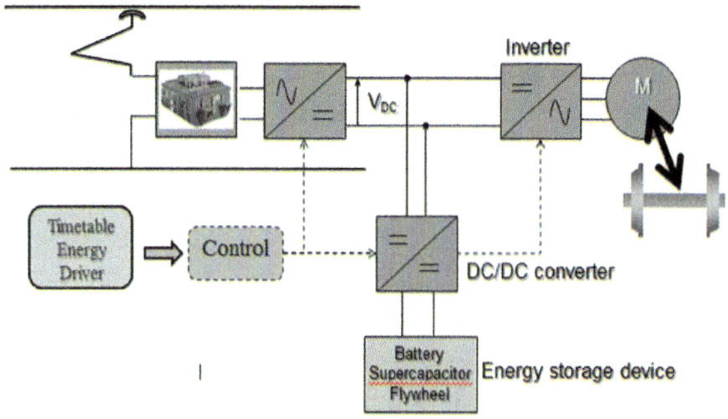

Fig. 5 Traction system schematic

Power and Propulsion

Electric trains are a mature technology with a significant majority of vehicles employing this form of traction. This is an instance where the automotive industry is catching up with the established practice of the rail industry. A fully electrified railway is in a unique position to make the most of the decarbonisation of the energy system.

Figure 5 is a schematic of a typical traction system that collects energy from an overhead AC electrical supply, with the addition of energy storage devices. The latter is seen as a key part of any new drivetrain with the previously lost braking energy being reused in the accelerative traction process. This is, as of 2021, an underutilised method, due to the complexity of the control methods.

Active Suspension Systems

Active control of suspension systems has been proposed for the last half century, but most of the suspension systems on rail vehicles are still purely passive. The exception to this is the use of active vehicle body tilt control, which is now an accepted mature technology.

Railway vehicle suspension systems are complex, operate in multiple directions and control the steering characteristics. Most of the research work into active methods has focused on the vertical suspension aspect, either on the primary or secondary element. As such this has little impact on the operational concept of the vehicle (it is still fundamentally passively guided) and the business case is not clear for implementation. Passenger's perception of improved ride quality is mixed and very often not a driver in how they choose a service. Tilt control is part of this in that it influences passenger's perception of cornering, not the vehicle's ability to traverse curves. Therefore, the business case is difficult to build. Away from tilt control, active secondary lateral control of the body has been deployed in Japan on the E2 and E3 class of Shinkansen [17].

Mechatronics however can fundamentally be used to change how a vehicle steers and interacts with the infrastructure. This is covered in more detail below, with the business case being more convincing when system benefits can be shown.

Braking Systems

Most rail braking systems are based upon pneumatic actuation and friction materials, these can be used on either a disk attached to the wheelset or directly onto the tread itself. Other methods are to apply friction material directly to the track or make use of eddy current induction, again in the track.

The downside to the pneumatically actuated methods is the production, movement, and storage of the compressed air. Removal of this would greatly simplify the construction of vehicles and the move to electrically actuated friction brakes has been predicted for many years. But to date, a satisfactory solution to the fail-safe power-off application of brakes, balanced against not requiring constant energy flow to hold brakes off under operation, has not been reached.

The use of the electric drivetrain has increasingly become an important part of the braking power. The generation of electrical energy from braking requires dissipation, and this is achieved through resistor banks or moving electrical energy back into the power distribution infrastructure.

The ideal would be no mechanical brakes down to zero speed and re-use of as much of the energy as possible. This is a large area of focus.

More or All Electric Trains

The current trend in aerospace and automotive is to remove hydraulic and pneumatic energy systems and replace these with electrical systems. Modern airliners such as the Boeing 787 and the Airbus A350 are indicators of this, with engine starting, de-icing, air-conditioning, cabin pressure control, brakes, and hydraulic pumps moving to be electrically driven. In the automotive sphere drivetrains, braking systems, and steering system are transitioning to electric drive.

Where could rail move? The main targets are no pneumatic system, removing air-spring based suspension, and electric braking solely through the traction system.

Rail Mechatronics

As the previous sections highlight there is a strong argument for the rail industry to evaluate its long-term development strategy to be bolder and more inclusive of technical developments to meet its own ambitious targets for future performance.

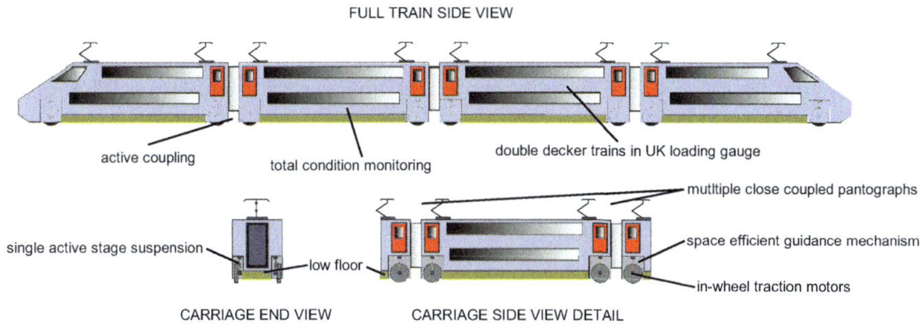

Fig. 6 Integrated mechatronic vehicle

Mechatronics offers the opportunity to rethink how vehicles operate. However, there is no real incentive to make changes to a vehicle alone unless there is a perceived system-wide benefit.

Fully Integrated Vehicle

Over many decades there has been extensive academic work looking at adding mechatronic aspects to existing railway vehicles [17], but very little looking at the vehicle as a system. The exception is the DLR 'Next Generation Train' (NGT) project [18] and the European Mechatronic Train project [19].

Work by the authors [20] looked at what a vehicle combining these insights would look like, and this is shown in Fig. 6. This encapsulated active guidance bogie-less operation, active suspension, independently rotating wheels with in-wheel traction motors, active pantographs, active coupling, total condition monitoring, and autonomous operation.

However, as mentioned above, by itself a vehicle like this would most likely not be adopted. This is due to a combination of issues: many of the benefits would appear for the infrastructure owner (reduced track damage, operational flexibility), rather than for the owner and operator of the vehicle. What would be the case for adoption?

Systems Perspective: What Can We Achieve with Control?

Railways are an integrated system, and any benefits can only truly be quantified as such. Researchers engaged in mechatronics of rail vehicle guidance have focussed on immediate tangible benefits. These tend to form around:

- Mechanical simplicity through 'bogie-less' operation
  Current bogie systems are highly complex, with multiple stages of suspension to cope with the mechanical stability trade-off inherent with wheelsets. The mechanism is much simplified here with the guidance being provided synthetically rather than passively. Reducing heavy engineering maintenance burdens.
- Improved functionality through removing curving versus stability trade-off
  Passive suspension systems are a trade-off in terms of straight-line stability and curving. In simple terms the choice is either straight-line high-speed stability or being able to go slowly around tight curves. This is due to the governing factor being the resistance of a wheelset to yaw, i.e. yaw stiffness. Active steering removes this trade-off and can be adapted to any running condition, meaning vehicles do not have to be specifically designed for a mode of operation.
- Reduced track damage
  The yaw stiffness that provides the guidance mechanism stability for a passive railway vehicle puts significant strain on the infrastructure. Essentially the yaw stiffness provides a damping mechanism by sliding the wheel across the railhead, removing material from each. The longitudinal creep forces associated with guidance can be removed almost entirely with mechatronic guidance, reducing the wear and tear on the infrastructure.

These aspects are often dismissed by railway people as being inconsistent with railway finance models that are 'vertically segregated' and therefore provide no justification for implementation. As outlined above the true benefits of mechatronics would come with what transformative action could be taken. The following are several aspects that could build a business case for adoption:

- New modes of operation
  Currently vehicles are under the control of the infrastructure. Switch and crossings and signalling system determining where and when vehicles move around a network constricting capacity. Vehicles that guide themselves could lead to autonomous decentralised control that is governed by passenger demand and not by timetable planning.
- Simplified infrastructure assets
  Switch and crossing could become purely passive items where the vehicle decides where it wants to go, rather than being driven by the position of the switch.
- Decentralised signalling system
  Self-guided vehicle could mean centralised signalling systems could be removed and vehicles could communicate peer-to-peer to locally optimise capacity; see Fig. 7.

All of this potentially adds up to a complete shift in how a railway operates, reducing carbon from the vehicle use, reducing the carbon locked up in infrastructure and making the proposition for passengers much more attractive. It would take true

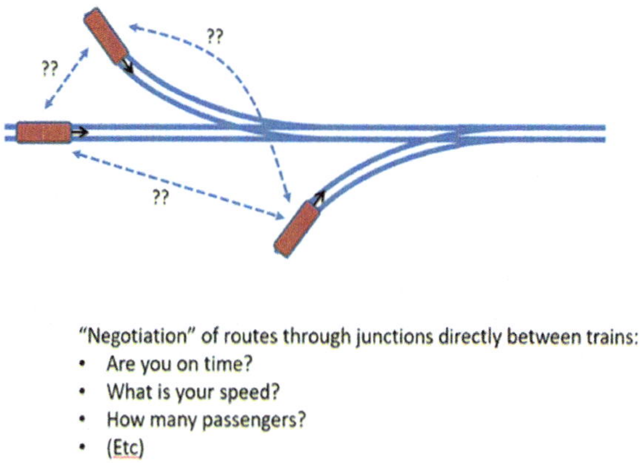

Fig. 7 Operational route negotiation paradigm

vision on the side of the industry to implement this, with the drivers coming from a push–pull in terms of legislative drivers, coupled with passengers demanding a step change improvement in the user experience. The current funding model, in the UK at least, would require a fundamental re-think to let this happen.

Examples of 'Enabling Technology' Projects

The following section highlights projects that have taken steps along the path to quantifying the benefits of and implementing the vision outlined above. These started as fundamental explorations of the technologies but have gradually transitioned to the operational aspects of railway.

Vehicle Mechatronics

The fundamentals of vehicle guidance have remained the same since the inception of the railways in the early 1800s: a wheelset (two conical wheels fixed on a single axle) running on a pair of shaped steel rails.

The basic principles of guidance can be found in [1], but the underlying kinematics are that a lateral displacement of a wheelset (as would happen through curving or due to track irregularities) results in a difference of rolling radius causing the wheelset to yaw and return to the centre of the track, at which point the rolling radius difference returns to zero.

An unconstrained wheelset however is a mechanical oscillator and due to a lack of damping in the movement described above results in a phenomenon called 'hunting'. This 'hunting' is constrained through the suspension system. Counter-intuitively, adding 'yaw stiffness', stabilises the wheelset, up to a critical speed, where the hunting returns. This yaw constraint restricts the movement of the wheelset and 'scrubs' the wheelset across the railhead, creating energy dissipation or 'damping', but at the cost of wearing the railhead and the wheel tread. This means that the first stage of suspension is required to be very stiff to constrain the wheelset and if directly connected to the vehicle body would restrict cornering ability (rail vehicles are typically 20+ m in length and the radial angles the wheelset would need to adopt would be large for the tighter radius curves that can be below 500 m) and be unacceptable to the passengers in terms of ride comfort. Therefore, short-wheelbase trucks called bogies are used with a pair of wheelsets, which are connected to the vehicle body through a soft vertical suspension, allowing the wheelset pairs to yaw and shielding the passengers from track borne vibrations. However, the 'wheelset hunting' can pass through to create 'bogie hunting' that requires dissipating through a secondary level yaw damper.

There is a fundamental issue in that this complicated mechanical arrangement presents a compromise and a critical choice for a designer, in simplistic terms:

- A vehicle that can go fast on relatively straight track then does not 'curve' very well
  Design fundamentals: Low conicity; high primary yaw stiffness; and large amounts of secondary yaw damping
- A vehicle that can curve well is then limited in terms of the maximum speed
  Design fundamentals: High conicity value: low primary yaw stiffness; and (occasionally) without secondary yaw damping

This ultimately means that a vehicle must be designed specifically for the types of routes on a network and cannot be generally used everywhere, increasing cost, and limiting operational flexibility. Mechatronics is a method of removing this compromise and allowing flexibility in the way (and where) vehicles are operated.

Vehicle Guidance and Vehicle-Based Switching

The principles of mechatronic guidance are covered in more detail in [21]. There are several possible configurations, as seen in Fig. 8.

- **Secondary yaw control (SYC).**
  Design fundamentals: Standard solid axle wheelsets and bogies; secondary yaw dampers replaced with linear actuators.
  - The simplest mechatronic form mechanically, with the guidance still provided by the wheelset. The primary yaw stiffness can be lowered considerably to allow better curving performance. The ensuing lower critical speed can be regained by the secondary active damper [22].
  - The disadvantage is that this is still relatively complex mechanically and the wheelsets demonstrate similar wear properties.

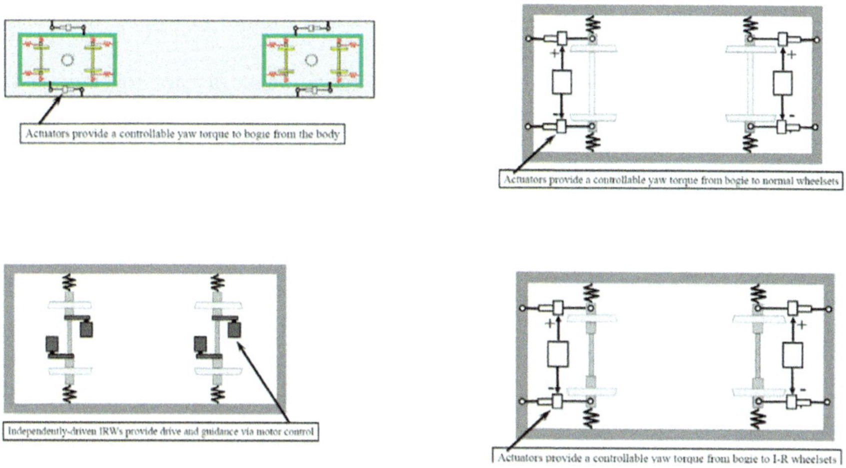

Fig. 8 Mechatronic vehicle configurations

- **Actuated solid axle wheelsets (ASW).**
  Design fundamentals: Retaining the solid axle wheelsets, but now providing yaw actuation at the primary suspension level.
  - The control algorithm balances longitudinal creep forces across the axle, achieving what is known as 'perfect curving'.
  - The wheelset itself is a 'closed loop system', and the issue can be the actuated control loop 'fighting' the steering tendency of the wheelset. Large actuation forces are required because of this and the compression required of the primary suspension.

- **Actuated independently rotating wheelsets (AIRW).**
  Design fundamentals: The wheels on a 'wheelset' are free to rotate independently, with yaw actuation provided at the primary suspension level.
  - Allowing the wheels to rotate independently on a wheelset removes the natural self-centring and curving ability of the wheelset. It also removes the longitudinal creep forces associated with curving, reducing track and wheel damage. The actuator and control algorithm are required to provide the cornering performance.
  - The downside with this approach is the requirement to estimate lateral displacement, which is a nontrivial exercise.

- **In-wheel motor wheelset (IWM).**
  Design fundamentals: IRW wheelsets with in-wheel motors for traction and guidance purposes.

- Like the AIRW system, but with the traction motors providing differential torque to steer the vehicle.
- Control algorithms require careful consideration so as not to saturate the available torque purely for steering.

Of the methods shown, the AIRW and IWM provide the maximum flexibility and provide the possibility of structural modifications to the vehicle. As illustrated, they no longer required the complicated structure of bogies and could lead to 'two-axle' vehicles, freeing space within the cabin for passengers.

Of major interest is the possibility of true autonomy as the 'vehicle' provides guidance through a track switch. This could ultimately lead to 'passive' switches removing one of the key failure modes of any rail system [23]. In the short term mechatronic vehicles can provide a huge reduction in the damage to current 'active' track switches, 25% of the current level, through steering [24].

SET Wheel Motor Project

To date, most of the vehicle mechatronics projects have purely been in the simulation environment. The exception has been the along the lines of demonstration of vertical suspension mechatronics [25] and a solid axle wheelset stabilisation method on a full-scale roller rig [26].

A UK projects led by SET in Derby, with Loughborough University and The University of Huddersfield, sponsored by RSSB, demonstrated the use of in wheel motors for guidance and traction in a full-scale rail vehicle (Fig. 9) [27]. This put beyond doubt the feasibility of active guidance and traction. Of note is the control design requirements and the ability to trade-off curving performance against straight line stability and the level of traction torque used [28]. It also introduced the concept of braking down to zero speed purely through the motors, further simplifying vehicle systems and removing the need for compressed air in the future.

Fig. 9 Class 230 vehicle used for the 'Actiwheel' project (Chris Ward)

Condition Monitoring Projects

As railways are considered safety critical a great deal of effort is spent on maintenance of the vehicles and associated systems. At the date of writing, most maintenance activities are calendar, rather than condition based. This can lead to systems being 'over-serviced' or in some cases 'under-serviced' where an item may have failed or be in a degraded state in service without this being noticed.

Moving to a more mechatronic railway leads to a greater proliferation of sensing, and therefore, more opportunity to do intelligent condition monitoring of assets. Vehicle manufacturers have been pushing forward this trend for over a decade, for example, Bombardier (now part of Alstom) took the Rolls Royce 'power by the hour' concept [29] to create their Orbita vehicle monitoring system [30].

There is still room for expansion in these concepts, incorporating the latest processing and artificial intelligence methods. A key aim of a project, again sponsored by RSSB, was to move from thinking about condition monitoring of specific components and move to assessing the systems health and fitness to run. If key operational metrics were not reached, to look at the health of specific components [31].

Moving away from monitoring the vehicle, there are multiple opportunities to explore the systematic health. A great deal of effort has been spent on development of monitoring of the rail infrastructure from on board the vehicle [32]. Of ongoing interest is monitoring of low-adhesion events that periodically occur, particularly during the autumnal season. These can cause vehicle to slide un-controllably through red signals and fail to stop at stations. Work from the simulation domain [33] is now moving to full scale demonstration of the concept. In the long term continuous monitoring from service vehicles should give a much clearer picture in terms of the variations in adhesion and validate correlations back to climate and seasons.

Infrastructure Mechatronics

In addition to the vehicles, there are aspects of infrastructure that are not only safety critical, operationally critical, but also due to their historic routes have multiple single points of failure; track switches being a prime example.

Repoint

The Repoint switch system began as a thought experiment to reduce congestion at network nodes (stations, junctions etc.). The nodes, rather than the plain line, define the capacity of the railway network. Increasing capacity at the nodes will increase the capacity of the system as a whole. Following discussion with industry insiders, the project focussed on the capacity constraints baked into the method of signalling trains on approach to points, by not allowing a train to approach until the points are detected as correctly set. If the switch were reliable enough that it was certain the switch would complete its movement, then the train could continue to approach at full speed while the switch is in motion.

To improve reliability, the Repoint switch introduces two novel features, redundant actuation and a bi-stable "passive locking" mechanism [23].

Redundant actuation is a concept used across safety critical industries. Multiple actuators are used, any one of which is sufficient to complete the necessary action. So long as one of them remains operational, the others can fail, and the system as a whole remains functional (Fig. 10).

Passive locking is necessary in this application. Conventional switch machines use an active locking mechanism; however, active locking precludes the use of multiple actuators. If the locking mechanism fails locked, then any number of duplicate actuators cannot move the switch. The Repoint system introduces semi-circular motion. The mechanism first lifts the rail out of a locking detent, then moves the rails laterally, before lowering the rails into the locking detent in the opposite position. Lateral movement of the switch rails is prevented when in the fully down position; however, any one of the actuators can lift the mechanism out of its locked position. A side benefit of this mechanism is that it is bi-stable. Should the power fail during actuation, the mechanism will fall under gravity into one of the locked positions (Fig. 11).

Fig. 10 Repoint mechanism, showing the semi-circular motion with the locking features highlighted in red (Loughborough University)

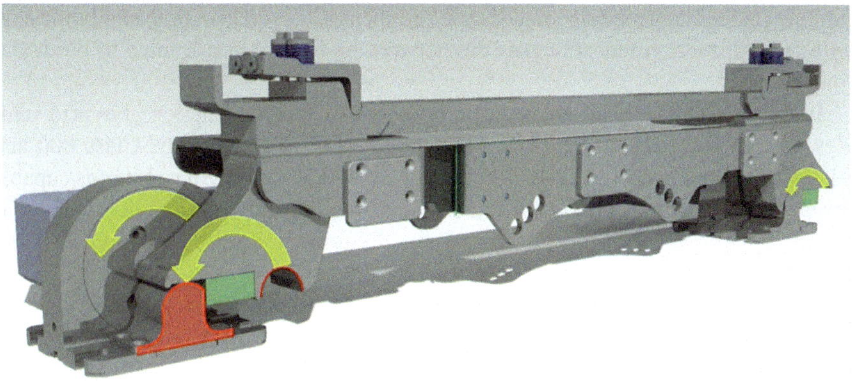

Fig. 11 Repoint switch in-track, showing the three actuator bearers (Loughborough University)

Development passed through several iterations, beginning with a lab-based demonstrator that proved the actuation and locking concept [34]. Subsequently, a full-scale switch system was designed, built, and tested in-track [35]. There are 2 motors in each of the actuator bearers. The protype system has three actuator bearers to achieve the redundancy aims. All 6 of the motors are controlled and synchronised by a controller located in the lineside equipment cabinet. Development continues as a part of the In2Track series of projects led by Network Rail.

Systems Mechatronics

Background

Technology is sometimes applied to transportation simply for technology's sake. At the extreme, we have 'futuristic' systems like Hyperloop [36] and the Las Vegas Convention Centre Loop [37]. These rely on large capital outlays to build infrastructure (vacuum-sealed tubes or underground tunnels) to move small volumes of passengers (in air-tight pods or electric cars) between stations, which makes their economic and environmental value questionable. As outlaid above, the metal-on-metal running contact of traditional rail transportation has many benefits, and there is huge scope for system wide benefits from mechatronics.

Applications to the Rail Network

Electrification is the best way to decarbonise high-speed intercity rail, yet only around 38% of Great Britain's rail network was electrified as of 2019 [38]. This leads to some routes being partially electrified which means that there are still diesel trains running under electrified overhead line equipment (OLE). This is especially the case with freight traction that traverses the network and is not constrained to particular routes.

In the near-term this has been partially solved by the introduction of hybrid vehicles such as the Hitachi AT300 product family. Of interest are the Class 800 and Class 802 trains, which are 5- or 9-car bi-mode (25 kV electric/diesel) trains capable of operating at up to 225 km/h [39]. These bi-mode trains run on OLE when it is available and change to their on-board diesel engine-generators when OLE is not available.

This vehicle is an excellent example of mechatronics engineering. It includes mechanical systems (the suspension and bogies), electrical systems (the generator, 25 kV rectifier and power electronics for the electric machines), and software systems (for controlling the diesel engine-generator and the electrical systems).

However, more can be done to use mechatronics to reduce the environmental impact of this rail transport system. On the vehicle side, research has shown that a feedforward control system can reduce the emissions of the Class 800/802 trains by 19% over a representative route [14].

The addition of batteries to the train could further reduce its emissions by allowing the vehicle to run in 'hybrid mode' when operating on the diesel engine. A 'deployable pantograph' with integrated contact sensing, similar to that developed by Brecknell-Willis [40] would allow the pantograph to be raised and lowered at speed and the reduction of wear at the pantograph-OLE interface.

Combining these technologies (advanced controllers, battery systems and deployable pantographs) will allow rapid decarbonisation of the rail transport system through 'intermittent electrification'. Intermittent electrification involves electrification of discrete parts of a rail route, rather than electrifying continuously from one end to the other. By reducing the energy that trains use (through advanced controllers), efficiently storing this energy for use when not under OLE (in batteries) and allowing the train to connect to the OLE at speed (using mechatronic pantographs), intermittent electrification is an excellent example of mechatronics engineering allowing complete (operational) decarbonisation of a transport system.[1]

Further Improvements

Integrating improvements such as mechatronic switches (e.g. REPOINT), vehicle-based switching, in-cab signalling [41] and moving-block signalling [42] into future railway systems will enable them to become truly mechatronic transportation systems.

---

[1] 'Operational emissions' are those from day-to-day operation of an asset/system. There are also 'embodied emissions' which are the emissions generated during the manufacture/installation of the system. Thankfully there are many research projects underway to reduce the emissions generated by building railways [43].

Such a system would have lower operational carbon emissions and be more reliable than a standard electrified railway. Mechatronic railway switches are more reliable than their standard mechanical counterparts, allowing higher throughput of rail traffic. Vehicle-based switching allows for intelligent re-routing of trains to reduce their energy consumption and provide better services to passengers or freight operators. In-cab signalling and moving-block signalling allow intelligent management of rail vehicles to maximise the number of trains through the network without compromising on safety.

Conclusions

Rail based transportation is at cross-roads. It is one of the few industries that can claim to be capable of having a zero-carbon output at the point of use, as it is easily capable of being fully electrified. But it is constrained by a running system that limits capacity and growth potential.

The industry has been reluctant to consider technologies that have a large impact on these, instead preferring an incremental improvement path. The authors believe this is reaching its limits and if the industry wants to fulfil its vision of making rail more attractive to the passenger, then mechatronics seems an obvious route.

Research and development projects have shown the potential short and long term improvement in running dynamics, energy consumption and flexibility with mechatronics. There is still a great deal to do in this area, not least of which will be convincing the decision makers and investors that fundamental change is needed. There is no technical reason not to, but the very nature of how the industry is constructed precludes change. Environmental consideration may be the only catalyst to make the changes that are required.

Acknowledgements We would like to thank our sponsors over the last decade (RSSB, Network Rail, Transport for London, EPSRC, EC), the many partners we have worked with (University of Birmingham, University of Huddersfield, University of Sheffield, University of Salford, University of Nottingham, Nihon University, Politecnico di Milano, Southern Rail, Telent, SET Derby, Hitachi Rail, Perpetuum) and the many researchers past and present who have contributed to the work.

References

1. Wickens A (1998) The dynamics of railway vehicles-from Stephenson to Carter. IMechE Part F J Rail Rapid Transit 212(3):209–217
2. Anon, "Capacity–helping reduce overcrowding," HS2 Ltd. https://www.hs2.org.uk/why/capacity/. Accessed 5 July 2021
3. Shirres D, Keemor G, Dolphin N, Hooper P (2021) "Why rail electrification?" Report. Rail Industry Association, London
4. Wolmar C (2009) Fire and steam: a new history of the railways in Britain. Atlantic Books, London

5. Connor, P (2017) Basic railway Signalling. (7 May 2017). http://www.railway-technical.com/signalling/infopaper-6-basic-railway.pdf. Accessed 3 July 2021
6. Anon (2021) Measuring and reducing embodied carbon in rail infrastructure, RSSB. https://www.rssb.co.uk/en/sustainability/carbon. Accessed 5 July 2021
7. Anon (2021) The future railway-the industry's rail technical strategy 2012, TSLG. https://www.sparkrail.org/Lists/Records/DispForm.aspx?ID=2899. Accessed 5 July 2021
8. Anon (2021) Rail technical strategy. https://railtechnicalstrategy.co.uk/. Accessed 5 July 2021
9. Anon (2021) Shift2Rail. European Commission. https://shift2rail.org/. Accessed 5 July 2021
10. Anon (2021) Horizon Europe. European Commission. https://ec.europa.eu/info/research-and-innovation/funding/funding-opportunities/funding-programmes-and-open-calls/horizon-europe_en. Accessed 5 July 2021
11. Tischler M (1996) Advances in aircraft flight control. Routledge, London
12. E. Commission (2021) Intelligent transport systems, vehicle safety systems. https://ec.europa.eu/transport/themes/its/road/application_areas/vehicle_safety_systems_en. Accessed 3 Aug 2021
13. Gilchrist, AO (2009) History of engineering research on British railways. Institute of Railway Studies, York
14. Harrison T, Midgley WJB, Goodall RM, Ward CP (2021) Development and control of a rail vehicle model to reduce energy consumption and carbon dioxide emissions. Proc Inst Mech Eng Part F J Rail Rapid Transit
15. Anon (2013) EN 15380–4:2013 Railway applications. Classification system for railway vehicles. Function groups, European Standards
16. Goodall R, Ward C (2014) Rolling stock technology for the future. In: International conference on high speed rail, Birmingham
17. Fu B, Giossi RL, Persson R, Stichel S, Bruni S, Goodall R (2020) Active suspension in railway vehicles: a literature survey. Railway Eng Sci 28:3–35
18. DLR, RTR Special (2011) NGT-Next Generation Train, Hamburg: Eurailpress
19. European Commission (2021) Mechatronic technologies for trains of the future. https://cordis.europa.eu/project/id/BRPR970527. Accessed 3 Aug 2021
20. Ward C, Mei T, Hubbard P, Mirzapour M (2012) Railway vehicle optimisation using the concept of "Design for Control". In: Railways 2014: the second international conference on railway technology: research, development and maintenance, Ajaccio
21. Farhat N, Ward C, Goodall R, Dixon R (2018) The benefits of mechatronically-guided railway vehicles: a multi-body physics simulation study. Mechatronics 51:115–126
22. Goodall RM, Ward CP, Prandi D, Bruni S (2016) Railway bogie stability control from secondary yaw actuators. In: 24th Symposium of the international association for vehicle system dynamics (IAVSD 2015), Graz
23. Bemment SD, Goodall RM, Dixon R, Ward CP (2018) Improving the reliability and availability of railway track switching by analysing historical failure data and introducing functionally redundant subsystems. Proc Inst Mech Eng Part F J Rail and Rapid Transit 232(5):1407–1424
24. Farhat N, Ward CP, Dixon R, Goodall R (2020) Benefits of mechatronically guided vehicles on railway track switches. Proc Inst Mech Eng Part F J Rail Rapid Transit 234(3):276–288
25. Qazizadeh A, Persson R, Stichel S (2015) On-track tests of active vertical suspension on a passenger train. Veh Syst Dyn Int J Veh Mech Mobility 53(6):798–811
26. Pearson JT, Goodall RM, Mei TX, Himmelstein G (2002) Assessment of active stability control strategies for a high speed bogie. IFAC Proc Vol 35(2):743–748
27. Dobell M (2019) Steering a course to the future. Rail engineer, pp 28–30. (29 Nov 2019)
28. Farhat N, Ward C, Shaebi O, Crosbee D, Stow J, Wang R, Goodall R, Whitley M (2019) Controlling a rail vehicle with independently-rotating wheels. In: 26th Symposium of the international association of vehicle system dynamics, IAVSD 2019, Gothenburg.
29. Anon (2021) Aftermarket services. Rolls Royce. https://www.rolls-royce.com/products-and-services/civil-aerospace/aftermarket-services.aspx. Accessed 9 Aug 2021
30. Anon (2021) Services. Bombardier Rail. https://rail.bombardier.com/en/solutions-and-technologies/all-services.html#Operations_and_maintenance. Accessed 9 Aug 2021

31. Anon (2018) Railway operator challenge: integrated bogie monitoring & prognostics system. RSSB. https://www.sparkrail.org/Lists/Records/DispForm.aspx?ID=24727. Accessed 28 Oct 2021
32. Ward CP, Weston PF, Stewart EJC, Li H, Goodall RM, Roberts C, Mei TX, Charles G, Dixon R (2001) Condition monitoring opportunities using vehicle-based sensors. Proc IMechE Part F J Rail Rapid Transit 225(2):202–218
33. Hubbard PD, Ward CP, Dixon R, Goodall RM (2014) Models for estimation of creep forces in the wheel/rail contact under varying adhesion levels. In: Vehicle system dynamics: international journal of vehicle mechanics and mobility, vol Sup 1, no IAVSD proceedings supplement, pp 370–386
34. Bemment SD, Ebinger E, Goodall RM, Ward CD, Dixon R (2017) Rethinking rail track switches for fault tolerance and enhanced performance. Proc Inst Mech Eng Part F J Rail Rapid Transit 231:1048–1065
35. Olaby O, Dutta S, Harrison T, Ward CP, Dixon R (2021) Realisation of a novel functionally redundant actuation system for a railway track-switch. J Appl Sci
36. Ross PE (2016) Hyperloop: no pressure. IEEE Spectr 53(1):51–54
37. The Boring Company (2020) LVCC. https://www.boringcompany.com/lvcc. Accessed 14 July 2020
38. Office of Rail and Road (2020) Rail infrastructure, assets and environmental 2018–19 Annual Statistical Release. https://dataportal.orr.gov.uk/statistics/infrastructure-and-emissions/rail-infrastructure-and-assets/. Accessed 22 April 2020
39. Rogers A, Robinson C, Agatsuma K, Iwasaki M, Inarida S, Yamamoto T, Konishi K, Mochida T (2014) Development of class 800/801 high-speed rolling stock for. Hitachi Rev 63(10):646–654
40. Brun L (2016) Closed loop pantograph for improved current collection and condition monitoring
41. Institution of Mechanical Engineers (2016) 'In-cab' signalling system to improve London train service. (4 Aug 2016). https://www.imeche.org/news/news-article/advanced-in-cab-signalling-system-to-improve-london-train-service. Accessed 03 Aug 2021
42. Basile D, ter Beek M, Ferrari A, Legay A (2019) Modelling and analysing ERTMS L3 Moving block railway signalling with simulink and uppaal SMC. In: Formal methods for industrial critical systems. FMICS 2019. Lecture notes in computer science, vol 11687
43. Network Rail (2019) Case study: Electrification embodied carbon savings. RSSB (16 May 2019). https://www.rssb.co.uk/sustainability/Electrification-Embodied-Carbon-Savings. Accessed 15 July 2020

Sustainable Mechatronic Solution for Agricultural Precision Farming Inspired by Space Robotics Technologies

Cong Niu, Youhua Li, and Xiu-Tian Yan

Abstract Despite recent developments in satellite-based agricultural crop monitoring, agricultural food production is still very much a traditional process in which the monitoring of the farming fields is still a manual process. In addition, the application of crop nutrients and control of pest and disease remains at best a very much hard automated process in respect of spreading fertilisers and pest control agents. There is a strong desire from the farming community and at governmental level to accelerate the advancement of smart faming, which requires more frequent soil condition measurement, precision fertiliser application and more optimal pest agent control and application. Mechatronic systems as the outcome of an academic discipline have much to offer to smart farming, and sustainable and secure food production. This chapter looks at the recent development of mechatronic systems for farming applications and focuses on a sustainable approach with more energy efficient ways of designing a mechatronic solution for farm soil monitoring and selective harvesting. It is intended that a more sustainable paradigm can be followed and strengthened for more sustainable and energy efficient food production.

Introduction

World population continues to grow, and it is estimated that it will reach 9 billion by 2050 [1], an increase of 1.18 billion or 15% on 2020. As part of this growth, the proportion of what may be considered middle-class is expected to increase from 3.77 billion in 2020 to 5.4 billion in 2030 by adding the number for each region of the world from an interactive chart called Forecast of the global middle-class population from 2015 to 2030 by region [2], a 43.6% increase, and this trend is expected to continue. These population growth trends and middle-class population

C. Niu · Y. Li · X.-T. Yan (✉)
Robotics and Autonomous Systems Group, The University of Strathclyde, Glasgow G1 1XJ, UK
e-mail: x.yan@strath.ac.uk

C. Niu
e-mail: cong.niu@strath.ac.uk

P. Hehenberger et al. (eds.), *EcoMechatronics*,
https://doi.org/10.1007/978-3-031-07555-1_11

and demographic changes are putting significant pressure on the current global food production system. Current food production approaches are unsustainable and can be exemplified by over fertilisation of crops and over-fishing. Current food production is also increasingly negatively influenced by more frequent extreme weather conditions, such as droughts and floods.

Sustainable and secure food production hence becomes perhaps the only technological long-term solution to feeding this growing population and providing better food quality to meet the middle-class demands. Sustainable development is thus becoming one of the most important elements for future societal development and as without the consideration of sustainability this will become a "*contemporary issue*" as Klarin noted [3].

Similar to the food production challenge, energy is another scare resource. This is a complex challenge, especially in the context of societal pressure to reduce greenhouse gas levels to achieve net-zero emissions. Energy is however critical for sustainable development in many industrial sectors. Even though coal consumption for generating electricity has decreased from 38.4% in 2003 to 22.4% in 2019, electricity generated from fossil fuel still represents 53.8% of all electricity generated according to the International Energy Organisation (IEA), extracted from the chart of the share of OECD gross electricity production by source, 1974-2020p [4].

These energy development trends point to the need for system designs that in future hence consider energy efficiency more carefully [5]. In the context of agricultural equipment design, especially with the commitment of many global governments to the gradual removal of fossil fuel powered vehicles for both transportation of goods and commuting [6], electrification becomes a popular technological approach which will help to achieve net-zero emissions in 2050 [7]. It is hence increasingly important to design agricultural equipment with as high an electricity efficiency as possible to achieve overall energy efficiency [8], assuming that is that electricity will be the main form of energy available for agricultural equipment.

It is clear that traditional engineering design approaches face challenges in meeting the world's future population and demographic changes. Scarce energy resources demand more novel solutions to minimise energy wastage in order to optimise energy usage in food production. This overall energy saving will lead to better protection of our Blue Planet from irreversible climate change [9]. It is time to think more radically about different technological solutions to address the above challenges.

Mechatronic Approach With a Focus on Energy Consumption of The Systems

Mechatronics as an academic discipline is concerned with the exploitation and maximisation of potential benefits for society by searching and exploring a technological solution from a spectrum of traditional academic disciplines such as mechanical, electrical, control, computing and more recently artificial intelligence. This exploration often leads to novel solutions to address complex problems which themselves are challenging, dynamically evolving, and often demanding from multiple perspectives with conflicting requirements. In the context of sustainable development, in particular for sustainable food production [10], mechatronics is considered to be unique in enabling engineering designers to explore a vastly increased design solution space, flexible in coping with conflicting demands, expandable as more cross-disciplinary solutions can be incorporated and integrated through typically modular structure. A systematic approach through an appropriate design methodology is considered to be important for any of such mechatronic system to be designed to meet the demanding requirements.

Design Methodology

Before generating an energy efficient mechatronic design solution, it is necessary to investigate a generic methodology considering all important aspects of the system and environmental factors so that such a methodology would be applicable across more applications. In this chapter, a purpose-generated design methodology is presented for developing an energy efficient system, focusing on path planning of mobile robotics designed for agricultural apparitions. An energy optimisation algorithm is then derived for detailed modelling and path optimisation to improve energy efficiency and long-distance operations.

An initial design process model presented by Yan et al. [11] consists of three pillars, namely an information repository pillar, a design process pillar and an energy pillar. The information repository pillar is concerned with the representation of the design information expansion process in which as more design decisions are made, more design parameters values are committed, and hence more information determined. The design process pillar represents the conventional process of designing mechatronic systems such as a mobile rover. These processes flow from top to bottom and help a designer to progress to a satisfactory design. Building on this work, an extended design methodology is proposed which introduces a new pillar which focuses on the consideration of energy conversion and the external impact factors of field robots with a focus on agricultural applications. Specifically, the methodology now includes the consideration of the dynamically changing environment and its impact on the energy consumption of a mobile robotic system. These factors include the elevation of the terrain, the surface unevenness of the terrain and the hardness of the surfaces over which the mobile rover needs to travel. With these considerations in the design process model, it is believed that a full and comprehensive consideration for a systematic design can be achieved to ensure that the mobile robot is energy efficient by design.

In the information repository pillar on the left of Fig. 1, the design information for the mechatronic system design process is shown and is available to support the whole design process. It gradually expands or grows as system design decisions are made and the relationships between each interconnecting part are clarified and decided.

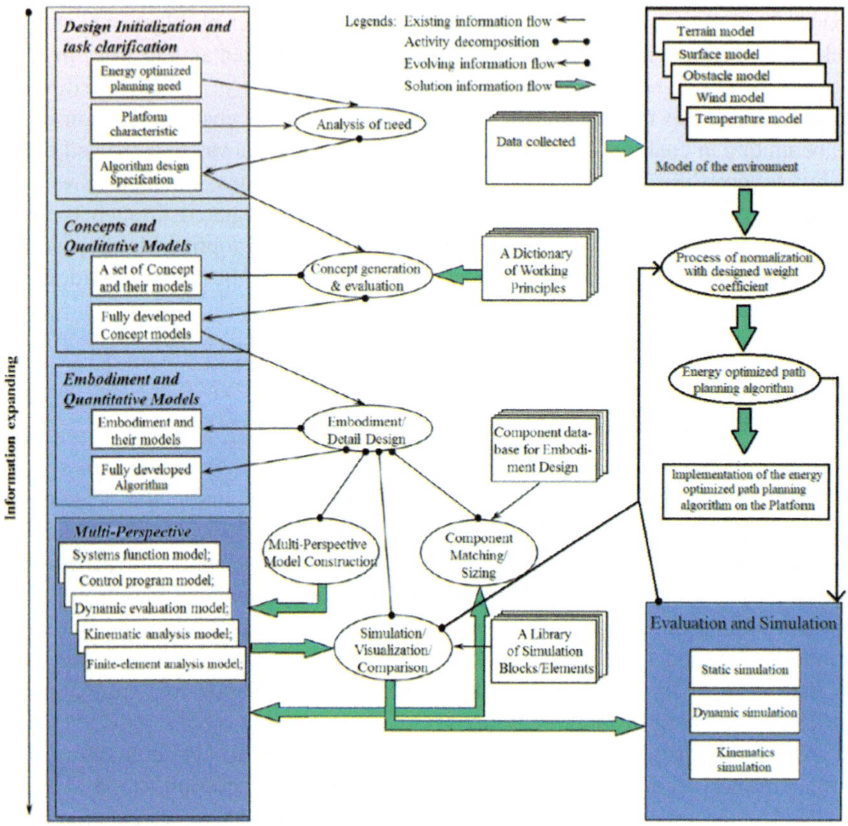

Fig. 1 A three-pillar mechatronic design methodology focusing on energy

This evolving process is shown on the left of the figure with the Evolving Information flow arrow. This evolving information can be categorised into four groups with different level of details from the top to the bottom of the pillar. These include information on design initialisation and task clarification, which can be further divided into the description of the system level needs of customers, specifications through analysis, also the market intelligence for commercialising a mobile robot for a researched market. The information becomes richer as shown in Fig. 1 by the darkening colour on the robot design specification. This evolving information flow is enriched in the form of the existing design information for the next stage. Ultimately, there results in a design solution in the form of a mixture of statements of the customers need and benchmark research results, conceptual design solution models, embodiment design and importantly a set of design models from multiple perspectives, and a control design model. All of these help in make decisions in designing the robot.

The middle pillar, from the top to the bottom in Fig. 1, shows the mechatronic system design process and activities in generating a final design solution. It involves both concept generation and qualitative modelling. In this step, concepts are generated and evaluated according to specific working principles and with the existing information of design specifications produced earlier.

The information of a concept as generated and evaluated is passed to the next step of embodiment design and quantitative modelling for evaluation. Embodiment and detail design are first undertaken so that all components are specified, then the evolving information is passed to the embodiment models, along with the fully developed solution models. These can then be simulated and evaluated for quantitative evaluation.

Finally, the last stage of the mechatronic system design process in the middle pillar is about multi-perspective modelling and simulation, in which a mobile robot solution is fully modelled and simulated from several perspectives as in the last element of the information repository pillar. Together, these multi-perspective models provide a full representation of a mobile system for intensive evaluation. There are more intertwined relationships and trades-offs required and considered at this stage to achieve an optimal solution. In this step, many models are designed as shown in the bottom left and top right of Fig. 1. Building on the library of simulation models and simulation and visualisation, a mechatronic system can be fully generated, evaluated and validated in this virtual modelling world. A fully developed concept model is created for further evaluation at the end of this process.

From the energy optimisation point of view, the design process model of Fig. 1 considers the following: environment, path planning and terrain. Environment covers the wind, temperature and lighting conditions which have impact on navigation and actuation. Terrain consideration includes surface modelling of the field from, typically, satellite images and the difference between it and the actual terrain as experienced by a mobile platform. For the path planning, on top of the left pillar a clear set of requirements for this need to be formulated, including how the planning algorithm focuses on the ratio between shortest path and lowest energy cost and their tradeoffs. The platform characteristics needs to be considered, depending on the difference between each platform's characteristics and different path planning methods to be utilised. After the first stage of Analysis of Need, the design can proceed to stage 2 and concept generation and evaluation. At this stage, the energy critical design parameters such as weight and friction coefficients are considered. This allows energy optimisation when a path is generated. Such a path can be virtually implemented on a target mobile platform, on which a series of simulation covering the static, kinematic and dynamic behaviour of the system. These provide a full and in-depth understanding of the energy behaviour of the design system through this qualitative conceptual design and quantitative design and simulation at the embodiment design stage.

Finally, on the right of Fig. 1 is the design of the path planning algorithm with from the top the data collected by various methods including on-board and offboard sensors, and weather stations. Satellite images are also used to model the terrain of the target field, according to the requirements of the applications which could include air-based ground-based water-based autonomous systems. After the modelling of the environment, there is a processing algorithm for each of the environmental models that provides different priorities according to the objectives of the specific platform.

These can include a higher rating on the requirement of maximum distance travelled for a task, or faster coverage of a full field, or exploration of an unknow farming environment and so on. Based on the importance of these possible objectives, each of the environmental models will be selected, processed and used for simulation accordingly. With these selected inputs of priorities, the energy optimisation algorithm is used independently to generate an energy efficient path for the specific platform. Then the result of the path generated is simulated again and the parameters of the planning algorithm can be tuned to improve the performance of the energy optimised path planning. When the performance is determined and satisfactory, the algorithm can be implemented on an individual platform and the field test can be performed.

Energy Focused Mechatronic Modelling

For energy efficient mobile rover design, a list of considerations has been compiled based on a literature review and field studies. These can then be used to help to build comprehensive representations of energy models for a mobile rover design. It is aimed that the energy model will provide a comprehensive representation of the key energy outputs required for a field robotic system.

A total energy model is developed, considering all critical energy used in a typical rover system and the energy in consideration is represented in the form of power in this case and can be derived by the following equation (Eq. 1):

$$\int_{t_1}^{t_2} \left(P_m + P_s + P_f + P_h + P_e\right) \cdot dt \tag{1}$$

where:

- P_m is the summation of the power on each wheel and at any given time, $P_m = F \cdot V_l$.
- P_s is then the summation of the power consumed by the steering system, and is calculated as $P_s = \tau \cdot \omega$.
- To overcome the friction between the rover and the contact surface the term P_f is introduced.
- During operation, there will be heat generated due to mechanical friction and electrical energy to mechanical rotation conversion. The power associated with these heat losses in the propulsion driving system can be represented by P_h. This portion of the power is proportional to the power that is used by the rover for its motion and can be further integrated by considering its sub-systems.
- Finally, P_e represents the power required for the on-board electronic equipment to perform varying tasks at different stages. This can be a significant part of energy consumption if computationally intensive calculations are performed for tasks such as navigation and path re-planning.

AgriRover Design as an Example

AgriRover [11] (Fig. 2) is designed as a multipurpose autonomous robotic mobile platform specifically for agricultural applications, aiming to address the pressing needs of monitoring soil condition and harvesting for overall development of smart farming. Specifically, the design is focused on functions such as soil sampling, fruit and vegetable harvesting and field mapping. This multipurpose agricultural robotic mobile platform is designed with the inspiration of Interplanetary Rovers. In particular, cost minimisation is a key driver of the design and this leads to the consideration of operational cost, including energy usage from on-board power sources in order to maximise the economic benefits to farmers. The key energy components of AgriRover utilise a set of rechargeable lithium batteries as the energy source and drive electrical actuators and motors as the mobility system. The AgriRover is designed with the focus of eco-friendly and sustainable development as it is important both environmentally for the wellbeing of ourselves and financially and makes sense in long term. One of the design perspectives is to make the AgriRover energy efficient during its operation and this can be achieved by following an energy efficient pathway using an energy model.

Perception of Agricultural Environment Through Vision

Perception of the farming environment is of significance in understanding the surroundings of an agriculture robot. A full mapping of the obstacles, including buildings, tractors, trees, fancies and other agricultural objects, helps a robot to plan

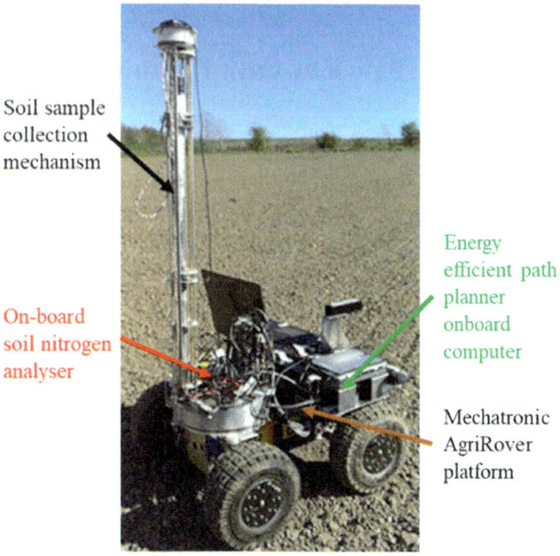

Fig. 2 AgriRover prototype performing soil sampling and monitoring

(a) (b)

Fig. 3 AgriRover's vision-based perception system and processing **a** Original image and **b** Processed image for recognition

its path more efficiently. In the context of energy perspective modelling, this critical location information of these objects can help a mobile robot to generate an energy efficient path for a particular agriculture task. The exact location of the subject is stored in the navigation map and used to build the occupancy grid for the map. Based on this occupancy grid, a suitable navigation path can be generated, and the overall distance travelled is calculated to optimise energy consumption. Whenever possible, the shortest path is generated for the rover to reach the next via-point and perform the next operations.

A visual image of a farmhouse is captured and shown in Fig. 3a. This is a typical farming environment within which these objects exist, and it is important for a robot to understand this environment. To train a robot to understand these objects, these objects are segmented for training purposes and the resulting segmented image is as shown in Fig. 3b and is used for developing an obstacle recognition capability of commonly found surroundings.

An Energy Efficient Path Planning for a Field Trial

Through further discussion with partners in Beijing, it became clear that the initial selected field has been divided into 32 small blocks and each of these blocks have had applied a different fertiliser for experimental purpose over several years. It is therefore desirable to sample each of these blocks so that a comparison and link to the yield could be made. A number of path plan concepts have therefore been developed.

Plan 1 shown in Fig. 4 is designed to cover all 32 points in the centre of each block in a sequential manner following the vertical path. This plan requires the rover to turn 14 times and it never passes any sampling point twice. The total distance the rover travels is calculated to be 530 m. The advantage of this plan is that it travels the least distance.

Plan 2 shown in Fig. 5 is also designed to cover all 32 points in the centre of each block in a slightly more complicated pattern travelling along the diagonal direction between the blocks. This plan requires the rover to turn 11 times, and it passes twice

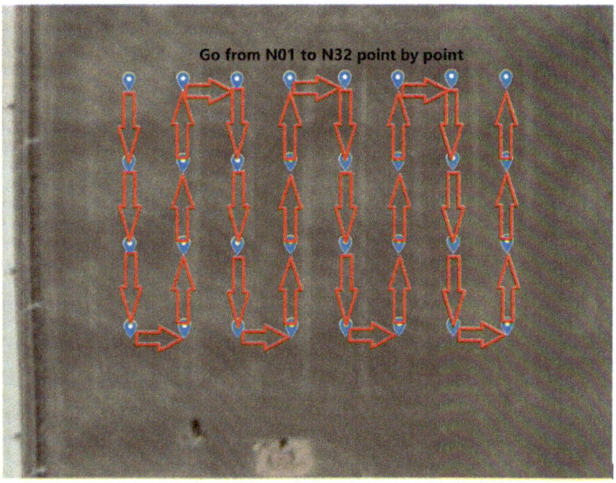

Fig. 4 A complete parallel robotic soil sampling path plan for the selected farm at NERCITA

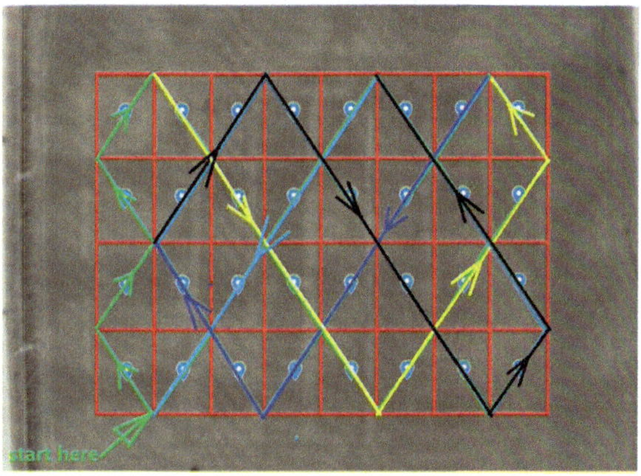

Fig. 5 An alternative diagonal robotic soil sampling path plan for the same selected farm at NERCITA

two sampling points at the bottom left corner. The total travel distance is 576.9, 46.7 m or 8.8% more than Plan 1. Based on the above, it was decided Plan 1 offers an optimal solution and was chosen for the field trial to minimise the energy consumption.

Dynamic Energy Modelling and Simulation

Robots working in a field typically experience significant variations in energy consumption due to the nature of unstructured environment such as hilly terrain, uneven surfaces, and soil clay soils. It is therefore desirable to investigate the dynamic nature of energy consumption in such environments. One of the approaches used is to perform a dynamic cooperative simulation to assess the energy consumed. This simulation can also be used to verify the results of the planned path. Dynamic cooperative simulation uses two models from MATLAB/Simulink and Adams. The former is used as a control module and the ADAMS model then provides a 3-dimensional kinematic and dynamic representation including detailed modelling of the terrain and the Rover used for the dynamic simulation. The two systems work together collaboratively to generate a solution to evaluate the energy performance with systematic and detailed dynamic modelling.

In this co-simulation the path generated with the path planning algorithm has the goal set to lowest energy consumption. The start point is indicated as a green circle. For a given terrain, it is necessary to optimise a path and a result is shown in Fig. 6(1). From the co-simulation, the total energy consumption of the planned path shown in Fig. 6(2) is lower than the straight-line path one shown in Fig. 6(5) and the height change of planned path has a lower slop shown in Fig. 6(2) compared to that for the straight-line path in Fig. 6(6).

Figure 6(1), (2) and (3) show the simulation results for the planned path, energy consumption and height profile for the energy optimised path. The total run time of this simulation is 305 s and the total energy equivalent cost shown is $6.277 \times 10^7 \mathrm{N \cdot mm \cdot s}$ by each wheel, this intermediate variable is due to the limitation of the software. This value is an integral of $\tau \times v$ with respect to time, where the speed v

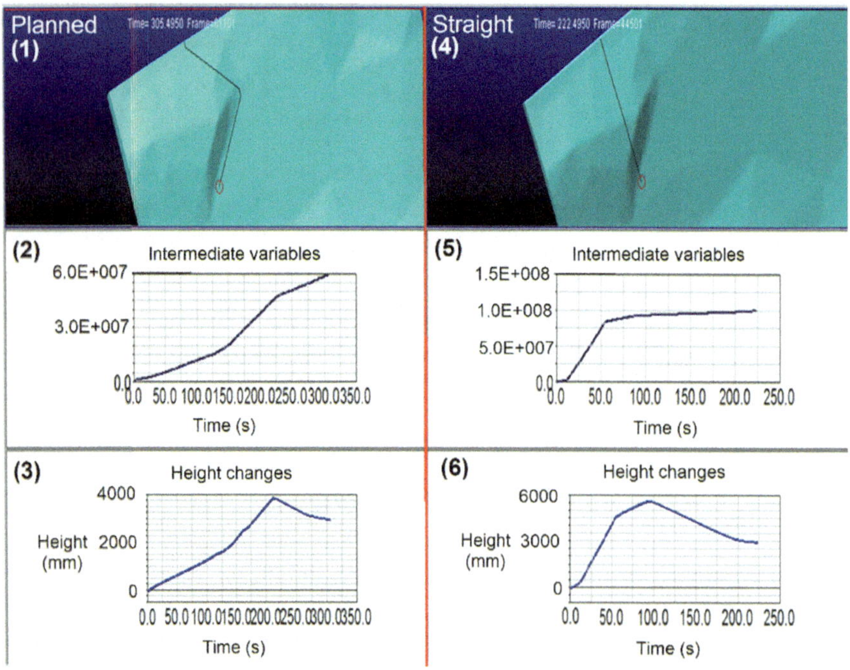

Fig. 6 Co-simulation results of an optimal path in comparison with a standard path: (1), an optimised path, (2) energy used for the path in (1), (3) the height profile of the optimised path (4) standard or normal path, (5) energy used for the path in (4), (6) the height profile of the straight path

is in unit of mm/s as the ADAMS uses this intermediate variable as the output. So to calculate the final total energy consumption in joules, Eq. 2 is used to represent the total work done in the period from t_1 to t_2. by integrating the rotational work.

$$Work = \int_{t_1}^{t_2} \tau\omega dt = \int_{t_1}^{t_2} \tau \times \frac{v}{r} dt \tag{2}$$

As mentioned the intermediate variable is $\tau \times v$ therefore the radius needs to be divided to fit the Eq. 2. The radius of the AgriRover wheel is 80 mm. As the AgriRover has two driving wheels when simulated, and the length units were in mm with the intermediate variable, the final energy cost is calculated as $6.28 \times 10^7/80/1000 \times 2$ and the total energy required is 1,567.5 J.

As shown in Fig. 6(4), 6(5), 6(6) for a straight path, the total run time is 223 s and this is shorter than 290 s for the planned path as shown in Fig. 6(1), 6(2) and 6(3). Despite the total length of the path is less with the straight path, the total energy cost shown is $1.01 \times 10^8 \text{N} \cdot \text{mm} \cdot \text{s}$ which is higher due to the work done by the AgriRover against gravity is higher. Using Eq. 3 following the same procedure as before, the energy cost is 2,525 J.

Field Tests

The first field test was conducted at a farmland field near Glasgow in the UK as shown in Fig. 7 and two groups of tests have been set up and undertaken. The first group of tests focused on the energy optimised planning. The second set of field tests took place in Beijing and focused on multipoint planning from an energy perspective. Due to technical problems with the AgriRover, full results were not obtained based on the original plan.

Fig. 7 Field tests at a farm near Glasgow

The first set of tests is designed to perform the following tasks.

1. Demonstrate the capability of the planning algorithm in identifying an optimal path for any given points in the field, e.g. from point A at the bottom to point B at the top of other locations.
2. Utilising the energy optimised path planning to establish alternative, and more energy efficient, paths from point A to point B.
3. Obtain relevant comparison data:
   (a) The first test is to capture data on the energy consumption between a direct straight or shortest path between set points,
   (b) The second test is to obtain data energy consumption on a path that has been generated by the energy optimised planning algorithm.

These results of two types of planning are compared in Table 1.

The energy optimised path planning algorithm generates an energy optimised path from any point A to point B taking into consideration of field terrain using a potential field path planning algorithm. The waypoint then is converted from National Grid (British National Grid System) to GPS coordinates which can be used by the AgriRover. The conversion has been done by utilising a MATLAB function named '*OS2LL*'. All the waypoints for the field test are converted into GPS coordinates and the results are shown in Table 1. These points are plotted on a satellite map as shown in Fig. 8.

When a straight path is followed by the Rover. The height change data plotted has a higher interval of height change that requires the rover to overcome a climb of a 5.63 m hilly slope from the lowest 78.85–84.48 m above the sea level. This is 1.62 m higher than optimised path in which a peak of 5.63 m on the straight path is avoided.

Table 1 Details of energy optimised way points set

Points	Latitude	Longitude		Points	Latitude	Longitude
1	55.90361	−4.1826		12	55.90395	−4.1835
2	55.90365	−4.18268		13	55.9039	−4.18358
3	55.90369	−4.18277		14	55.90386	−4.18365
4	55.90374	−4.18285		15	55.90381	−4.18373
5	55.90378	−4.18293		16	55.90376	−4.18381
6	55.90382	−4.18301		17	55.90372	−4.18389
7	55.90387	−4.1831		18	55.90367	−4.18396
8	55.90391	−4.18318		19	55.90362	−4.18404
9	55.90395	−4.18326		20	55.90358	−4.18412
10	55.904	−4.18334		21	55.90353	−4.1842
11	55.904	−4.18342		22	55.90353	−4.18428

Fig. 8 Energy optimised way point set plotted on the satellite map

Figure 9 shows the 3D view of the energy optimised path, starting from green circle to destination shown in magenta diamond which is generated by the algorithm and focused on energy optimization. The elevation change data is plotted with a height change that requires a climb of 4.01 m from lowest 78.85 m to a highest point of 82.86 m as shown in Fig. 9(2).

When a straight path is followed by the AgriRover. The height change data plotted has a higher interval of height change that requires the rover to overcome a climb of a 5.63 m hilly slope from the lowest 78.85–84.48 m above the sea level. This is 1.62 m higher than optimised path in which a peak of 5.63 m on the straight path is avoided.

As shown in Fig. 6(3), the height change that the rover needs to overcome with the Energy optimised path is 4.01 m starting from the starting point as 0 m high, compared to the straight path which is 5.63 m as shown in Fig. 6(6). This is over 28% height reduction in climbing and reaching the same point, during which unnecessary energy is used and which can be avoided.

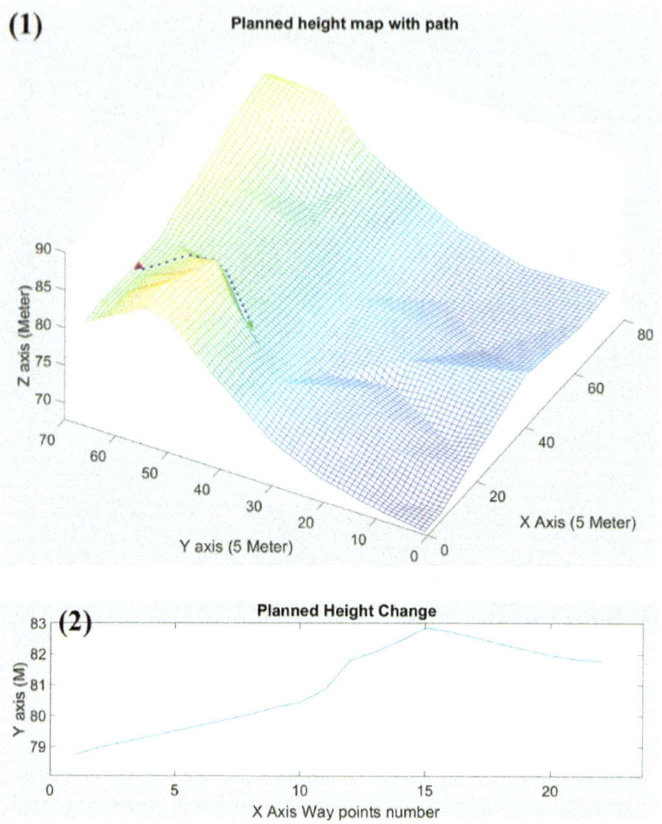

Fig. 9 Energy optimised path in 3D map (1) and the height change of the path (2)

Comparison of Simulation and Field Test Results

Comparing the straight path and the optimised path, the simulation shows the latter represents a saving of 957.5 J or 37.9% in total energy cost. This is achieved from a total energy cost saving of 1567.5 J for the optimised path in comparison with 2525 J for the straight path. The actual energy measured during the field tests differ as expected. The energy in the field test shows an energy consumption of 3640.5 J for the planed path and 4628.3 J for the straight path, resulting in a 21.3% of the energy saving for optimised path. This is slightly less than the 37.9% energy saving derived from the simulation run.

There are several reasons for the discrepancies between simulation and field test results and these are explained below. Firstly, the test field is far more complex in its terrain surface conditions and the presence of small agricultural obstacles. The field had stalks remaining after harvesting, and these are clearly seen in Fig. 7. This additional resistance in moving AgriRover forward was not considered in the simulation as this condition results in additional resistance as denser stalks increase the resistance, and hence requires more energy to drive the AgriRover. Furthermore, the farmland has variable rolling resistance due to the condition of the soil, high moisture content and low temperature. Some parts of the field are muddy and some parts had ice coverage, all conditions which increase the resistance and also introduce more randomness to the whole system.

Secondly, as the simulation of the system is based on a static rolling resistance factor, energy consumption will be different to the real world. It appears that the rolling resistance set in the simulation is lower overall in comparison to the real world, hence the overall energy consumption in simulation when compared to the field test is lower. Due to very high calculation resource requirement by implementing the ADAMS and MATLAB collaborative simulation, the randomly generated rolling resistant coefficient matrix was not implemented. The current simulation takes three days to complete, using a computer that is equipped with a 4 core 8 thread processor and 32 Gigabyte of RAM. By adding more complexity, the simulation has a higher rate of crashing which makes it hard to generate useful results in a reasonable time frame.

In addition, it was raining for 2 days before the test, so the farmland around those areas with lower elevation was muddy and some parts were frozen because of the low temperature during the test. All of these make the AgriRover experience a harder time in traversing the field due to the increased friction and slippage from the terrain.

Future EcoMechatronic Development

Smart farming is only at its infancy. This new way of farming can be potentially enabled by advancements in a range of technologies, including robotics, and satellite enabled and unmanned aerial vehicle-based remote sensing along with advances in the genomic studies of crops along with crops management. Many of these technologies are enabled and supported by a variety of electronic components devices which could also be found in systems such as robots, satellites and unmanned aerial vehicles. It is therefore argued that electronics system solutions or ever wider information scientific systems do provide promising enablers to support sustainable development and contribute significantly to an emerging EcoMechatronic discipline. Specifically, this chapter demonstrates a contribution to sustainable development from two perspectives as set out in the following sections.

AgriRover Enabled New Energy Efficient and Sustainable Approach to Food Production

The AgriRover is designed to be sustainable from three perspectives. Firstly, instead of using typically large and energy hungry farming equipment such as tractors, AgriRover is more efficient and has a significantly smaller footprint both in the field and in terms of energy consumption. Secondly, through energy optimization in path planning, a high proportion of energy can be saved as demonstrated in the chapter. Finally, the soil quality measurement capability embodied in the AgriRover enables timely precision measurement of the fertilisers required for the current growing season. In contrast to this real time and in situ soil fertility and quality measurement, traditional practises typically rely on out of date data for farmers to determine the amount of fertiliser to be applied in the field. This often results in over fertilisation and consequently has a negative impact on the environment and in long term on sustainable food production. The AgriRover work has therefore demonstrated a new way of smart farming.

An Energy Optimization Method in Path Planning for Multiple Robots

The path learning algorithm utilises an artificial potential field as the means of pathfinding, therefore it is possible to introduce an additional matrix field containing the real-time position of multiple Rovers. As evaluated before, the path planning algorithm that utilising the artificial potential field uses less computational power when compared to other path planning algorithms such as those based on machine learning and genetic algorithms, and therefore makes it possible to be implanted on the on-board computer of the Rover which then runs the path planning algorithm in real time.

Such architectures can be configured as a centralised or decentralised cluster depending on the application. For a centralised cluster architecture there will be a master and slaves structurer in which the master can be a ground-station or a particular Rover which all the other Rovers report to. A current position matrix can be generated and broadcasted to all Rovers. For a decentralised cluster architecture, each Rover will have their own position sent via a data link to neighbouring Rovers or Rovers within a specified radius which will then have the position data relative to themselves, which is enough for the task of collision avoidance.

Both of the architectures had their benefits and limitations. For the centralised cluster the planning algorithm will have a real-time position for each Rover and therefore is more controllable for planning a task, the limitation is that communications distance is limited and therefore the performance of the task solving capability of the cluster will degrade when the size of the cluster or the distance between rovers increases. For the decentralised cluster architecture this problem is avoided because each member of the cluster only needs to communicate with a certain radius, but the capability of centralised planning will be harder to achieve. To solve this problem the architecture of the Rovers can be operated in a hybrid mode where the control centre sends the tasks to the Rovers within the communication radius, then the task is then passed to all the Rovers via data-link, and the position of the Rovers returned. Although this will increase the load on each of the Rovers because of the additional data that needs to be repeated, and which may require a hardware upgrade, this is a solution for a collaborative multi rover architecture that can be investigated in the future.

Conclusions

The chapter has described a new approach to sustainable and energy efficient food production by deploying technologies such as AgriRover. This type of mechatronic solution brings along with it many advantages. It enables farmers to have available the most recent data when required and the example in the chapter shows that AgriRover is capable of providing autonomous navigation in a field in an energy efficient manner. The chapter also shows that the total energy costs of same planning task when using the energy optimised path planning algorithm with the AgriRover when running in an undulating terrain environment can save energy in comparison with a standard planning for straight-line path. The dynamic collaboratively simulation results illustrate a 37.9% total energy cost saving with 1567.5 J for the planned path in comparison to 2,525 J for the straight path. This is confirmed and verified through a field test, showing 3,640.5 J for the optimised and planned path against 4,628.3 J for the straight path, resulting 21.3% of the energy saving. The energy consumption reduction rate with unit distance travelled is 45.5%, which proves the energy optimised planned path has successfully reduced the energy consumption by avoiding the unnecessary terrain overcrossing. There is a difference of 16.6% in the energy consumption between the simulation results of 37.9% and field test results at 21.3%, this is explained by the variations of condition parameters used in the simulation.

A new energy optimised path planning method is proposed which is capable of generating a path for the AgriRover whilst simultaneously avoiding the problem of unnecessary terrain overcrossing when running in terrain rich environment. The chapter demonstrates that with an application of energy optimisation, AgriRover can achieve a 21.3% total energy saving. This can be applied to other mechatronic systems designed for agricultural applications and collectively they can result in a significant saving of energy.

By implementing electrically powered mobile platforms such as AgriRover instead of fossil fuel, this also represents a significant step towards an improvement for the environment. Currently, electrical energy is mostly generated using fossil fuels, and in fact this amounts to 53.8% of all electricity generated according to the International Energy Organisation in 2020. This trend will however change as increasingly the public and governments around the world realise the potentially irreversible damage that increasing temperature will do to the fragile earth environment. The Paris Agreement of the United Nations Framework Convention on Climate Change specifically sets target of maintaining a global temperature rise this century well below 2 °C above pre-industrial levels. The mechatronic solution presented in this chapter surely provides a feasible option for engineers, designers and policy-makers to pursue for a more sustainable and energy efficient food production. Further improvements can be made by deploying solutions such as the proposed path planning method and other mechatronic technologies and moving a big step forward towards more eco-friendly agriculture powered by advanced EcoMechatronics.

Acknowledgements The authors of the chapter would like to thank Innovate UK and UK Space Agency for providing funding to develop an AgriRover for trial under SmartFarm project. *The authors would also like to acknowledge the significant contribution towards AgriRover prototyping by colleagues from Robotics Group of RAL Space.* In addition, the authors would like to thank colleagues at the National Engineering Research Center for Information Technology in Agriculture (NERCITA), China for providing information of their farm for trial planning reported in this chapter.

References

1. World Population to Reach 9.9 Billion by 2050. News. Retrieved November 15, 2021, from https://sdg.iisd.org/news/world-population-to-reach-9-9-billion-by-2050/
2. Forecast of the global middle class population from 2015 to 2030, by region (in millions). Statista. Retrieved November 15, 2021, from www.statista.com/statistics/255591/forecast-on-the-worldwide-middle-class-population-by-region
3. Klarin T (2018) The concept of sustainable development: from its beginning to the contemporary issues. Zagreb Int Rev Econ Bus 21(1):67–94
4. Electricity Information: Overview. Retrieved November 15, 2021, from www.iea.org/reports/electricity-information-overview
5. Gerstlberger W, Præst Knudsen M, Stampe I (2014) Sustainable development strategies for product innovation and energy efficiency. Bus Strateg Environ 23(2):131–144
6. Hosseini SE, Butler B (2020) An overview of development and challenges in hydrogen powered vehicles. Int J Green Energy 17(1):13–37
7. Walker P, Mason R, Carrington D (2019) Theresa may commits to net zero UK carbon emissions by 2050. The Guardian 11(6):19. Retrieved November 14, 2021, from www.theguardian.com/environment/2019/jun/11/theresa-may-commits-to-net-zero-uk-carbon-emissions-by-2050
8. Gellings CW (2009) Energy efficiency in agricultural equipment. EOLSS Publishers, Efficient Use and Conservation of Energy, pp 106–122
9. Hunt ND, Liebman M, Thakrar SK, Hill JD (2020) Fossil energy use, climate change impacts, and air quality-related human health damages of conventional and diversified cropping systems in Iowa USA. Environ Scie Technol 54(18):11002–11014
10. Negrete JC (2017) Mechatronics and precision livestock farming in Mexican animal production. Anim Rev 4(1):1–7
11. Yan X-T, Bianco A, Niu C, Palazzetti R, Henry G, Li YH, Tubby W, Kisdi A, Irshad R, Sanders S, Scott R (2020) The AgriRover: a reinvented mechatronic platform from space robotics for precision farming. In Reinventing Mechatronics. Springer, pp 55–73

The Achievement of Sustainability in the Built Environment

Crina Oltean-Dumbrava and Margherita Finamore

Abstract The built environment presents particular challenges in relation to the achievement of sustainability, not least because of the number of building types and the uses to which they are put. Design issues thus include ensuring that buildings are energy efficient with effective ventilation in relation to use while enabling the maximum utilization of shared spaces and flexibility of use. In this context, many of the underlying technologies, as for instance those involved in energy management and use, are mechatronic in nature while developments such as Building Digital Twins are supporting real-time responses to changing conditions. The chapter examines the background to these issues and places them into the context of the developing approach to building construction, operation and management. In doing so, it draws on as an exemplar the award winning, including at COP26, design and development of the Antonio Brancati School in Pesaro in Italy to illustrate how innovative approaches to design and the use of technology have enabled the creation not only of a place of learning, but also a shared resource for the local community.

Introduction

As suggested by White [1], perhaps the most commonly referenced definition of sustainability is that proposed by the 1987 World Commission on Environment and Development chaired by Gro Harlem Bruntland, the then Prime Minister of Norway, as being: "... *development which meets the needs of the present without compromising the ability of future generations to meet their own needs*" [2]. A more socially oriented definition is that put forward in 1989 by Pearce, Makandia and Barbier [3] that "*Sustainable development involves devising a social and economic system, which ensures that these goals are sustained, i.e., that real incomes rise, that educational*

C. Oltean-Dumbrava (✉)
University of Bradford, Bradford, UK
e-mail: m.c.a.oltean-dumbrava@bradford.ac.uk

M. Finamore
Comune Pesaro Municipality, Italy
e-mail: m.finamore@comune.pesaro.pu.it

P. Hehenberger et al. (eds.), *EcoMechatronics*,
https://doi.org/10.1007/978-3-031-07555-1_12

standards increase, that the health of the nation improves, that the general quality of life is advanced." More recently, the United Nations has published a series of Sustainable Development Goals [4] of which Goal 11 has as its stated objective to "*make cities and human settlements inclusive, safe, resilient and sustainable*". This includes specific reference to aspects of service provision which may in turn be considered in relation to [5, 6]:

Infrastructure Services—Encompassing water and sanitation, waste collection and management, transport and energy.

Social Services—Encompassing education, health, housing and elderly and child care.

Quality of Life Services—Encompassing public safety, urban planning, culture and entertainment, sport and public spaces.

In the context of the built environment, the term *Sustainability Framework* has been defined as *"The structure used to select and organize criteria, indicators and benchmarks"*. Projects often have associated with them a number of sustainability frameworks for the assessment of environmental, economic and social factors. However, few address the technical elements separately.

Considering the built environment as a whole, issues associated with achieving sustainability include

- Building design for material recovery and reuse and associated construction methods and techniques.
- Development of a low carbon society.
- Effective utilization of shared spaces and recreational facilities.
- Energy systems and energy efficiency including local energy sources and smart grid infrastructures.
- Impact on climate and climate change.
- Management of change, including innovation.
- Management of the building stock.
- Materials recycling and waste reduction where the construction industry is one of the biggest sources of non-recyclable waste.
- Refurbishment and reuse of existing building stock as compared to demolition and rebuild.
- Resource management with an emphasis on the use of local resources.
- The design of new build to support easy change of function and use.
- The role of Information Technologies (IT) in supporting change of use and sustainability.

Many of the technologies associated with achieving an efficient and sustainable performance are associated with mechatronics systems and subsystems, for instance, in respect of environmental control and energy management. The EcoMechatronic concept, in which these systems are increasingly integrated with building operation and use, is therefore likely to play an increasingly significant role in building design and operation as part of a whole lifetime approach which includes recovery, recycling and reuse at the end of life.

Technologies

These include:

Construction technologies—The use of traditional technologies does not necessarily prevent the achievement of optimal results and design concepts increasingly aim to maximize the envelope efficiency and find natural solutions providing free gains. Thus, shape, position and orientation are chosen to best exploit solar radiation and optimize natural lighting and shade structured around the collection and interpretation of related climate change data.

Alternative and novel construction technologies such as the 3D printing of buildings [7] then afford opportunities to develop new strategies for the management of such structures over their lifetime. Similarly, by designing elements such as digital communications as part of the construction process, the ability to create smart buildings capable of reconfiguration is increased.

Smart building technologies—A Building Energy Management System (BEMS) is an integral part of a building wide system which introduces artificial intelligence into building operation and enables innovative modes of automation, management and the remote control of plant [8, 9]. The hardware elements of such plant are often essentially mechatronic in nature and use intelligent sensors to define system operation along with cloud-based software platforms to optimize, in real-time, behaviours structured around Internet of Things (IoT) solutions to increase comfort and reduce energy consumption while transforming each building into a smart building with zero emissions.

Systems are also monitored, controlled and regulated remotely to ensure maximum control of the internal conditions and at the same time define and implement energy strategy and manage operating costs in real time. BEMS thus allows the user not only to remotely control all managed systems, but also to enable

- Access to environmental and energy certifications (e.g. ISO50001 and ISO14001).
- Control of indoor and outdoor lighting as well as security and fire alarms.
- Effective monitoring of energy consumption.
- Effective utilization of solar radiation, including the operation of shading systems.
- Greater reliability and durability of plant.
- Maintenance scheduling and support.
- Optimal values of comfort and the best quality of natural and artificial light.
- Real-time monitoring of electricity consumption from the grid and of energy production by any photovoltaic system.
- Real-time monitoring of water consumption.
- Reduction of maintenance costs and greater effectiveness in the planning of ordinary and extraordinary maintenance.

Building Information Modelling (BIM)—The ISO standard 19,650:2019 defines BIM as "*The use of a shared digital representation of a built asset to facilitate design, construction and operation processes to form a reliable basis for decisions.*" In contrast with conventional building design practices, BIM introduces a range of additional information such as time, costs and asset management as well as sustainability, in order to enable the evaluation of the building from planning through to construction, its operational life and ultimately disposal. The information model created in BIM integrates individual stakeholder models to support coordination and development along with decision-making [10].

User interactions—The ways in which users interact with buildings and the ways in which they utilize the space is a major design factor which impacts sustainability. This requires that the design team takes into all stakeholder views and integrates these with the building function, including external issues such as landscaping.

Environment—The key issue here is generally user comfort which involves factors such as

- Customized temperature control throughout the occupied space.
- Integrated control of both natural and artificial light.
- High acoustic performance obtained to create the best environment for the intended use.
- Monitoring and control of indoor air quality with an air exchange cycle of 5 times per hour.
- Use of natural ventilation.
- Parking for bicycles, electric charging stations for cars.

Communication and Networking—The ability to create both internal and external networks in support of Internet of Things configurations as part of a smart building infrastructure is of increasing importance. In an EcoMechatronics context, this enables the individual systems to interact with each other, and hence respond in real time to changes in weather conditions and occupancy. In the wider context, the extension of the network from the building to the wider environment then supports the interaction both with users and with external systems such as transport.

Digital Twins—Digital Twins are dynamic digital models which use artificial intelligence, machine learning and data analytics to monitor and update in real time the status of their physical counterpart. In operation, the Digital Twin uses both real time and historic data from the physical system to learn about the behaviour of that system and hence to maintain itself as close to the real systems working condition or position as possible. The results can then be integrated to both update and improve the generic model and to identify areas and issues of potential concern. Digital Twins are therefore increasingly finding application in support of the optimization of the operation of a range of physical assets, including sustainable smart buildings. Here, the Digital Twin would be developed in parallel with the building, including the design and construction phases, to the point where it provides real-time support for the operation and maintenance of the building, for instance in respect of energy management and efficiency. Twins at the level of individual buildings can then be integrated as part of a network of twins within the wider context of a Smart City environment [11–13].

Antonio Brancati School

The Antonio Brancati School (Fig. 1) is a new school built in an abandoned urban area of Pesaro in Italy (Fig. 2) and has a very important role in terms of social sustainability for the area, as well as providing an example of green construction. The decision was made by the municipality to reuse a site with zero land consumption for the new middle school by redeveloping an abandoned urban area and replacing three army barracks built in the 1950s. The excellence of the resulting school building is the result of the ambitious commitment of Pesaro municipality to preserve the environment and tackle climate change through a refocusing of the mandate of local construction on efficiency, sustainability and responsibility. Then in 2017 it was decided to utilize the school design and construction as pilot project with the aim of encouraging a holistic approach within an ambition to accelerate the transition towards a circular economy [14]. This would enable construction firms to move beyond the current, and outdated, '*take-make-dispose*' model and contribute to a mainstreaming of the circular economy [15] concept within and outside the city.

In the case of the Antonio Brancati school, the approach to public procurement [16] has changed significantly. Instead of taking a traditional approach focusing on the construction of the building itself on the basis of the lowest tender price, the focus shifted towards achieving a high energy environmental performance [17] building. The building itself also has to take into consideration issues of environmental protection, health and comfort, well-being and social requirements based on public and user needs. The school will accommodate some 375 students in 15 classrooms along with

Fig. 1 The Antonio Brancati school

Fig. 2 Pesaro

6 classrooms for special activities and one for music, a teacher meeting room, a canteen, multipurpose spaces and a library also open to the local residents. Though it is strictly speaking a school building, it is designed to act as a catalyst for the whole neighbourhood and its residents.

An analysis of not only educational needs but also of public expectation was carried out to make the building more circular in terms of its functionality and make sure that it can be used by everyone all day, every day as a user-friendly community space. The aim was to transfer all the requirements into relevant technical elements to reduce the environmental impact of the building from the construction phase to its use. Thus, a great deal of significance is given to thermo-hygrometric comfort, acoustic comfort, correct use of natural and artificial light, optimization of the building envelope to maximize solar effects, appropriate use of water, choice of materials and the use of appropriate technologies to maximize the energy performance including adjustable solar shading and green roof designed not only to insulate but to mitigate water use and the heat island effect [18]. It is important not only to achieve a reduced energy consumption, [19] but also to enhance the indoor environment for students and staff through advanced acoustics, improved air quality and high levels of thermal comfort [20] to create a good working environment. All these considerations were part of the tender requirement, so construction firms had to respond to these as part of the bidding process.

This analysis of not only educational needs but also of citizens expectations established the central theme, how to make the building more flexible and adaptable in terms of its functionality and ensure that it can be used by everyone all day, every day. To respond to this need, there is a large entrance hall to encourage circulation and interaction with others. The hall has been planned as a mixed-use space and the design of the circulation pattern encourages people to sit down and use the space.

This is the first Italian school project to use a Green Public Procurement approach with such clear and ambitious conditions. Above all two criteria were selected as mandatory in respect of achieving the desired outcomes:

- High energy efficiency goal by obtaining a nearly zero energy building (nZEB) design for the school [21]
- Improved environmental quality by providing environmental energy certification using the LEED certification scheme with a minimum LEED Gold certification.

To ensure that companies could prepare appropriately, the tender documents [22] included technical specifications and provided measurables against which the tenders would be evaluated. Performance-based criteria describe the expected results and mandatory outputs were defined. Eleven core criteria were chosen from a range of 23 and grouped into three macro-areas: improvement to building envelope performance, improvement to energy performance and improvement to environmental performance as set out in Table 1. Each element was described in detail in the tender documents and linked to a reward system to lead companies towards a quality offer rather than minimum cost.

Referring to Table 1, the role of information, and hence of an EcoMechatronics approach, within the context of a smart building [23] in supporting both the design and operation of such buildings become clear, particularly in relation to the monitoring and control functions associated with achieving the desired energy performance, but

Table 1 Tender technical specifications

Building envelope	Thermo-hygrometric efficiency of transparent elements
	Thermo-hygrometric efficiency of the roof
	Thermo-hygrometric efficiency of outer walls
	Ventilated facade
	Solar screen
Energy performance	Automated lighting system
	Use of controlled mechanical ventilation and heat recovery systems
	Heat pumps
	Energy consumption monitoring system
Environmental performance	Waste materials management
	Environmental energy efficiency certification

also in relation to features such as ventilation and solar screening. Further, information gained on the operation of the building in areas such as the management of waste materials both during construction and in operation will support long-term decision-making, perhaps supported in future designs by reference to concepts such as Digital Twins. Such information flows also have the potential to enable the building to learn about its profile of use, and hence to autonomously adapt and adjust over time to better match users' requirements, for instance, in relation to lighting and comfort levels.

In this context, as bidders were asked to offer a radical improvement of the energy and environmental performances of the building, the use of the Building Information Modelling (BIM) [24] was not required in the tender. Nevertheless, the Architectural, Engineering and Construction firms (AEC) involved decided voluntarily to use BIM from the planning stage in order to provide a more effective construction cost estimation with a focus on higher value factors such as identifying construction activities and risks.

The experiment was successful as it achieved the goal of a building that performs exceptionally in every sustainable LEED evaluation area with a score of 88/110 points as confirmed by the assessment records. It is the first school building in Europe to achieve this high score. The tendering process can be replicated by other municipalities to obtain not only energy-efficient buildings, but above all to reduce the buildings' carbon footprint.

This project thus provides the first concrete example of mainstreaming circular practices in public procurement. The innovation of this project is in putting into practice circular principles within public procurement and ensuring that the procedure runs efficiently and achieves the expected results based on environmental and social requirements. The project also demonstrates that the Green Public Procurement process in which environmental requirements are made mandatory within the tendering process by using predetermined award criteria is essential for delivering environmental social value, influencing the market towards sustainable materials and products and improving new skills on green jobs.

The environmental impacts were improved by the tendering process to build a low energy concept school using good design to support not only passive heating and cooling technologies, but relevant active mechanisms as needed. Moreover, considering how green requirements will affect the procurement process for works and implementing them in line with the legal obligation, materials and methods of production and provision are considered when defining appropriate selection criteria.

Concerning the tendering process and the requirements asked of the bidders, a thorough analysis of the costs and of the possibility of finding materials and building components in the market was carried out to check the sustainability of the '*Business Plan*' and its consequent operational effectiveness. This analysis was carried out to make effective the requirements in the call for tender and to enable the bidder to be able to put forward their best proposal.

As a result, the tender includes technical specification on energy efficiency, product lifecycle, waste management, water efficiency with attention paid to internal comfort to create the best indoor environmental conditions for the students and teachers. It is relevant not only to the energy consumption reduction but also to the indoor quality for the students and staff through acoustics, air quality and thermal comfort requirements. These results are guaranteed by the environmental energy certification made mandatory in the tender notice and measured in accordance with the Minimum Environmental Criteria (MEC).

As a consequence of this approach, the Antonio Brancati school has received:

- The highest score in the school's energy environmental certification in Europe [25, 26].
- The third highest score in the school's energy environmental certification in the world out of 1,733 such certifications using the LEED system [27–29].
- The highest score in v4 school's certification in the world [30].

Construction and Environmental Issues

When planning the building, emphasis is given to thermo-hygrometric comfort, acoustic comfort, correct use of natural and artificial light, optimisation of the building envelope to maximize free benefits of the sun, appropriate use of water, choice of materials, use of appropriate technologies to maximize the energy performance with adjustable solar shading and green roof designed not only to insulate but also to mitigate water use and the heat island effect. All these features were part of the requirements for tender, so companies had to respond to these criteria as part of the bidding process. The coloured façade of the building is composed of a modular system of porcelain stoneware and finished with earthy colours introduce a smoothly playful dash of colour in the streetscape. The characteristic colour provides a strong and recognizable identity and establishes the school as a playful, eye-catching beacon for the vibrant neighbourhood.

A study of heat gains and losses was carried out to optimize the thermal capacity of the building envelope, which has been achieved by using a green roof, automated solar shadings, and the ventilated wall. All the elements which make up the envelope contribute to its efficiency:

- Integrated and motorized shutters with sensors for natural light and overheating control.
- Insulation is rock wool with 120/150 kg/m^3, free of petroleum derivatives and recyclable.
- A green roof made of Mediterranean plants with low water need to moderate the impact of high temperatures, capture stormwater, abate pollution and act as carbon sinks, while enhancing biodiversity.
- A ventilated wall made of modular porcelain stoneware avoids overheating, minimizes losses and reduces also the maintenance costs of the façade.
- High-performance wooden windows with $U = 0.74$ also certified Forest Stewardship Council-Chain of Custody (FSC-CoC)

The ventilated wall, made of porcelain stoneware, reduces the maintenance costs of the facade and with the green roof and solar shading supports minimization of energy demand. As a result, the mechatronic elements such as the PV system are simpler and less expensive in terms of initial costs and maintenance.

The school is equipped with a mechanical filtered ventilation and climate control system to guarantee higher levels of indoor air quality with a rate of exchange equivalent to 5 times the enclosed volume per hour. To obtain the energy and environmental certification, it was mandatory to take into account the following benchmarks:

- Global warming potential (greenhouse gases).
- Depletion of the stratospheric ozone.
- Acidification of land and water sources.
- Eutrophication.
- Formation of tropospheric ozone (photochemical oxidant formation).
- Depletion of non-renewable energy resources.

In relation to the four main phases of the building life, namely,

- Construction materials.
- Transportation to site.
- Maintenance and material replacement.
- End of life.

Acoustic quality was another crucial goal requiring compliance Italian standards UNI 11,367: 2010, UNI 11,534–1: 2018 and UNI 11,534–2: 2020. Usually, a report is provided with the evaluation of external noise soundproofing by the material data sheets and rarely a campaign of in situ measurements is carried out at the end of works to verify the real quality of acoustic performance.

As it is a school building, it was crucial to verify the acoustic standard to achieve the best quality of sound transmission so that teachers are not forced to raise their voices to be heard and make equivalent all student locations inside the classroom.

To ensure the best level of acoustic comfort two sets of in situ measurements were carried out. Based on the first set of results it was necessary to replace a small area of the suspended ceilings in the ground floor classrooms and add some elements to the walls on the first floor to correct the reverberation levels. This demonstrated the importance of carrying out an in situ verification.

The second series of tests was carried out after these remedial actions to verify the acoustic standard achieved and it gave the following results:

- Speech transmission index in classroom is ≥ 0.55.
- The clarity C50 (500–1000-2000 Hz) is ≥ 2 dB.
- Reverberation time is compliant with formulas and graphs from DIN 18,041, the optimal reverberation time (depending on volume) is referred to the furnished and occupied state at 80% of capacity.

Table 2 Antonio Brancati school energy system

Heating system	Condensing gas boiler
	Heat pump
	Low-temperature floor heating
	Variable Air Volume system (VAV) System
Hot water system	Heat pump
Cooling system	Reversible heat pump
	VAV System
	Floor cooling
Ventilation system	Double flow heat exchanger
Renewable systems	Solar photovoltaic
	Heat pump

Smart Building Technologies

A BEMS system was installed in the building for the control, operation and regulation of thermomechanical systems according to need, including:

- Continuous monitoring of weather conditions.
- Detection and regulation of thermo-hygrometric comfort parameters.
- Hygienic replacement of air according to indoor CO_2 level.
- Operation of thermo-refrigerators, UTA, networks and terminals.

In particular, the motorized actuators of the installed valves allow for a finer regulation of the operation of the radiant floor circuits (0 to 100% in continuous mode), depending on the temperature of the individual room, with the possibility of minimum attenuation (2–3 °C), in case of short periods of occupant absence, and '*energy saving*' (4–5 °C), in case of closure of the building.

During the construction phase, the maintenance service was involved in meeting all building and school requirements with a particular attention to the BEMS features.[1]

Heating, Lighting and Ventilation

The energy system provides 57% of renewable energy production and is made up of the components of Table 2.

The renewable energy system as installed has a significant impact on the performance and overall quality of the services provided. It features a high-efficiency

[1] It should be noted that this community engagement process is not usual in Italy or in this municipality, so a great effort was made to involve people who are not usually involved at any level.

reversible air/water heat pump, rated at 135 kwt[2] and 113.5 kWf.[3] The heat pump can produce hot water at up to 55 °C with outdoor air temperatures of −6 °C, and up to 40 °C with an outdoor temperature of −15 °C,[4] so it can also work for pre-heating of the ACS upstream of the relevant boiler. In summer, water can be produced at 7 °C with outside temperature up to 46 °C. This system leads to a significant increase in the energy efficiency of the generator even when operating at partial loads. Despite the extension to the entire building of the mechanical ventilation system, it also presents a significant step forward in relation to primary energy systems.

In addition to the air used for heat exchange, the electrical input necessary for operation comes from green energy, including the deployment of photovoltaic panels on the roof of the school and grid electricity from certified renewable sources.

From an environmental point of view, the main advantage is in the significant reduction of the consumption of fossil energy sources and hence CO_2 emissions. In detail:

- 0 ppm NOx (−100%);–0 ppm CO_2 (−100%);
- Expected annual emissions of 2,420 kgco2 (−86%);
- Expected annual economic savings of EUR 1,052 (−84%).

From the economic point of view, it also results in a significant reduction in operating costs, further increased by the installation of an additional heat pump of the same type and size, to ensure continuity of service and increase the useful life of system components.

With particular regard to renewable energy production, the building is equipped with a photovoltaic system of 96 monocrystalline silicon panels each with a rating of 300 Wp located on the roof for a specific power of 28.8 kwp. The PV system is installed on the central part of the roof of the school building. Performance details are:

- Photovoltaic installation of 158.40 m^2.
- 96 photovoltaic modules of 300 W and 1.96 m^2 each.
- Photovoltaic power ratings 28.8 kW continuous.
- SolarEdge inverter 27.6 kW alternating current.
- 113.5 kW heat pump.
- Heating system–Floor heating and VAV system (variable air volume)
- Cooling system: Floor cooling and VAV system (variable air volume)

Lighting

The school is equipped with high luminous efficiency LED lamps that will allow optimization of the energy consumption of the building without limiting the visual

[2] Water to the condenser at 30–35 °C and external air at 7 °C b.s., or 6 °C b.u.

[3] Water to the evaporator at 12–7 °C and external air at 35 °C.

[4] The winter design temperature for Pesaro is −2 °C.

comfort produced. It was also decided to equip the lighting systems for the classrooms with a system for the automatic regulation of the luminous flux according to the levels of daylight. This results in an important optimization of the energy consumption of the building. The is managed by the Digital Addressable Lighting Interface DALI protocol controlled by a supervision system in KNX standard (EN 50,090–EN 13,321–1 and ISO/IEC 14,543).[5] The system also controls the automated shutters to optimize the natural light and integrates via a Bacnet interface with the mechanical plant management system.

The call for tender highlighted that the lighting system must meet the minimum requirements imposed by the standards UNI EN 12,464–1 '*Lighting of indoor workplaces*' and the D.M. of 11 January 2017, in point 2.4.2.11 '*Lighting systems for indoor and outdoor*' and emergency lighting standards. Lighting analysis of all rooms was required, including both Isolux curves and the average illumination value within each room. This must meet the minimum values set by UNI EN 12,464–1 according to the intended use of each environment. Only offers that took into account the use of LED technology were evaluated. According to Italian standard UNI 10,840, the daylight factor is $> = 3\%$ in each classroom. Furthermore, according to the EN 12,464 standard, an illuminance value of not less than 500 lx is guaranteed in each classroom through a sensor that integrates daylight with LED light.

In the call for tender bidders were asked to provide a specific lighting simulation and light efficiency measurements and the lighting analysis of the entire building developed by the specific reports referring to each room. Each report had to be accompanied by a plan of the classrooms and multifunctionality spaces indicating the Isolux curves, the average, maximum and minimum illumination values on worktables as required by the regulations, number and type of equipment provided,[6] their arrangement and their installation height. The reports had to demonstrate that the minimum values imposed by the regulations based on the intended use of the premises are respected (average illumination, UGR glare index, colour rendering Ra, etc.).

Ventilation

The air conditioning system consists of

- A heat pump generator, installed on the roof, with an additional methane boiler located in the technology plant room.
- Thermostatic storage upstream of the fluid distribution networks, which provides the radiant heating system and the primary air plant with heat recovery.

[5] KNX comes from Konnex, an open standard (EN 50,090, ISO/IEC 14,543) for commercial and building automation.

[6] Defined by brand and type.

The school is equipped with a mechanically filtered ventilation climate system HVAC with a CO_2 monitor able to guarantee higher levels of indoor air quality with the air exchange of 5 volumes per hour. The individual air networks branch off to all classrooms and spaces to ensure constantly renewed air of high quality without any air exchange among the classrooms themselves. The energy supply for the HVAC system is provided by photovoltaic (PV) panels installed on the roof in compliance with the nZEB standard.

Instead of using galvanized sheet metal for the distribution channels as usual, to ensure the best health conditions and the minimum maintenance costs those installed are made with sandwich panels composed of

- Rigid polyurethane foam as a structural element.
- Coating on with zeolite-treated aluminium foil to be antimicrobial. The nanostructured coating retains a surface in contact with the extremely smooth airflow that, also thanks to the absence of sharp edges, gives the channels a self-cleaning property, reducing the need for manual cleaning of the channels.

Also, to the benefit of air quality, the surfaces of flexible ducts are treated with the technology effective against 600 strains of micro-organisms and certified for the reduction of the proliferation of pathogens. The plenums are equipped with linear vents installed on the wall or in special lowered ceilings inside some rooms. The exhaust air extraction system from the toilets has also been extended to the service rooms.

The mechanical ventilation and heat recovery system optimize energy performance, with an improvement of 42% of the performance of the building compared to the reference project. The design of the ventilation system has been carried out with a BIM system, in order to solve in advance any interference. It is installed a thermal generator with a backup function of the heat pump with a 150 kwt boiler (T sent 40 °C) with modulating burner, that provides complete backup for the winter period, allowing not only to support the operation of the heat pump in periods of low outside temperatures, but also to face the entire thermal demand of the building. The increase in power is also accompanied by an increase in efficiency, from 104.6 (50–30 °C) to 109.4 (40–30 °C) and numerous other advantages, including

- Low start-up emissions due to modulating burner and high-water content of circuits.
- Minimum climate-altering and polluting emissions from the surface burner.
- Increase in thermal recovery from condensation (up to 6%).
- Rapid adjustment of operating times.
- Simplification of the methane distribution system, with a consequent reduction in safety risks.
- High thermal comfort thanks to the detection of external temperature and solar radiation (by weather forecast).
- Control of management costs.
- Low maintenance costs due to low emission combustion and self-cleaning Al/Fe heat exchanger.

The installation of the most powerful boiler did not involve changes to the layout of the technology plant, nor to the layout of the fluid distribution networks that, before reaching the terminals in the environment, are stored in an inertial accumulation (1,000 l). This has also reduced the consumption of electricity by the circulation groups, due to the high-water content of the circuits.

Green Solution Awards

The Green Solutions Awards are an international competition run by the Construction21 [31] network and supported by the UN environment programme, Global Alliance for Buildings and Construction [32] in partnership among others with the World Green Building Council, BUILD UP, the European portal for energy efficiency in building, FEDARENE, ICLEI Local Governments for Sustainability.

The competition includes a national phase, where candidates first compete within their own country, and an international phase which brings together the national winners. Originally a competition dedicated to green buildings, the Green Solutions Awards have in recent years expanded their scope to districts and infrastructures in order to take into account a wider variety of sustainable city solutions and actors. The competition is the concrete embodiment of all statements on ecological transition; it aims to promote already existing solutions, integrated in buildings, districts and infrastructures.

The purpose of the competition is, as Christian Brodhag the President of Construction21 [33], has stated, to showcase innovations in order to turn them into standard practices on a global scale. The competition is open to all professionals to demonstrate their thinking and technology and aims to contribute to the prevention of climate change by becoming a source of inspiration through the worldwide communication plan provided by the Construction21 network.

In the 2020–2021 competition, 192 candidates participated in the contest at the global level, 148 of which were buildings from 15 countries including Australia, Austria, Benin, Canada, India, Italy, Moldavia, the Netherlands, Slovenia, Sri Lanka, Switzerland, Turkey, the United Arab Emirates, the United Kingdom and the USA. From these, 52 finalists were selected including 39 buildings, mostly in the Low Carbon and Energy & Temperate Climates categories.

Out of 192 buildings, districts and infrastructures, only 24 went forward to the international phase of the competition and the Brancati school project, after winning from among 9 finalists at the national level, was among the 8 overall international winners [34], receiving the 2020–21 Energy and Temperate Climates award.

The merit of the project was to establish not only a friendly environment favourable to teaching and learning in a high-energy-efficient building but also to develop a replicable model on a large scale to make it possible to reduce carbon emissions. The model is based on the dialogue between project holders and professionals seeking to adopt innovative solutions, and thanks to this strong collaboration among all the people involved it was possible to achieve valuable results.

Issues and Concerns

Future Trajectories

The built environment comprises residential and other buildings and related infrastructure elements such as roads. In terms of waste generation, construction and demolition is by far the greatest contributor [19], implying that there needs to be a radical shift in the way in which these industries operate, both to reduce waste and to increase the proportion of recovered and recyclable materials. The World Economic Forum in its report on the Infrastructure and Urban Development (IU) industry considered three primary scenarios as suggested in Table 3 [35]

A key component of each of these scenarios is that they rely on the close integration of the physical elements with an information infrastructure to provide the connectivity. This same information infrastructure is then likely to form the basis for operational issues such as might be associated with smart city concepts as well as providing the information link with individuals within the physical and information environments.

Table 3 Possible future scenarios

Scenario		
Building a virtual world	Factories run the world	A green reboot
• Requires increased residential, recreational and passenger mobility assets • Interconnected intelligent systems and robots run the industry • Software organizations gain power • New businesses emerge around data and services	• Prefabrication and modularization create cost-efficient structures • Emphasis on industrial and commercial buildings and infrastructure • Value chain emphasizes prefabrication, lean processes and mass customization • Suppliers benefit • Businesses focus on integrated systems and logistics	• Emphasis on environmental-friendly construction methods and sustainable materials • Increased conflict over scarce resources and climate change • Sustainability is the main decision criterion • Innovative technologies, new materials and sensor-based surveillance to ensure low environmental impact • Participants with a deep knowledge of materials and local brownfield portfolios thrive • Opportunities centre around environmental-focused services and materials recycling

Smart Cities and Big Data

The concept of a smart city is generally structured around the deployment and use of information technologies and the Internet of Things in order to:

1. Enable, enhance, the effective use of the physical infrastructure in support of economic, social and cultural development.
2. Enhance the involvement and engagement of people in local governance through strategies of e-participation and e-governance.
3. Support adaption and innovation in respond efficiently, effectively and promptly to changing circumstances.

This implies empowering the population through access to open platforms and the introduction of city-wide information networks to collect, gather and transmit data in real time. To be effective, such data must have both value and context at the level of the individual, and be accessible as required [36–38].

The use of Big Data analysis raises all sorts of issues, but also links into the self-management of local communities empowered by mobile technology to bypass otherwise restrictive practices, for instance, in the formation of farm co-operatives to sell produce directly rather than through middlemen. There are also issues regarding healthcare and health management [39, 40].

In a Smart City context, the nature of the data and its context is of primary importance in establishing viability with users. What works in one urban community may not have value in another. For instance, information on taxi availability as used in Singapore [41] may have limited value in the city of Jos in central Nigeria where the emphasis might be on (say) rubbish collection. With a limited resource budget, the impact of the unit spent needs to be maximized and this implies that

1. There is an understanding of the impact of decisions made in relation to the short, medium and long terms.
2. That the data gathering and analysis processes are neutral.

The Internet of Things has a significant potential for impacting on quality of life, but is structured around an ability to access and utilize robust data and information in a timely and appropriate manner, with in some cases such data and information becoming a commodity to be traded. Additionally, there needs to be in place mechanisms to properly understand the nature of the technology being employed in order to avoid issues of technological lock-in and the '*dumping*' of a technology (from tractors to televisions) on a society to essentially force them to adopt, and become locked-in to, a specific technology.

People

It acknowledged that "*we know very little about what interventions are most effective in reducing vulnerability, saving lives and rebuilding livelihoods in urban contexts*"

[42]. Thus, there is an overwhelming need to target interventions to maximize the security, prosperity and well-being of urban populations while recognizing that these are not discrete challenges but are interconnected with each other and with a range of other factors. This is indicative of a need to train and educate the local population to operate and maintain the resulting systems and technologies, and education thus needs to be incorporated into the project evaluation and assessment process. Technical considerations then include:

- The ways in which infrastructure projects are linked to sustainability, resilience and climate change.
- The identification of their impact on sustainability and related factors.
- The ways in which data is gathered and integrated in relation to issues such as Smart Cities.

Conclusions

To comply with the concepts of sustainability in the built environment while applying circular economy principles the main lessons learnt from this project are as set out below.

Firstly, the EU Sustainable Product Policy sets out that "*EcoDesign is a core tool in the matrix of approaches that enables the Circular Economy*" and the Green Public Procurement is then an instrument with a key role to play in the EU's efforts to become a more resource-efficient economy. This is then a key driver for decision-making. Thus when considering the building as a product along with its complexity, it is necessary to use defined sustainable design concepts in respect of the creation of products and services that have properly and fully considered the environmental, social and economic impacts from the initial phase through to the end of life.

As a consequence, these elements and issues are a crucial aspect of the thinking about a multiuse building not only for educational purposes but for whole community use. This includes the incorporation of technical features which will allow for its use all year and all day, despite the fact that in Italy schools are currently closed from the beginning of June till the middle of September and indeed are usually used only in the mornings.

By making informed decisions conforming with citizen's expectations and needs and integrating these with climate data collection, designers can reduce environmental impact, resulting in less greenhouse gas emissions and a more efficient use of natural resources. Climate data collection is thus key to designing an efficient operating envelope to minimize energy demand and make the plant less expensive as well as integrating a cooling functionality which is equally crucial with respect to energy saving. Thanks to the operating envelope efficiency, the PV panels provide all the energy needed for the school at a cost of only 36.000€.

Another crucial point is the IT-based strategy designed to reduce operational and maintenance costs and lead to higher profitability as the monitoring process during the construction phase results in the saving of both time and materials.

Finally, to build trust in the project, an integrated and multidisciplinary approach as used to share the vision with the community to obtain a sustainable building is extremely important element in achieving the task in a proper and timely manner.

References

1. White MA (2013) Sustainability: i know it when i see it. Ecol Econ 86:213–217
2. Keeble R (1988) The Brundtland report: 'Our common future'. Med War 4(1):17–25. https://doi.org/10.1080/07488008808408783
3. Pearce D, Makandia A, Barbier E (1989) Blueprint for a green economy. Earthscan Publications
4. Retrieved July 14, 2021, from sdgs.un.org/goals
5. Perkins R (2013) Sustainable development and the making and unmaking of a developing world. Eviron Plann C Gov Policy 31(6):1003–1022
6. Collier MJ et al (2013) Transitioning to resilience and sustainability in urban communities. Cities 32:S21–S28
7. Shahrubudin N, Lee TC, Ramlan R (2019) An overview on 3D printing technology: technological, materials, and applications. Proc Manufact 35:1286–1296
8. Manic M, Wijayasekara D, Amarasinghe K, Rodriguez-Andina JJ (2016) Building energy management systems: the age of intelligent and adaptive buildings. IEEE Ind Electron Mag 10(1):25–39
9. Froufe MM, Chinelli CK, Guedes ALA, Haddad AN, Hammad AW, Soare CAP (2020) Smart buildings: systems and drivers. Buildings 10(9):153–161
10. Gao H, Koch C, Wu Y (2019) Building information modelling based building energy modelling: a review. Appl Energy 238:320–343
11. Jones D, Snider C, Nassehi A, Yon J, Hicks B (2020) Characterising the digital twin: a systematic literature review. CIRP J Manufact Sci and Technol 29:36–52
12. Farsi M, Daneshkhah MA, Hosseinian-Far A, Jahankhani H (eds) (2020) Digital twin technologies and smart cities. Springer
13. Ketzler B, Naserentin V, Latino F, Zangelidis FC, Thuvander L, Logg A (2020) Digital twins for cities: a state of the art review. Built Environ 46(4):547–573
14. Merli R, Preziosi M, Acampor A (2018) How do scholars approach the circular economy? A systematic literature review. J Clean Prod 178:703–722
15. European Commission (2014) Communication from the commission to the European parliament, the council, the european economic and social committee and the committee of the regions towards a circular economy: a zero waste programme for Europe. Retrieved March 4, 2021, from ec.europa.eu/environment/circular-economy/pdf/circular-economy-communication.pdf
16. European Commission (2008) Communication from the Commission to the European Parliament, the Council, the European Economic and social committee and the committee of the regions: public procurement for a better environment. Retrieved March 4, 2021, from ec.europa.eu/transparency/regdoc/rep/1/2008/EN/1-2008-400-EN-F2-1.pdf. EUR-Lex - 52008DC0400 - EN - EUR-Lex (europa.eu)
17. European Commission (2016) EU GPP criteria for office building design, construction and management. Retrieved March 4, 2021, from ec.europa.eu/environment/gpp/pdf/swd_2016_180.pdf
18. Li X, Zhou Y, Yu S, Jia G, Li H, Li W (2019) Urban heat island impacts on building energy consumption: a review of approaches and findings. Energy 174:407–419
19. Italian Ministry of Economic Development (2015) Application of methodologies for calculating energy performance and definition of minimum requirements for buildings. Retrieved March 4, 2021, from www.gazzettaufficiale.it/eli/id/2015/07/15/15A05198/sg

20. Italian Ministry of the Environment (2017) Minimum environmental criteria for the award of design services and works for the new construction, renovation and maintenance of public buildings. Retrieved March 5, 2021, from www.gazzettaufficiale.it/eli/id/2017/11/06/17A07439/sg
21. Attia S, Eleftheriou P, Xeni F, Morlot R, Ménézo C, Kostopoulos V, Betsi M, Kalaitzoglou I, Pagliano L, Cellura M, Almeida M (2017) Overview and future challenges of nearly zero energy buildings (nZEB) design in Southern Europe. Energy Build 155:439–458
22. Italian Government (2016) Implementation of Directives 2014/23/EU, 2014/24/EU and 2014/25/EU on the award of concession contracts, public contracts and procurement procedures of entities operating in the water sector, energy, transport and postal services, as well as for the reorganization of the current regulations on public contracts relating to works, services and supplies. Legislative Decree 2016 18 April 2016, n. 50 Italy. Retrieved March 5, 2021, from www.gazzettaufficiale.it/sommario/codici/contrattiPubblici
23. Sinopoli JM (2009) Smart buildings systems for architects, owners and builders. Butterworth-Heinemann
24. Shirowzhan S, Sepasgozar SM, Edwards DJ, Li H, Wang C (2020) BIM compatibility and its differentiation with interoperability challenges as an innovation factor. Autom Constr 112:103086
25. Goldman D, Ayalon O, Baum D, Weis B (2018) Influence of 'green school certification 'on students' environmental literacy and adoption of sustainable practice by schools. J Clean Prod 183:1300–1313
26. Pereira LD, Raimondo D, Corgnati SP, Da Silva MG (2014) Energy consumption in schools–a review paper. Renew Sustain Energy Rev 40:911–922
27. Reeder L (2010) Guide to green building rating systems: understanding LEED, green globes, energy star, the national green building standard, and more. John Wiley & Sons
28. Uğur LO, Leblebici N (2018) An examination of the LEED green building certification system in terms of construction costs. Renew Sustain Energy Rev 81:1476–1483
29. Retrieved March 19, 2021, from www.usgbc.org/articles/explore-30-leedcertified-schools-2019
30. Retrieved March 19, 2021, from www.qualhub.co.uk/qualification-search/qualification-detail/level-4-certificate-for-the-advanced-practitioner-in-schools-and-colleges-552
31. Retrieved November 29, 2021, from www.construction21.org/static/Green-Solutions-Awards.html
32. Retrieved November 29, 2021, from globalabc.org
33. Retrieved November 29, 2021, from www.construction21.org/member/44/christian-brodhag.html
34. Retrieved November 29, 2021, from www.construction21.org/contest/h/green-solutions-awards-2020-2021.html
35. World Economic Forum (2018) Shaping the future of construction–future scenarios and implications for the industry. Retrieved March 1, 2021, from www.weforum.org/docs/Future_Scenarios_Implications_Industry_report_2018.pdf
36. Perera A, Zaslavsky A, Christen P, Georgakopoulos D (2014) Sensing as a service model for smart cities supported by internet of things. Trans Emerg Telecommun Technol 25(1):81–93
37. Rathore MM, Ahmad A, Paul A, Rho S (2016) Urban planning and building smart cities based on the internet of things using big data analytics. Comput Netw 101:63–80
38. Helbing D, Frey BS, Gigerenzer G, Hafen E, Hagner M, Hofstetter Y, Van Den Hoven J, Zicari RV, Zwitter A (2019) Will democracy survive big data and artificial intelligence? Springer, Towards Digital Enlightenment, pp 73–98

39. Feroz A, Perveen S, Aftab W (2017) Role of mHealth applications for improving antenatal and postnatal care in low and middle income countries: a systematic review. BMC Health Serv Res 17(1):704–714
40. Iribarren SJ, Brown W III, Giguere R, Stone P, Schnall R, Staggers N, Carballo-Dieguez A (2017) Scoping review and evaluation of SMS/text messaging platforms for mHealth projects or clinical interventions. Intl J Med Inform 101:28–40
41. Retrieved March 1, 2021, from senseable.mit.edu/livesingapore/visualizations.html
42. Nussbaum T, Zorbas E, Koros M (2012) A new deal for engagement in fragile states. Confl Secur Dev 12(5):559–587

Eco-Design of Mechatronic Hydropower Device in River

Jun Chen, C. Greg Jensen, and Haiyan H. Zhang

Abstract The off-grid energy provided by renewable resources is urgently needed for modern life, economic growth, and even medical care in many remote communities. Very often near the remote communities, there are rivers with hydropower potential. As an affordable and available solution for off-grid energy, the hydropower device can generate electricity, which can meet a wide range of market needs. For this purpose, we present an eco-design of mechatronic modular river current energy converter (MRCEC), using an onboard intelligent mechanism to control the cross-flow cycloid turbine to operate optimally under various river conditions in a year, and the hydraulic-based transmission system transfers the mechanical energy from the underwater turbine to a generator on the platform floating on the surface of water.

Introduction

In the developing countries, extending the power grid to remote areas may be a costly option. In fact, a small energy system that can provide electric power of 6kWh per month can greatly improve people's living conditions and meet their education needs. Even in the United States, in the first decade of twenty-first century, at least 180,000 households were living outside the power grid, and this number is increasing year by year due to the pursuit of environmental protection and a return to nature lifestyle [4]. Consulting company Accenture further predicts that by 2035, 12% of American households and 11% of European households will be off the grid. Off-grid solutions will become a real need of the market in a decade.

For off-grid power solutions of utilizing different renewable resources, specific technical treatments for varying factors, such as the strong day and night changes of solar radiation or wind, are needed. In addition, attention should be paid to improvement of system efficiency over the technical barriers and the reduction of installation, operation and maintenance costs.

J. Chen · C. G. Jensen · H. H. Zhang (✉)
Purdue University, West Lafayette, USA
e-mail: hhzhang@purdue.edu

P. Hehenberger et al. (eds.), *EcoMechatronics*,
https://doi.org/10.1007/978-3-031-07555-1_13

The proposal of off-grid solutions for remote communities should not ignore the basic fact that most of the domestic and international remote communities are located in nearby locations with rivers of hydropower potential. Per IRENA reports, hydropower-based minigrids have globally supported the population of 6.4 million in 2016, which doubled the population served in 2007 [5]. In the United States, the undeveloped hydropower carried by the rivers has capability over one trillion watt-hours (TWh/yr) per year to support millions of people with locally sourced, clean and reliable energy.

Centuries of engineering practice have shown that the use of hydropower to generate electricity is an efficient and reliable way. It is very attractive to design a river energy converter (RCEC), suitable for low/zero water head hydropower, whose rotating water turbine driven by the river currents harvests hydropower for electric generation in the rural and remote areas. It neither produces greenhouse gases nor negatively affects the environment. Ref. [7] pointed out that it has become a major interest in Europe, and an important option for rural electrification in underdeveloped countries.

However, RCEC has several issues to be resolved: (1) low efficiency: because its turbine in low-head river usually runs at much lower efficiency than the theoretical Betz limit (59%) without pipe/diffuser enhancement; (2) high mounting cost: the installation of hydropower turbines, whether it is bottom structure mounting (BSM), near-surface structure mounting (NSM) or floating structure mounting (FSM), the design of the mountings are more complicated than the hydropower device itself, and cost of the structure is very expensive; (3) high maintenance cost: Because the turbine and its power system (transmission and generator) are underwater, and the lubrication and replacement of faulty parts are performed outside the water, the maintenance cost is high, especially at remote sites. The underwater powertrain is sealed and waterproof, further increasing the complexity of the design; (4) high operation costs: For the seasonal flow changes of most rivers, the traditional design of turbines lacks the flexibility to work effectively under different inflow conditions. Since the individual modules in the multi-module array have their own independent systems, their generators, control electronics, etc. are redundant, which further increases operating costs. In order to maximally harvest the river hydropower, further development is needed for RCEC, whose integration and design while ensuring its structural stability and reliability under different operating conditions must be properly treated for the next generation of technology.

We present an approach of modular river current energy converter (MRCEC) as a stable, affordable, and environmental-friendly off-grid energy solution for residential households and small businesses in remote areas. MRCEC integrates high-fidelity hydrodynamic analysis, optimized mechatronics control, and high efficiency transmission system to maximize efficiency while minimizing installation, operation and maintenance costs. The design specifications of each MRCEC module are: to generate at least 10KW of electricity at a water flow speed of 2.0 m/s; to improve turbine hydrodynamic efficiency up to 50% in comparison to 20–30% of other existing designs; to have smart onboard control optimizes the system operation at seasonal variation;

to be environmentally-friendly, less disturbance to the eco-system, at the same time affordable; to meet various power needs using the modular design that can formulate individual or array applications.

Cycloidal Turbine with Optimal Blade Pitch Control

Cycloidal turbine is a subset of the H-bar type turbines, and it can maximize the harvested energy by varying the blade pitch angle during the revolution. Figure 1 presents the power coefficient of different wind turbines as a function of tip speed ratio (TSR, $\lambda = \Omega r / V_{\infty}$) and the same analysis can guide the analysis of their hydrokinetic counterparts. The traditional fixed-pitch Darrieus turbine's blades often fail to extract more power or stall during each revolution. These attenuate power-harvesting, and the oscillatory loads cause undesirable damages, and prohibit self-start at a low rotational speed. Cycloidal turbines can overcome these difficulties. However, the proper selection of blade pitching scheme is nontrivial, and most of the previous efforts were based on a certain trial-and-error approach.

Our approach incorporates the concept of a horizontally deployed cycloidal turbine (cross-flow cycloidal turbine) for harvesting the mechanical energy from the river current. (see Fig. 2) The optimal turbine blade design and optimal blade pitching adjustment are explored. We have applied this approach to study the loading of a 10-m diameter wind turbine [10]. Recently, we investigated the design of the cycloidal turbines and proposed a model, Fluxline Theory, to optimize the pitch adjustment scheme [1]. The experiments on a three-blade vertical axis wind turbine demonstrated an efficiency of 40–55% when the TSR ranges from 1.5 to 2.5 [2], significantly higher than that with other pitch-control schemes, even when the pitch scheme is applied to off-design conditions (Fig. 3). However, from that test, it was

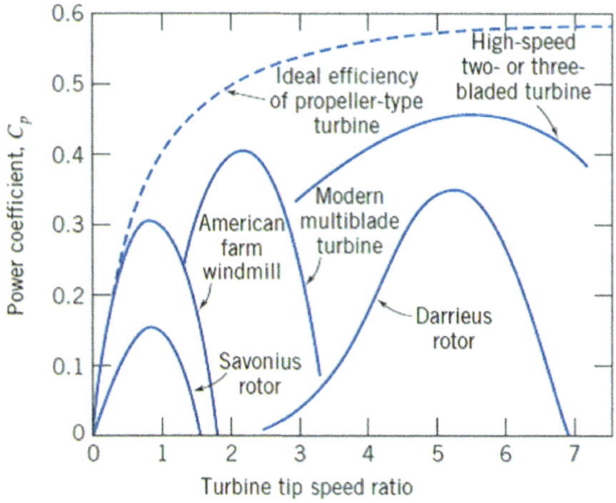

Fig. 1 Efficiency trends of wind turbine types versus tip-speed ratio [8]

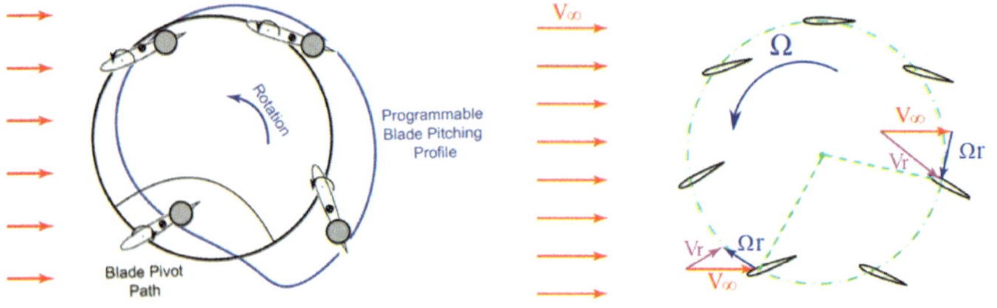

Fig. 2 Schematic of pitching adjusting scheme of a cycloidal turbine blade (left) and velocity vectors (right) at different cyclic

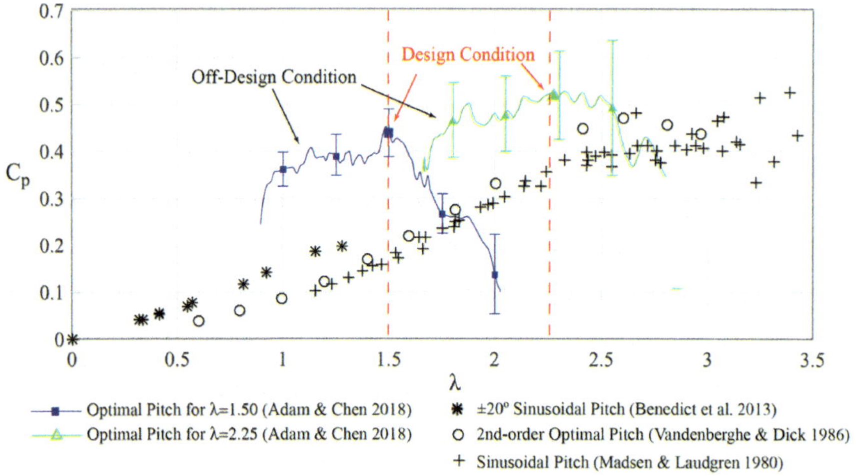

Fig. 3 Comparison of measured power coefficient (efficiency) of a three-blade cylcodiao turbine with results from other cycloturbine experiments. More details can be found in Adam and Chen [2]. Two optimal pitching schemes are applied for two design conditions ($\lambda = 1.5$ and 2.25), showing the performance at both design and off-design conditions. Error bars represent a 95% confidence interval

also identified that the single section straight blades result in unwanted oscillating loads on the turbine shaft, which may cripple the duty life of the system.

By simulating both the turbine revolution at a chosen TSR and rotation of individual turbine due to pitch adjustment at different phase angle, we are able to understand the evolution of the different hydrodynamic parameters, e.g, the power coefficient, as the turbine/blade rotates. This approach has been proven effective by comparing to the experimental results. Properly controlling the TSR in the vicinity of the optimum, the turbine system can perform optimally at efficiency much higher than the existing technologies.

In order to maximize the power coefficient, a pitching mechanism algorithm is developed to determine the pitch motion of the hydrofoil, which is the pitch angle of

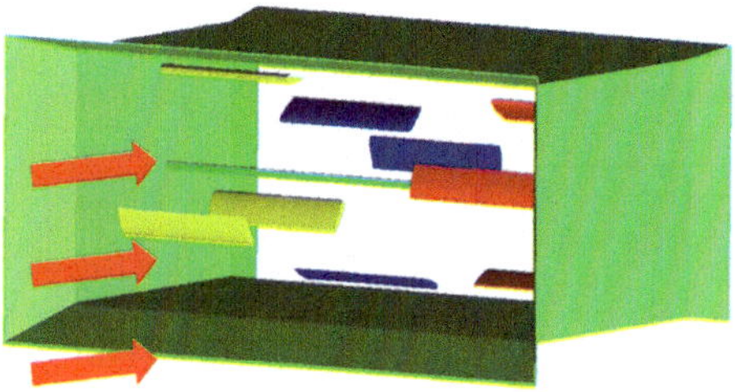

Fig. 4 Schematic of three-section turbine installed inside the inlet-diffuser duct

the hydrofoil in a revolution. Based on the following Eq. (1), for a given freestream velocity and rotation speed, the power coefficient is proportional to the torque. Thus, the pitch angle of the hydrofoil is determined by maximizing the generated torque on the rotation shaft at each azimuthal angle,

$$\theta(\phi) = \text{argmax}(T(\theta, \phi)) \tag{1}$$

with the input of hydrofoil data, incoming flow velocity, and turbine dimensions.

The pitch motion of the hydrofoil is highly related to the tip speed ratio. By running the pitch analysis with different input of freestream velocity and rotation speed, the pitch motion for different tip speed ratio. At low rotation speed ($\lambda < 1$), the pitch angle exceeds 90° for some azimuthal angles, which is difficult to achieve physically. Since the turbine usually operates at a relatively higher rotation speed, a ±75° pitch angle limit is applied.

To determine the optimal TSR for the cycloidal turbine, the three-section model with the inlet-diffuser duct (Fig. 4) is studied at a TSR range from 0.25 to 2.5 via 3D CFD simulations. Figure 5 presents the pitch motion optimized for different TSRs. The torque generated by the turbine at different TSRs is presented in Fig. 6a. With

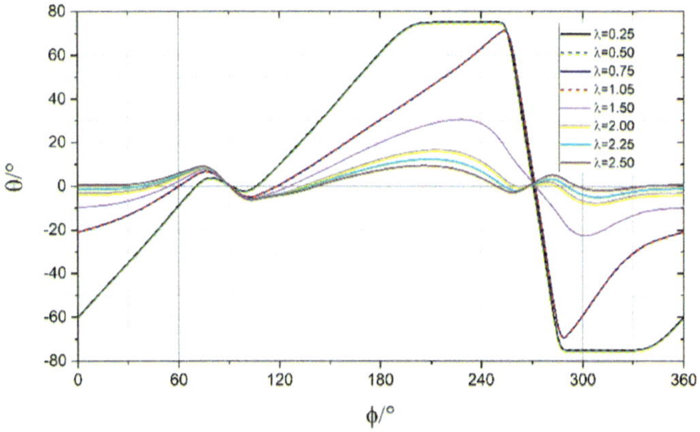

Fig. 5 Pitch schedule optimized for different TSRs

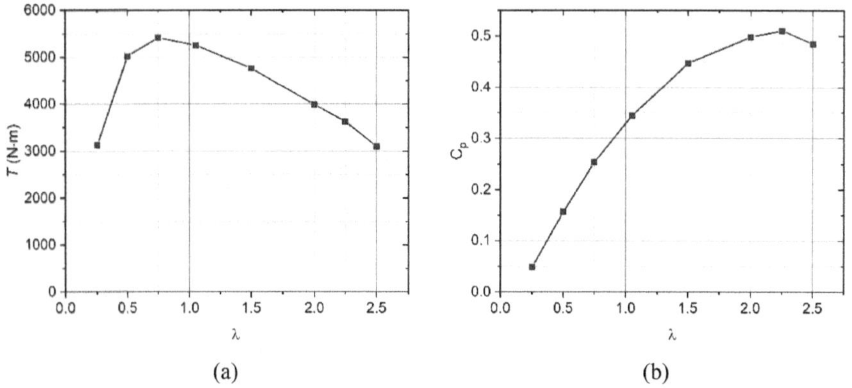

Fig. 6 **a** Torque at different TSRs and **b** averaged power coefficient at different TSRs

the increase of λ, the torque increases for $\lambda < 0.75$ and decreases for $\lambda > 0.75$. 5(b) illustrates the power coefficient at different TSRs. The maximum power coefficient, $C_p = 0.511$, occurs at $\lambda = 2.25$. Therefore, $\lambda = 2.25$ is chosen as the design condition for TSR.

Figure 7 illustrates the power coefficient of this design in one revolution. The power coefficient maintains a nearly constant value of 0.515, which is significantly higher than other existing designs. An innovative three-dimensional (3D) multilayer cam is designed to achieve the pitch motion to achieve an impressive extracted power coefficient and facilitate effective mode switches to adapt to the change in river flow velocity. As a critical component in achieving pitch motion, the cam profile is developed with the pitch variance mechanism for different modes: park mode, startup mode, and several operation modes for different tip speed ratios.

The theoretical equation to reflect the relationship between the cam radius and the pitch angle as shown in Fig. 8,

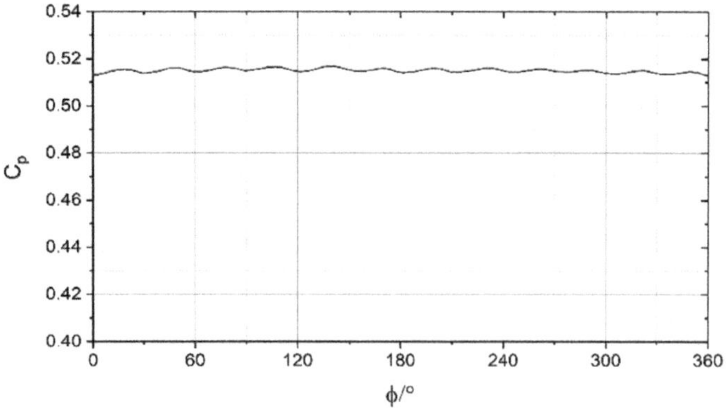

Fig. 7 Power coefficient obtained by the final design

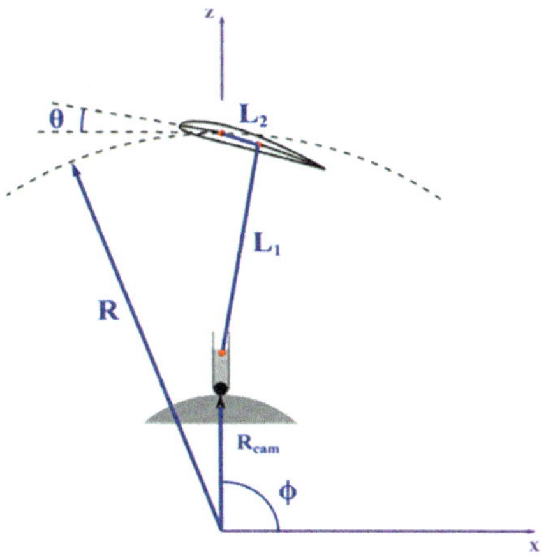

Fig. 8 Schematic of cam profile determination

$$r_{\text{cam}} = r_{\text{cam},\theta=0} + \sqrt{(L_1^2 - L_2^2)} - \left(\sqrt{L_1^2 - (L_2 cos\theta(\phi))^2}\right) + L_2 sin\theta(\phi) \quad (2)$$

where L_1, L_2 and $r_{\text{cam},\theta=0}$ are strategically selected constants for the pitch variance mechanism to aid in the ease manufacturing of these components.

By repeating the equation for a revolution of azimuthal angle, the cam profile can be further determined. For the operation modes, several cam profiles are generated with the same algorithm based on different pitch motion optimized at $\lambda = 1.5$, $\lambda = 2.0$, $\lambda = 2.25$, and $\lambda = 2.5$. For the startup mode, the pitch motion is specially developed based on a very low rotation speed (0.25 rad/s) with the cam profile generated using the algorithm discussed above. For the park mode, the cam profile is designed to set the angle of attack of the hydrofoil in a nearly always stalled position.

Hydraulic Transmission System

For our MRCEC, neither a gearbox nor a heavy generator with multiple pole pairs is an ideal solution for the cross-flow cycloidal turbine. A hydraulic transmission system with much less weight than a gearbox becomes the best choice to transfer energy from the turbine to the generator for the design of the MRCEC. The hydraulic transmission system is easy to be arranged on the turbine system, which can damper the energy fluctuation shocks from the incoming river streams. It also has good heat dissipation performance underwater and can integrate power from multiple modules of MRCEC together into one generating system, which leads to the great potential of reducing the cost of installation, operation and maintenance.

For the MRCEC, the hydraulic pump will be immersed in the river. If hydraulic oil is used as the medium in the hydraulic system, the oil leakage will pollute the

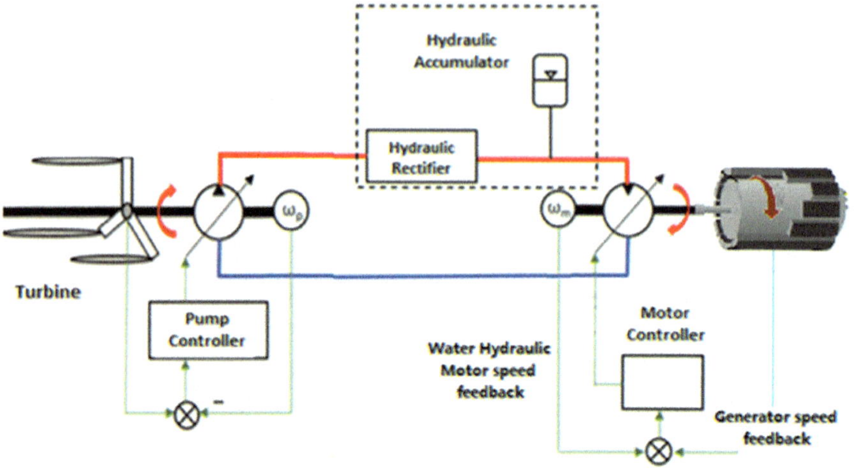

Fig. 9 Hydraulic transmission system of the MRCEC

river environment and significantly weaken the functionality of the power generation system. To ensure complete environmental friendliness, and to utilize the water from the river to replenish the leaked hydraulic medium without manual refilling of hydraulic fluids, a hydraulic transmission system based on water as the hydraulic medium is designed for the MRCEC.

The hydraulic transmission system consists of a radial piston pump, an axial piston motor, a bladder accumulator, control valves, and hoses. (see Fig. 9) To provide theoretical support for the product development stage of the MRCEC, mathematical models of the hydraulic components in the transmission system are derived, and the control strategy of the rotational speed regulation of the generator is developed as well. Also, special features of the radial piston pump are designed for the water medium directly taken from the river.

The radial piston pumps with large displacement are usually the choice for applications with high-torque but low rotational speed. The structure of the radial piston pump is shown in Fig. 10.

The mathematical modeling of the radial piston pump includes a steady-state equation for the operational states, and the dynamic equations for hydraulic machines as follows:

$$d\rho = \left(\frac{\partial \rho}{\partial p}\right)_{T_l} dp + \left(\frac{\partial \rho}{\partial T_l}\right)_p dT_l \tag{3}$$

$$\frac{v}{\beta_e}\dot{p} + \dot{V} = \frac{(w_{in} - w_{out})}{\rho} \tag{4}$$

$$q_p = D_p \omega_p \eta_{pv} \tag{5}$$

Fig. 10 Radial piston pump of the MRCEC

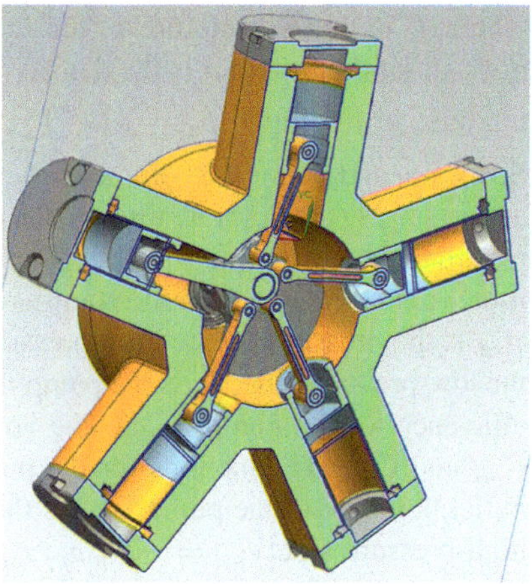

$$\eta_{pv} = 1 - \frac{q_{pv\_loss}}{D_p \omega_p} \tag{6}$$

$$T_p = \frac{D_p \left(p_{ph} - p_{low} \right)}{\eta_{pmh}} \tag{7}$$

$$\dot{V}_{ph} + \frac{V_{ph}}{\beta_e} \dot{p}_{ph} = -C_{pil}\left(p_{ph} - p_{low} \right) - C_{pel}\left(p_{ph} \right) - q_{ph} \tag{8}$$

$$C_{pil}\left(p_{ph} - p_{low} \right) = r_{ie} C_{pil}\left(p_{ph} \right) \tag{9}$$

$$J_p \dot{\omega}_p = T_d - B_p \omega_p - D_p \left(p_{ph} - p_{low} \right) \tag{10}$$

$$\dot{\theta}_p = \omega_p \tag{11}$$

$$J_p \dot{\omega}_p = \eta_{pmh} T_d - D_p \left(p_{ph} - p_{low} \right) \tag{12}$$

$$\eta_{mmh} D_m (p_{mh} - p_{low}) = J_m \dot{\omega_m} + T_l \tag{13}$$

$$q_m = D_m \omega_m + k_{m\_leak} (p_{mh} - p_{low}) \tag{14}$$

$$k_{m\_leak} = \frac{k_{m\_HP}}{v \rho} \tag{15}$$

$$k_{m\_HP} = \frac{V_{m\_max} \omega_{m_{nom}} (1 - \eta_{mv}) v_{m\_nom} \rho_{m\_nom}}{\Delta p_{m\_nom}} \tag{16}$$

where T_l is the temperature in the hydraulic liquid; ρ and p are the density and pressure of the liquid; ρ is the density of the hydraulic fluid; β_e is the effective bulk modulus of the liquid $\beta_e = \rho\left(\frac{\partial p}{\partial \rho}\right)_{T_l}$; $w_{in} = \rho q_{in}$, and $w_{out} = \rho q_{out}$ are the mass flows into and out of the volume of the pump with the volumetric fluid flow rates q_{in} and q_{out}; q_p is the flow rate through the pump; D_p is the pump displacement; ω_p is the rotation speed of the pump shaft; η_{pv} is the volumetric efficiency for the pump; q_{pv_loss} is the volumetric losses for the pump as a function of the pressure $\left(p_{ph} - p_{low}\right)$, where p_{ph} is the pressure on the high-pressure side for the pump, p_{low} is the pressure on the low-pressure side; T_p is the pump input torque; η_{pmh} is the mechanical hydraulic efficiency for the pump; V_{ph} is the volume on the high-pressure side of the pump; C_{pil} and C_{pel} are functions for the pump internal and external leakage; q_{ph} is the outlet flow rate of the pump; $\dot{V}_{ph}$ is the change rate of the hydraulic volume on the high-pressure side:$\dot{V}_{ph} = -D_p\omega_p$; r_{ie} is the ratio of the pump internal leakage to the external leakage, changing within the range 1–3; J_p is the moment of inertia for the pump; θ_p is the angle of rotation of the pump shaft; T_d is the driving torque on the pump shaft; B_p is the viscous damping coefficient; η_{mmh} is the motor mechanical hydraulic efficiency; D_m is the motor volumetric displacement; p_{mh} is the pressure on the high-pressure side of the motor; J_m is the moment of inertia for the motor; ω_m is the motor rotational speed; T_l is the torque produced by the generator load; q_m is the fluid flow rate at the motor side; k_{m_leak} is the leakage coefficient at the motor side; k_{m_HP} is the Hagen-Poiseuille coefficient for the motor; V_{m_max} is the maximum motor displacement; $\omega_{m_{nom}}$ is nominal motor angular speed; η_{mv} is motor volumetric efficiency; v_{m_nom} is nominal motor fluid kinematic viscosity; ρ_{m_nom} is nominal motor fluid density; Δp_{m_nom} is nominal motor differential pressure.

For the hydraulic system of the MRCEC, an accumulator with 1 L volume with a maximum pressure of 207 bar is selected in the market to damper the energy fluctuation from the incoming river streams. Ideal gas law is applied to the nitrogen in the bladder, and its isothermal model is as follows:

$$P_g V_g = nRT \tag{17}$$

$$P_g = \frac{P_0 {V_0}^n}{{V_g}^n} = \frac{P_0 {V_0}^n}{\left(V_0 + \int Q_a dt\right)^n} \tag{18}$$

$$Q_a = -\frac{V_0 {P_0}^{\frac{5}{7}}}{1.4 {P_g}^{\frac{12}{7}}} \frac{dP_g}{dt} \tag{19}$$

where P_g and V_g are the pressure and volume of the nitrogen gas in the accumulator; P_0 and V_0 are the pre-charge pressure and the initial volume of the nitrogen gas; Q_a is the hydraulic fluid flow into or out of the accumulator; $n = 1$ in this isothermal model.

Hoses are used to connect hydraulic components and carry hydraulic fluid in the transmission line. Neglecting the hose deformation, hydraulic pressure losses along the hose line are neglected. Therefore, the hose dynamics is as follows:

$$\frac{dP}{dt} = \frac{(Q_e - Q_b)\beta}{V} \tag{20}$$

where V is the fluid volume inside hoses; β is the bulk modulus of the fluid; Q_e is the flow rate at the end of the hose; and Q_b is the flow rate at the beginning of the hose.

Based on the mathematical model of all the components in the hydraulic transmission system, numeric simulation of the system operational performance is performed according to technical datasheets (see Fig. 11).

Giving the signal input of the turbine's designed operational speed (4.5 rad/s for the 2 m/s incoming river flow) with some turbulence in the input curve to simulate the sudden increase of the river flow speed, the reaction rotational speed curve of the hydraulic motor (generator) with electrical load is shown in Fig. 12. The actual overall efficiency of the hydraulic transmission system is calculated to be 74.6%.

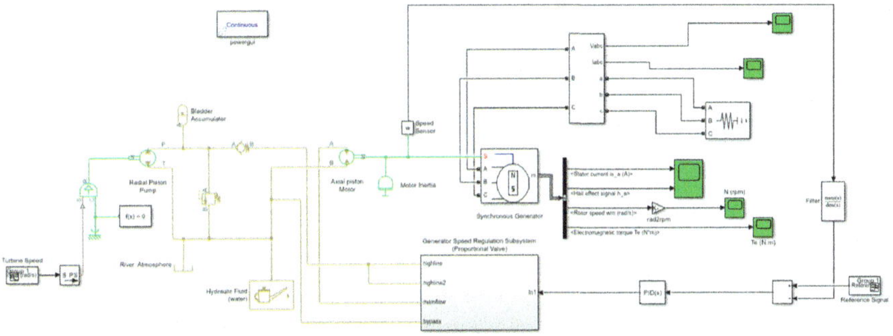

Fig. 11 Simulation blocks of the hydraulic transmission system

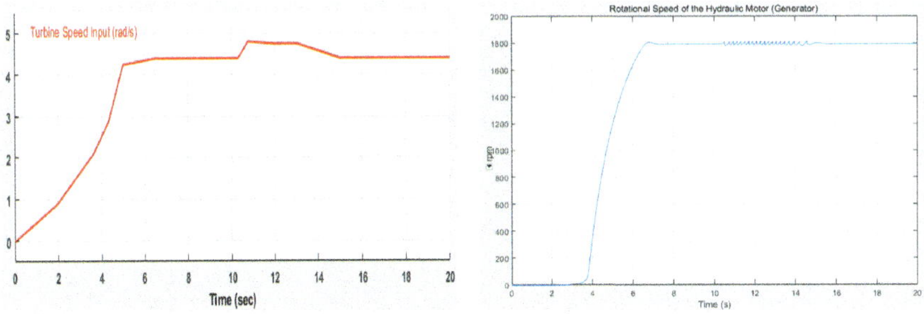

Fig. 12 **a** Input speed curve of the turbine; **b** Reaction speed curve of the hydraulic motor/generator

Mechatronic Control System

The blade pitch scheme is real time controlled by onboard sensing and smart mechanism. The cycloidal turbines are controlled by the mechatronic system. The pitch adjustment of the cycloidal turbines is realized with smart mechanisms and supporting structures, smart sensing techniques and data acquisition, precision actuators, and their operations, electrical/hydraulic power resources, embedded microcontrollers, control laws, and the drives for actuators, integrated circuits, controller as well as its programming. Linear actuator PA-17 and turbine speed sensor are integrated into a waterproofed sealing enclosure box, mounted on the MRCEC framing adapter, to drive the 3D cam, on which the smart mechanism adjusts the cycloidal turbine's blades according to the blade pitch scheme (see Fig. 13).

The sensors used in the MRCEC system include the following devices: a flow velocity sensor, a turbine rotary speed sensor, a 3D cam position sensor, and a battery level sensor. For the flow velocity, a pitot tube with a pressure transducer is employed, and its difference of total flow pressure P_t and static flow pressure P_s is transduced as a voltage signal to calculate the flow velocity. M393 speed sensor module is used to indicate the turbine rotary speed. The position sensor of the 3D cam is integrated with linear actuator PA-17, and the position signal will be utilized to calculate the linear actuator stroke. The battery level sensor monitors the battery level.

The working modes of MRCEC include startup mode, high efficiency mode, constant power mode, and park mode. The startup mode will be activated when the mechatronic control system of MRCEC is initiated. Then MRCEC will automatically enter other different working modes according to the river flow conditions. There are two working modes, i.e, high-efficiency mode and constant power mode. Under extreme circumstances, the control system will enter the park mode.

During the MRCEC's startup mode, the load from the generator is eliminated by the proportional valve being fully opened to bypass all the flow. At the same time, the river flow speed will be measured by the flow velocity sensor to determine further action. For the flow speed less than 0.5 m/s the generator will be idled because the flow does not have enough energy to harvest. For the flow speed between 0.5 to

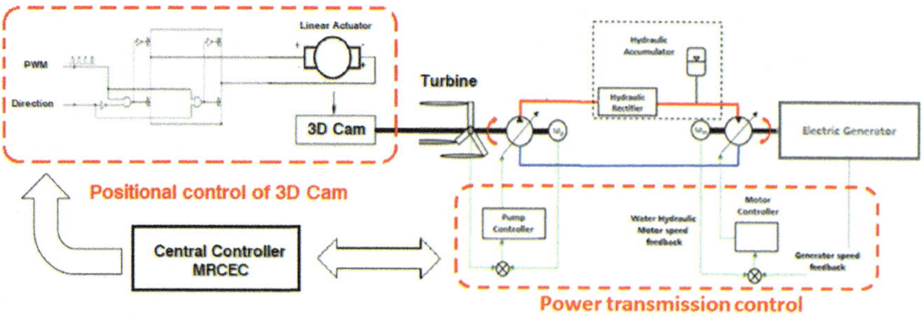

Fig. 13 Control analysis and simulation of MRCEC

3.5 m/s, the control system will operate in high-efficiency mode. For the flow speed higher than 3.5 m/s, the system will shift to park mode due to the extreme conditions.

For high-efficiency mode, the power harvested by the turbine will depend on the power coefficient under a given flow speed, as shown in Eq. (21):

$$P = 0.5\rho u_{\infty}^{3} S C_{p} \tag{21}$$

where u_{∞} is freestream velocity, ρ is density of water, and S is the blade surface. The turbine can keep running at maximum power coefficient $C_{p_{max}}$ as long as the tip-speed ratio of the turbine is $\lambda = 2.25$. By controlling the position of the 3D camshaft, if the turbine angle is adjusted to keep the optimal tip speed ratio, the best power coefficient remains.

In order to protect the MRCEC system under extreme conditions, the maximum output power $P_n = 11$ KW is determined due to mechanical strength, consumer electricity consumption, and other physical properties. Once the control system adopts constant power mode, keeping the output power at P_n rather than optimal tip speed ratio will be a priority.

To ensure the safety of the MRCEC system at extreme conditions, the system will shift to park mode when flow speed is higher than 3.0 m/s. In this mode, the turbine shaft will not rotate, the output power will be 0.

The control procedure of MRCEC is implemented as shown in Fig. 14. The entire MRCEC system is shown in Fig. 15.

The output ports of hydraulic systems of MRCEC units can be connected with flexible hoses to drive a single hydraulic motor. Thus, the generator can be placed on a deck above the water or directly on the nearby bank. The hydraulic transmission system is designed watertight. The hydraulic transmission system has features of a low-pass filter for smoothing the fluctuations of hydrodynamic power collected from a mechanical blade system so it contributes to lengthen turbines lifetime and thus further reduce maintenance cost, as shown in Fig. 16.

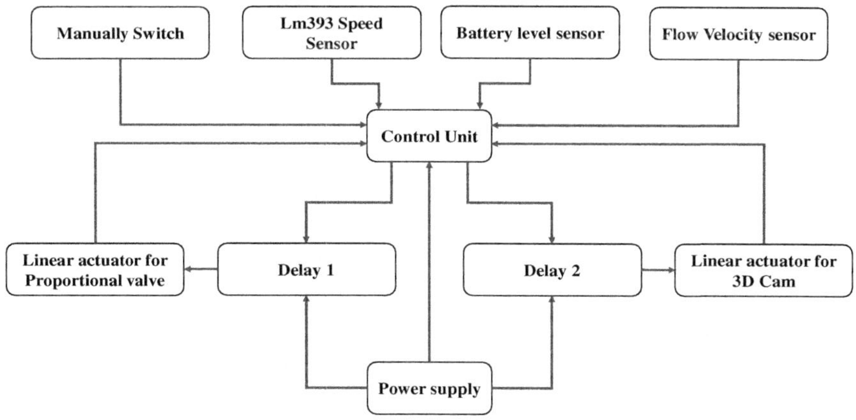

Fig. 14 Control system block diagram

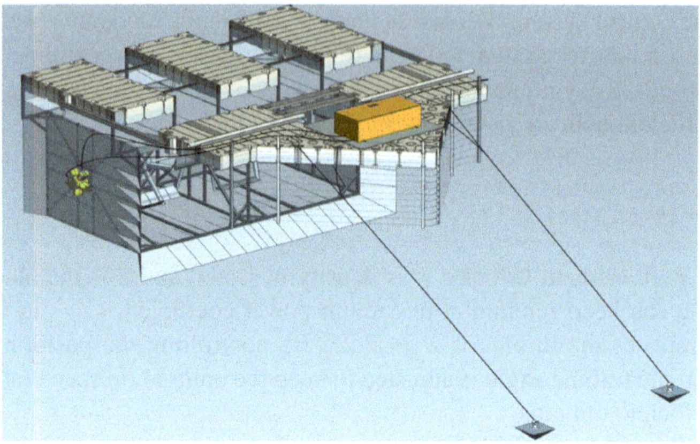

Fig. 15 MRCEC system

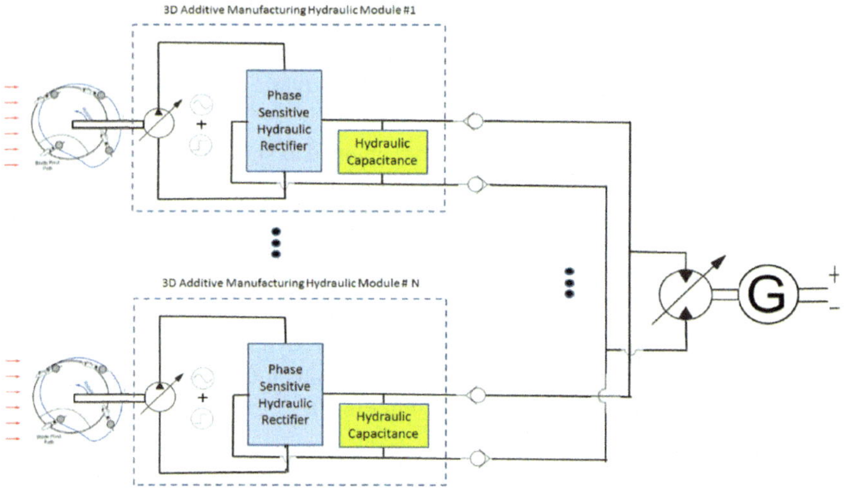

Fig. 16 Electricity generation of the multi-module array by a hydraulic manifold

Conclusion and Future Work

Constant reliable energy impacts nearly every aspect of life, standard of living and is a determinant factor of societal well-being, economic growth, healthcare, etc. Most of the off-grid energy solutions are frequently disturbed by the unstable nature of these renewable resources. Likewise, most of the off-grid remote communities, domestic and international, are located on sites with nearby hydropower potentials. Producing electricity from the riverine currents thus presents an ideal off-grid energy solution and presents a significant market opportunity.

In this work, our team summarizes the design of a new product, Modular River Current Energy Converter (MRCEC), which can effectively produce electricity from rivers with increased energy affordability and availability. The design features: (1) a cross-flow cycloidal turbine that yields a high power coefficient (51.5%); (2) an onboard intelligent mechanism controls the turbine functioning optimally over a range of river conditions; and (3) a hydraulic-based drivetrain feeds the harvested mechanical energy from the underwater turbine to the generator(s) placed above the water.

The hydrodynamic performance of the turbine at different operating conditions was analyzed using high-fidelity CFD simulation, which accounts for the varying of the blade pitch angle at different rotating phase. An optimal hydrofoil was designed for MRCEC. A four-section three-blade design was adopted for the turbine to obtain smooth power output and stable hydrodynamic loads. At different inflow conditions, the blade pitch can be controlled accordingly.

With the hydrofoil pitch variance, the turbine can produce efficient optimal power as the river flow varies across its annual cycle. A control system monitors the flow condition and turbine operation conditions by collecting signals from sensors installed on the turbine and system. The controller sends out control signals to a linear actuator that adjusts the position of the 3D cam accordingly. The turbine is enclosed by an inlet–outlet-duct to further improve its performance.

The drivetrain includes a radial piston pump, an axial piston motor, generator(s), and hydraulic lines. Water is used as the hydraulic fluid so it eliminates the contamination to rivers. Optimal design and model analysis report a power efficiency 74.6% for the entire drivetrain. The turbine system is supported by a floatation platform such that it harvests energy from the near-surface current that has a larger velocity than the bottom current. The self-contained design of MRCEC also significantly eases the installation, operation, and maintenance.

Ongoing and future work includes a prototype test to validate the performance of individual subsystem, as well the integrated one. This involves tests in well-controlled lab experiments (e.g, in water tunnels) and realistic open water tests that experience unavoidable flow variation. Reducing the manufacturing cost and operation cost is further explored by adopting new materials and manufacture procedures. Further study of MRCEC in an array setting will identify new need for elaborated mechatronic control that does not exist for individual MRCEC module. Moreover, incorporation of the mechatronic sensing and control into a digital twin for the MRCEC will add its merit for industrial application.

Acknowledgements This work was supported by the grant DOE-EERE Project Review EE0008950. We would like students Yijie Wang, Zhen Li, Pradip Murugan, Heming Liu, Ethan Guenther, Joohyun Lee, Mridul Rathi, Gunnar Rich, Lauren Sailor, Haoyang Shi, and Zack Zhang in School of Mechanical Engineering in College of Engineering, and in School of Engineering Technology in Polytechnic Institute at Purdue University.

References

1. Adams Z, Chen J (2017) Flux-line theory: a novel analytical model for cycloturbines. AIAA J 3851–3867
2. Adams Z, Chen J (2018) Optimization and validation of cycloturbine blade-pitching kinematics via flux-line theory. AIAA J 1894–1909
3. Benedict, M, Lakshminarayan, V, Pino, J, Chopra I (2013) Fundamental understanding of the physics of small-scale vertical axis wind turbine with dynamic blade pitching: an experimental and computational approach. In: 54th AIAA/ASME/ASCE/AHS/ASC structures, structural dynamics, and materials conference, AIAA Paper 2013–1553
4. Davidson P (2006) Off the grid or on, solar and wind power gain; incentives, savings push more families to renewable energy. USA Today, 13
5. International Renewable Energy Agency (2018) Off-grid renewable energy solutions: global and regional status and trends
6. Madsen H, Lundgren K (1980) The Voith–Schneider wind turbine: some theoretical and experimental results on the aerodynamics of the Voith–Schneider Wind Turbine, Tech Rept NP-2900722, Institue of Industrial Constructions and Energy Technology, Aslborg Univ. Centre, Aslborg, Denmark
7. Paish O (2002) Micro-hydropower: Status and prospects. Proc Inst Mech Eng, Part A: J Power Energy 216(1):31–40
8. Pritchard PJ, Mitchell JW (2016) Fox and McDonald's introduction to fluid mechanics. John Wiley & Sons
9. Vandenberghe D, Dick E (1986) A theoretical and experimental investigation into the straight bladed vertical axis wind turbine with second order harmonic pitch control. Wind Eng 10(3):22–138
10. Zhou N, Chen J, Adams DE, Fleeter S (2015) Influence of inflow conditions on turbine loading and wake structures predicted by large eddy simulations using exact geometry. Wind Energy

Micro/Nanopositioning Systems with Piezoelectric Actuators and Their Role in Sustainability and Ecosystems

Morteza Mohammadzaheri, Payam Soltani, and Mojtaba Ghodsi

Abstract Micro/nanopositioners based on piezoelectric actuators and their role to preserve the biodiversity of the ecosystem and achieve a sustainable manufacturing sector are presented in this chapter. These positioners are precise at micro/nanometre resolution and have improved and assisted reproduction and somatic cell nuclear transfer, playing an increasingly important role in preserving endangered species from extinction. Studies suggest these technologies are potentially key factors in our ability to decelerate the degradation of our nature. Further, piezo-actuated micro/nanopositioners are the foundation of add-on accuracy increasing systems, which can return outdated machine tools to service, with minor changes and at a performance level higher than new machines. This avoids waste of energy and materials, as the outdated machines or their major parts would otherwise be deposed of. In addition, piezo-actuated micropositioners play an essential role in vibration-assisted machining, which reduces energy consumption, increases product quality and extends machine lifetimes.

Introduction

Nano/micropositioning aims to control position at the nano/micrometre scale for applications such as fine machining [1], manipulation of biological cells [2], scanning probe microscopy [3] and robotic surgery [4]. Different types of actuators such as magnetostrictive actuators [5] and worm gears [6] have been used in nano/micropositioning. However, piezoelectric actuators are the least bulky, most precise and now the most common actuators for nano/micropositioning [7]. The technology also has a high potential to contribute to environmental sustainability.

M. Mohammadzaheri (✉) · P. Soltani
School of Engineering and the Built Environment, Birmingham City University, Birmingham, UK
e-mail: mmzahery@gmail.com; morteza.mohammadzaheri@bcu.ac.uk

M. Ghodsi
School of Energy and Electronic Engineering, University of Portsmouth, Portsmouth, UK

P. Hehenberger et al. (eds.), *EcoMechatronics*,
https://doi.org/10.1007/978-3-031-07555-1_14

Piezo-actuated micro/nanopositioning systems may include (i) a piezoelectric actuator, which deforms in relation to its input electrical voltage, (ii) a sensor to obtain the position of (a point or a surface on) the piezoelectric actuator and (iii) a control system which provides the required excitation voltage to realise the desired position. Of these three components, the actuator and the control system are essential and sensors may be excluded [8, 9]. Choice of the designer in respect of any of these components affects the performance, expense and operational requirements of the nano/micropositioner.

Role of Piezo-Actuated Micro/Nanopositioning Systems in Biodiversity Protection and Sustainable Manufacturing

Biodiversity Protection

Biodiversity and maintaining an intact biodiverse ecosystem have a direct link to health and sustainable development. According to recent research [10], more than one million species are currently at risk of extinction, and losing any takes our ecosystem one step closer to irreversible degradation. Recent developments in biotechnology, particularly in cell manipulation, have aided species preservation [11] with assisted reproductive techniques (ARTs) such as intracytoplasmic sperm injection (ICSI), in vitro fertilisation (IVF), cryopreservation of gametes, artificial insemination, and embryo transfer. ARTs are being used extensively to increase the population of endangered animal species [12].

Piezo-actuated micro/nanopositioning systems are employed in cell manipulation and play a crucial role in increasing the performance and success rate of ARTs [13]. For instance, in ICSI, a single spermatozoon of the species is injected into the cytoplasm of a mature oocyte as shown in Fig. 1. The use of a piezo-actuated nanopositioner (attached to the needle) for injection improves the performance of the system [14]. With the use of a piezo-actuated nanopositioner, wider pipes (with outer diameters up to 30μm) can be successfully driven into the plasma without a membrane break; this is almost impossible in the conventional ICSI, where needle size is restricted to only a few micrometres [14]. Piezo-actuated ICSI systems provide a precise and fast motion for the tip of the needle with speeds of nearly 40μm/s; this facilitates control of the procedure as the sperm tail oolemma is disrupted in a well-controlled manner rather than manually. Positive effects of using piezo-actuated nanopositioners on the outcome of ICSI procedures have been reported [14, 15]. As an instance, a recent study [16] clearly indicates the significant increase in fertilisation rates with the use of piezo-actuated micro/nanopositioning systems in ICSI when compared to conventional ICIS; and that piezo-ICSI reduces the rate of oocyte degeneration and abnormal fertilisation.

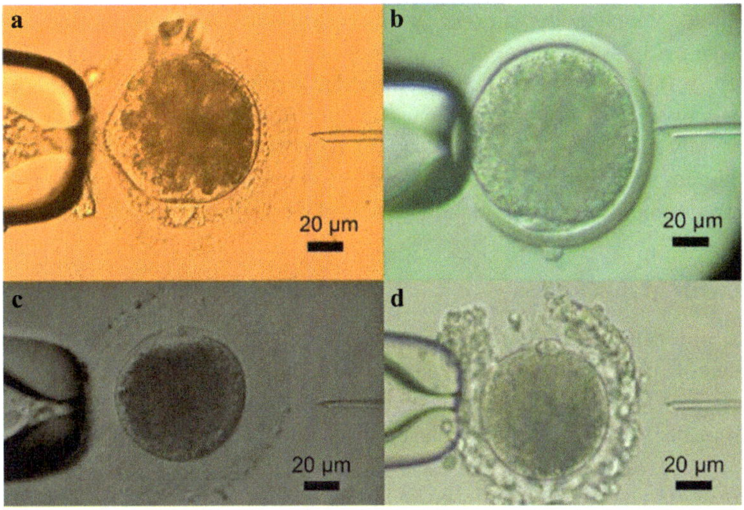

Fig. 1 ICIS setup and oocyte of different species: **a** Equine oocyte, **b** ovine oocyte, **c** domestic cat oocyte and **d** leopard oocyte. PB, polar body [15]

Somatic cell nuclear transfer (SCNT) is another emerging and promising technique for maintaining biodiversity and tackling the issue of the extinction and endangerment of many species [17]. SCNT could be used not only for saving the endangered species [18], but also as an effective tool for the reproduction of genetically valuable animals, generation of transgenic species, and manufacturing of modern medicines [18]. SCNT cloning also has a great potential to help in fighting new diseases like Covid-19 [19]. As depicted in Fig. 2, nuclear DNA is removed from an oocyte and replaced by a somatic cell nucleus. There are two main methods to place

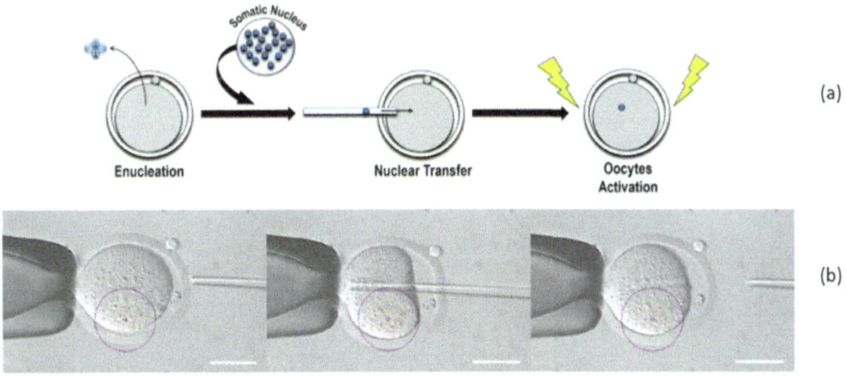

Fig. 2 **a** Different steps in somatic cell nuclear transfer (SCNT) [18], **b** donor cell injection process [22]

a somatic cell nucleus into an oocyte: (i) conventional electrofusion of the somatic cell to the cytoplasm of the oocyte and (ii) one-step micromanipulation (OSM) as depicted in Fig. 2. Research has revealed that the latter provides better efficiency and leads to a higher in vitro blastocyst development rate [20]. In OSM, the needle tip is usually driven by a piezo-actuated micromanipulator. Figure 3 shows a typical experimental setup for SCNT. A piezo-actuated micropositioner is mounted on the injection micropipettes for both enucleation of the somatic cell and nuclear transfer to oocytes [21].

Sustainable Manufacturing

Sustainable manufacturing and production systems also benefit from the technical attributes of piezo-actuated micro/nanopositioning systems. Machine tools significantly influence sustainability, and social, economic and environmental issues [23]. A third of the worldwide energy consumption and CO_2 emissions is in the production industry, where machine tools are major role players [24]. The use of outdated machine tools, common in developing nations, leads to a considerable waste of energy and raw materials [24]. The constant replacement or retrofitting of machine tools consumes material and energy resources, a threat to sustainable growth. According to the International Energy Agency (IEA), there exists significant potential for energy and raw material saving through the upgrade and reuse of these outdated machine tools without extensive modification or replacing their core components [25]. Upgrading existing machine tools with minor changes, as an alternative method

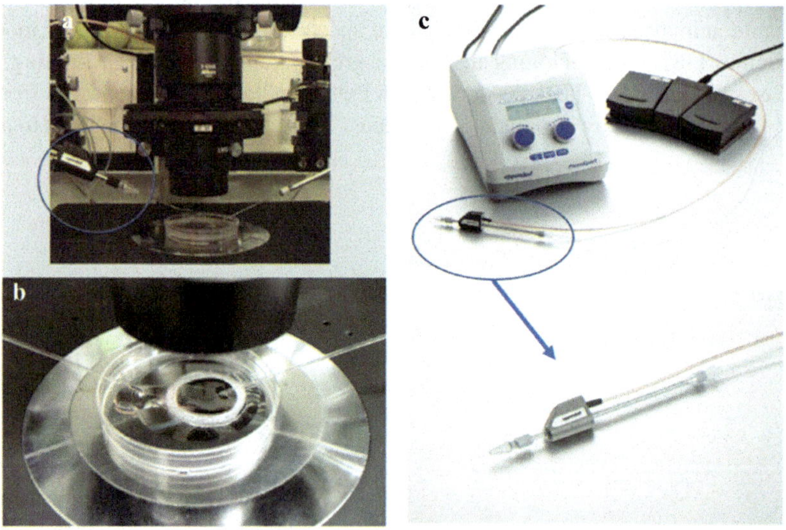

Fig. 3 **a** Micromanipulation chamber and setup for somatic cell nuclear transfer (SCNT) to oocytes including micropipettes **b** holding pipette (right) and enucleation or nuclear transfer micropipette (left) coupled to **c** piezoelectric micromanipulating system Eppendorf PiezoXpert [21]

to retrofitting and refurbishment, can considerably improve the sustainability aspect of manufacturing processes [26].

One of the major issues with outdated machine tools is insufficient positioning accuracy. In outdated machine tools, mispositioning errors yield low-quality machined products, the main reason for their replacement, retrofitting and major overhaul [25]. A sustainable solution to this problem is to use an add-on system to compensate for these positioning errors and such add-ons are often based on piezo-actuated micropositioners. As an example, a piezo-actuated add-on accuracy increasing system (AAIS), shown in Fig. 4b, was introduced [24], as a sustainable solution. This is a modular station with high accuracy in error compensation that can be easily mounted on the worktable of a machine to upgrade it and improve its functionality, as shown in Fig. 4a [23]. Figure 4c shows that an outdated machine with an AAIS may perform even better than a new machine tool, in terms of positioning accuracy. The add-on not only increases the precision of positioning but also acts as an active control system to diminish chatter and avoid stick–slip problems [24], increasing the life span of the machine and further sustainability. An AAIS consists of a series of sensors measuring kinetic (forces and torques) and kinematic (displacement, velocity and acceleration) errors of the machine during its operation. A control unit, separate from the main control unit of the machine, analyses these

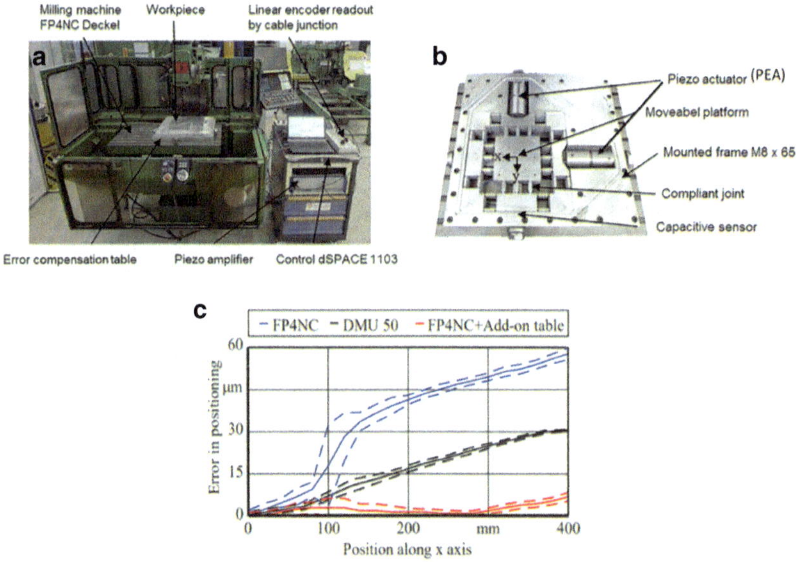

Fig. 4 **a** AAIS setup mounted on a milling machine FP4NC, FRIEDRICH DECKEL AG, München, Germany [23]. **b** Error compensation table with piezoelectric actuators in two directions (x,y) [24]. **c** Measured positioning error in the x-axis of the setup for FP4NC (old), upgraded FP4NC (FP4NC+wAAIS) and a new DMU 50 milling machine [24]

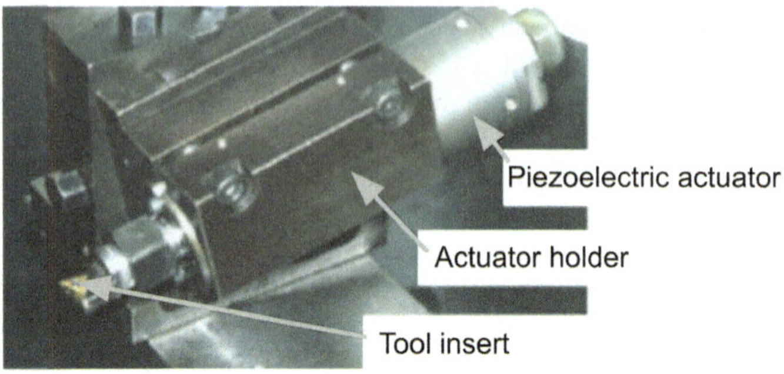

Fig. 5 Schematic of a piezo-actuated micropositioner/vibrator for VAM [29]

errors and commands a high stiffness two-dimensional piezo-actuated micropositioning system mounted on the AAIS table, as depicted in Fig. 4a. Add-ons, including piezo-actuated micropositioners, can be widely used to upgrade all different types of machine tools. This avoids replacing or retrofitting outdating machines without loss of product quality.

In addition to improved positioning accuracy and reduction of chatter, piezo-actuated micro/nanopositioning systems can enhance manufacturing sustainability with a realisation of vibration-assisted machining (VAM), as depicted in Fig. 5. In VAM, a precise low-amplitude, high-frequency vibration in the cutting direction is applied on the cutting tool [27] to create a reciprocating motion at the tip. For a specific combination of amplitude and frequency and feed velocity, the tip of the tool periodically gets detached from the workpiece, reducing machining forces, friction and heat, improving the surface finish and increasing the accuracy [27]. VAM also increases tool life span and leads to almost zero burrs [28]. For the frequencies (>5 kHz) and vibration amplitudes (<50 μm) required by VAM, piezo-actuated micropositioners attached to the tooltip or the worktable are the best candidates to generate the required vibrations.

Different Types of Piezoelectric Actuators in Micro/Nanopositioners

Piezoelectricity, the interaction of mechanical and electrical quantities in piezoelectric materials, was discovered in the 1880s for quartz by the French physicists, the Curie brothers [30]. Currently, quartz [31] and other crystals [32] are still used to produce piezoelectric materials. These materials are also produced with piezoceramics (e.g. barium titanate, $BaTio_3$), ferroelectric polycrystalline ceramics and most commonly with lead zirconate titanate (PZT) [30].

The piezoelectricity phenomenon is rooted in the asymmetrical distribution of electrons within ions of particular materials [33]. As a result of this, in piezoelectric

Fig. 6 Stacked piezoelectric actuators with the shape of a cylinder [41] (left), a rectangular prism (middle) and a ring [38] (right)

materials, application of either mechanical force or an electrical voltage results in the generation of a voltage or mechanical deformation and force. The generation of deformation or force due to an applied voltage is known as inverse piezoelectricity [34]. Piezoelectric materials intentionally produced to use inverse piezoelectricity are called piezoelectric actuators [35].

Piezoelectric actuators are the foremost actuators in nano/micropositioning due to their nanometre displacement resolution, large bandwidth, fast response and high stiffness [36]. There are three types of piezoelectric actuators commercially available as follows.

Piezoelectric Stacks

Stacks are composed of layers of a piezoelectric material [37] and are commonly coated with polymers [38]. Stacked piezoelectric actuators are stiff and easy to use for nano/micropositioning purposes [7]. Stacks with various shapes have been used including rectangular prism [39], cylinder [38] or ring [40] as shown in Fig. 6. As a rule of thumb, the maximum deflection of rectangular prism stacks is generally slightly >0.1% of their length [38]. The relationship between the applied voltage on a stack piezo actuator and its deflection has been repeatedly reported to be highly nonlinear [39].

Piezoelectric Tubes

Piezoelectric tubes are made of a metal cylinder covered by piezoelectric material internally and externally. The inner layer is continuous and grounded, as depicted in Fig. 7. The outer layer is divided into a number of (e.g. 4 [42] or 12 [43]) segments or electrodes. These actuators should be differentiated from the piezo stack with a long ring or tube shape [38]. Piezoelectric tubes can deflect in three directions. Nevertheless, they are quite fragile; thus, these actuators need more care and occasionally more complicated stands for operation [33]. Piezoelectric tubes are widely used for probe scanning microscopy [44], particularly in atomic force microscopy [45]. The first natural or resonance frequency of the tubes employed in atomic force microscopy is in the range of 200–1500 Hz [46], though relatively short piezo tubes

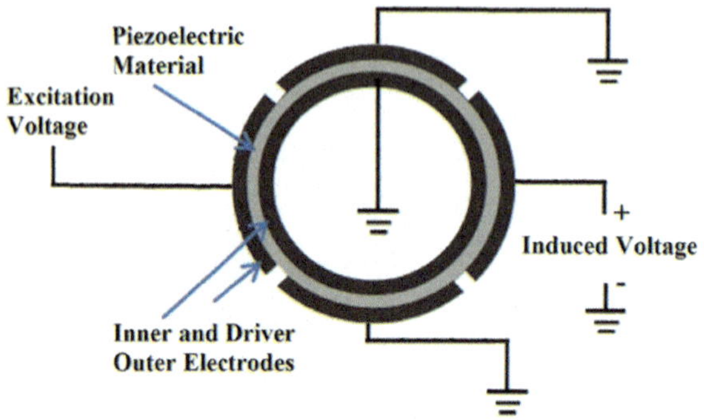

Fig. 7 Cross section and electrical connections of a piezoelectric tube actuator [48]

may exhibit first natural frequencies up to hundreds of kilohertz [47]. A large piezoelectric tube (e.g. 100 mm long and with an outer diameter of 15 mm) is likely to produce displacements of up to 100 μm. For some tube piezo actuators, the relationship between applied voltage and produced deflection has been reported to be slightly nonlinear [42] or nearly linear [44].

Piezoelectric Benders

Piezoelectric benders may be composed of one, two or multiple piezoelectric plate(s) and possibly thin layers of shim or metal [49]. Benders with rectangular [49] or circular [50] sections, depicted in Fig. 8, are commercially available. Piezoelectric benders with one (two) piezoelectric plate(s) are known as unimorph (bimorph) benders [35] and have been widely used in micro- and nanopositioning [35]. High deflection and low natural frequency, in comparison with tubes and stacks, are the two main characteristics of benders, i.e. a typical piezoelectric bender can have a maximum deflection of 2 mm [38] and the first natural frequency in the range of tens to hundreds of hertz [38].

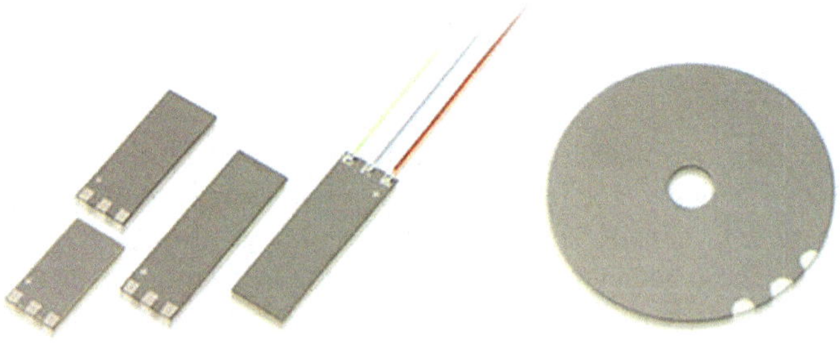

Fig. 8 Piezoelectric benders [50]

Different Types of Sensors Used in Micro/Nanopositioners

Two categories of sensors may be used in micro/nanopositioning: (1) sensors with mechanical contact and (2) contactless sensors.

Sensors with Mechanical Contact

Strain gauges and precise Linear Differential Variable Transformers (LVDTs) are suitable candidates for the measurement with mechanical contact, which are detailed in the following.

Strain Gauges

These inexpensive sensors originally measure strain; however, since the dimensions of the actuators are much larger than their deflections, they are widely used as displacement or position sensors [51]. The resolution of a precision strain gauge is approximately 1 μ (10^{-6}) [52]; thus for a piezoelectric stack with a length of 5 cm (5×10^{-2} m), a strain gauge can measure displacements as small as 50 nm ($10^{-6} \times 5 \times 10^{-2}$ m). Although the strain gauge can be bonded to the actuator surface, any misalignment between the longitudinal axis of the strain gauge with the strain direction of the piezoelectric actuator needs to be compensated. Furthermore, the strain gauges need a resistive bridge box, similar to the one in Fig. 9, and special amplifiers to provide measurable voltage [53]. In addition, dummy gauges are needed to compensate the effect of both environmental magnetic fields and thermal stress.

Linear Differential Variable Transformer (LVDT)

LVDT is one of the most reliable and robust displacement sensors. An LVDT, shown in Fig. 10a, consists of a primary coil, two secondary coils and a ferromagnetic core connected to a non-ferromagnetic stem [55]. To measure displacement, the tip of LVDT's plunger, Fig. 10b is attached to the target. The variation of the rectified output voltage of the secondary coil is proportional to the displacement of the target. LVDTs

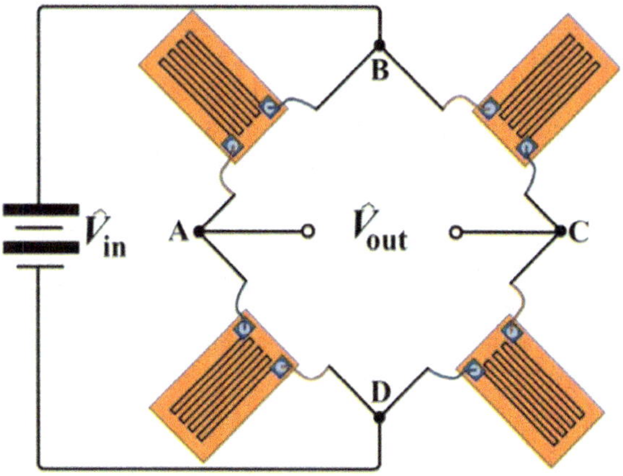

Fig. 9 A bridge strain gauge circuit [54]

Fig. 10 **a** Internal structure of LVDT. **b** Real photo of LVDT [57]

have a relatively low price (approximately US$ 1000) and a long life span; however, they are not recommended for motions with frequencies higher than 1 kHz [56].

Contactless Sensors

Contactless measurement can be conducted with optical sensors (e.g. laser triangular sensor, laser interferometer and fibre optic) and capacitive sensors, introduced in the following subsections.

Laser Triangulation Sensor

Laser triangulation sensors have a range of accuracies and resolutions. Resolutions of 1 nm at the sampling frequency of 392 kHz are achievable with a triangular laser sensor [58]. Compared to a laser interferometer, its reference distance (the required gap between the sensor tip and the target) is smaller; however, its lower cost (US$10,000) and easy initial setup make it attractive for many nano/micropositioning applications.

Laser Doppler Vibrometer

Laser Doppler Vibrometers (LDVs) are the most precise commercially available position sensors providing sub-nanometre resolution. However, they are expensive, costing at least US$ 20,000 and measuring velocity rather than position/displacement. To find the position, the velocity must be integrated which potentially leads to drift. Nevertheless, laser vibrometers offer a large reference distance and have been successfully employed in nanopositioning [59].

Fibre Optic Displacement Sensor

Fibre optic displacement/position sensors can reach resolutions as small as 8 nm [60] and have been reported to be successfully used in nanopositioning [8]. These sensors need to be calibrated before each use [61], itself a precise operation. A fibre optic sensor with its essential accessories, including a proper amplifier, costs around US$ 1000.

Capacitive Sensor

Capacitive sensors measure the capacitance between two metal surfaces and map the change of capacitance into displacement. They may provide resolutions as small as 1 nm [62]. This type of sensor is widely used in nanopositioning due to its high accuracy and reasonable price (e.g. below $2500) [42].

Control System of Micro/Nanopositioners

Figure 11 depicts a nano/micropositioning system in terms of control. The control system may receive three types of inputs:

(1) Desired position, normally as a function of time [42].
(2) Position signal from a position/displacement sensor, non-existent in some control systems.
(3) Voltage signals collected from the piezoelectric actuator, some or all of these signals are non-existent in some control systems. These signals are (i) the voltage across the piezoelectric actuator, known as the piezoelectric voltage [63] (V_P in Fig. 12), (ii) the voltage across an electrical element in series with the piezoelectric actuator, known as the sensing voltage [64] (V_S in Fig. 12) and (iii) the voltage induced on a non-excited segment or electrode of a piezoelectric tube actuator (unavailable in other types of piezo actuators), known as the induced voltage.

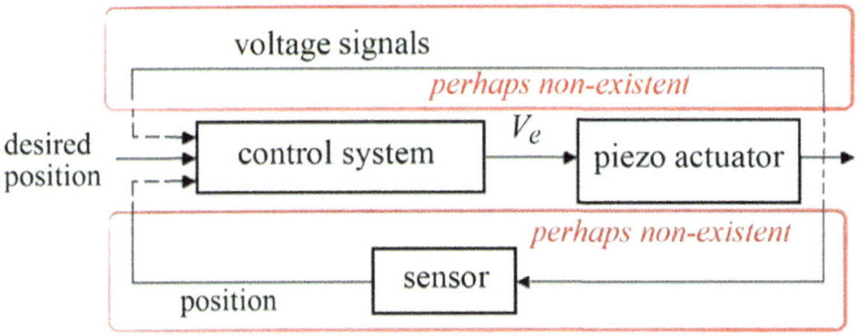

Fig. 11 A schematic of a piezo-actuated nano/micropositioning system

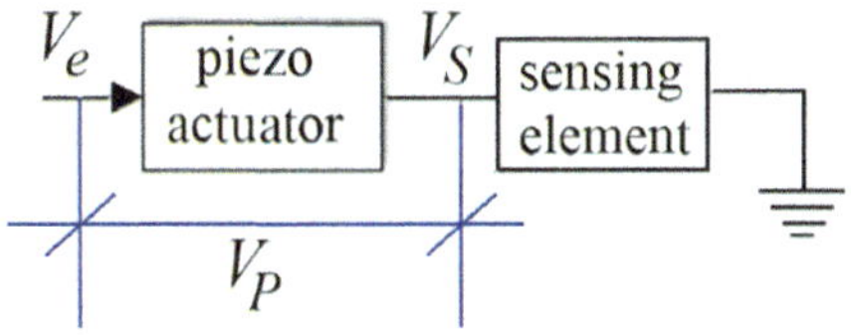

Fig. 12 Excitation, piezoelectric and sensing voltages (V_e, V_P and V_S) in a circuit. The sensing element, if exists, is a part of the control system

There are two major approaches to controlling piezo-actuated nano/micropositioners: (i) voltage-based and (ii) charge-based [7]. In the voltage-based approach (e.g. [42]), the position is directly controlled by the tuning of V_e. However, in a charge-based approach (e.g. [65]), the desired position is mapped into the desired charge, and then the charge is controlled by the tuning of V_e.

Voltage-Based Approach

Voltage-based control systems are widely used for piezo-actuated nano/micropositioners with different control architectures and controller types. Figures 13 and 14 depict two feedback voltage-based control systems with the use of a sensor (conventional feedback) and a position estimator, respectively. Nonlinear controllers [66] have been used in feedback control systems, similar to the ones of Figs. 13 and 14, due to their inherited nonlinearity. Figure 15 depicts a feedforward control system, which uses neither a measured position nor a voltage signal.

The development of position estimators for piezoelectric actuators is still a challenge; the reason is the complexity of the relationship between the voltage signals and position. Some researchers have tried to explain and model this relationship through three separate phenomena: vibration, hysteresis and creep [45]. Several physics-based [67] and data-driven [39, 42, 68–70] methods have been tried to estimate the position using one [70, 71] or a number of voltage signals [8, 65], and this area of research is still active [72]. Amongst the voltage signals, V_P has been tried most to estimate position, e.g. in [69–71, 73].

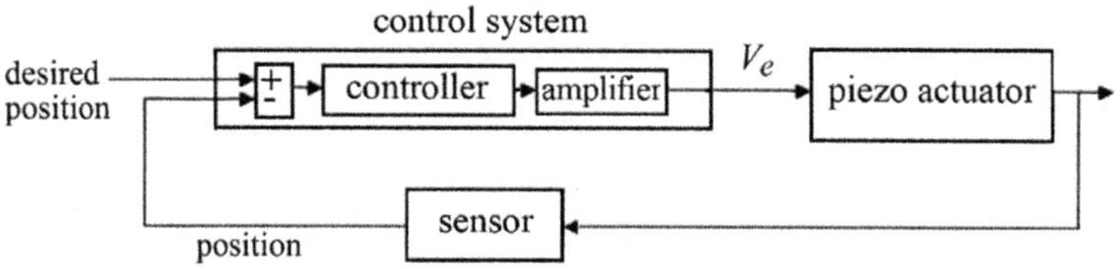

Fig. 13 A schematic of a conventional feedback voltage-based control system for a piezo-actuated nano/micropositioner

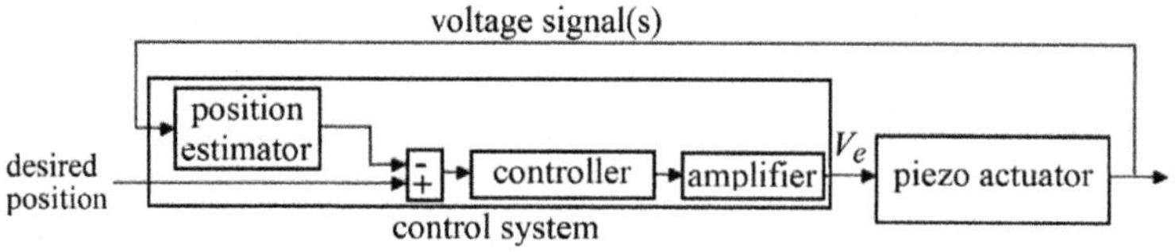

Fig. 14 A schematic of a feedback voltage-based control system for a piezo-actuated nano/micropositioner with use of estimated position

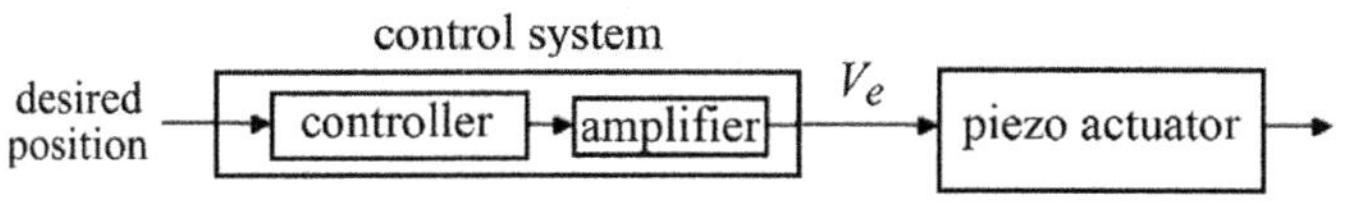

Fig. 15 A schematic of a feedforward voltage-based control system for a piezo-actuated nano/micropositioner

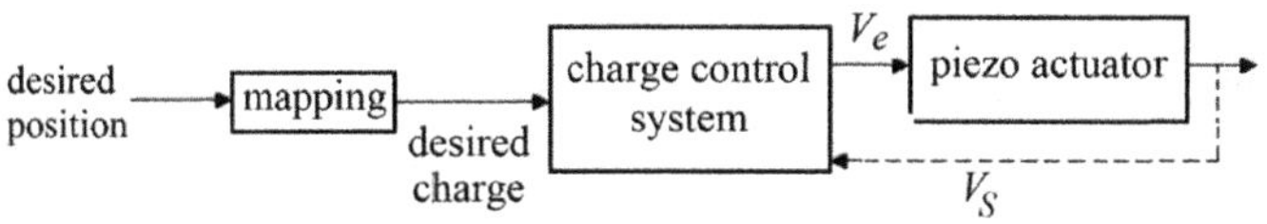

Fig. 16 A schematic of a charge-based piezo-actuated nano/micropositioning system

Charge-Based Approach

Figure 16 depicts the schematics of a charge-based control system for piezo-actuated nano/micropositioners. In this approach, position control is divided into two problems:

(i) Mapping position to piezoelectric electric charge;
(ii) Charge control by tuning V_e.

Regarding point (i), advantageously, charge and position are proportionally related in wide operating areas for many piezoelectric actuators. This simplifies charge–position mapping to identification/use of a simple gain.

Regarding point (ii), estimation and control of charge can be performed within an electrical circuit with no need to involve complexities of a voltage–position relationship, e.g. hysteresis and ... [74]. In this approach, V_S is used to estimate charge, rather than to deform the actuator. As a result, V_S is also called voltage drop.

Summary of Control Systems Section

Both voltage-based and charge-based control approaches have been successfully used for piezo-actuated nano/micropositioners. Each has its own advantages, as briefly listed in Table 1.

Charge-based control systems do not include a position sensor while retaining accuracy. This is often considered as an advantage, as position sensors are costly and

Table 1 Relative advantages of different approaches to piezo-actuated nano/micropositioners

Relative advantages of voltage-based	Relative advantages of charge-based
• No voltage drop in most cases	• Satisfactory sensorless performance
• No limits in operating area	• No need for complex nonlinear controllers or models • Implementable as an analog circuit

occasionally time-consuming to calibrate and operate, particularly in small spaces. Further, charge-based control systems, unlike voltage-based control systems, do not require complex nonlinear control to provide satisfactory performance. Such intricate control systems not only have challenging design procedures but also can only be implemented with digital processors, adding to implementation complexity. On the other hand, charge-based control systems advantageously can be implemented as analog circuits and still perform well.

However, hardware-wise, the potential of charge-based control systems for analog implementation does not necessarily mean more simplicity. In comparison with voltage-based control systems, charge-based ones need an extra sensing element e.g. the one shown in Fig.12 [75]. Moreover, if they are implemented as analog circuits, an extra voltage amplifier and, at least, an extra resistor parallel with the actuator are also needed [74]. In addition, a portion of excitation voltage is wasted in charge-based methods due to voltage drop. The most serious limitation of the charge-based approach is, however, its relatively limited operating areas. The operating areas, where the charge–position relationship for the piezo actuator is not proportional, are not considered appropriate to use charge-based methods. In addition, low-frequency operating areas (e.g. frequencies below 2 Hz in [64] and below 5 Hz in [63, 75]) do not suit charge-based methods. Voltage-based methods, on the other hand, can be designed and used for all operating areas.

Conclusions

This chapter introduced piezo-actuated micro/nanopositioners, the systems which exhibit controlled position at the micro/nanometre scale. In addition, the contribution of these systems to the biodiversity of our ecosystem and sustainability of the manufacturing sector was presented.

Piezo-actuated micro/nanopositioning, through the enhancement of cell manipulation, has created significant advancements in assisted reproduction and somatic cell nuclear transfer. These techniques are increasingly used to raise the population of threatened animal species and save them from extinction. It is believed that such technologies are essential to help species in extremis, and any future habitat restoration depends upon progress in these technologies.

In addition, piezo-actuated micro/nanopositioners are progressively employed as a core component of add-on accuracy increasing systems (AAISs). These systems, with minor changes, can considerably improve the performance of outdated machine tools. This is a substantial contribution to sustainable manufacturing, knowing that all existing alternatives lead to a huge waste of materials and energy. Without piezo-actuated AAISs, outdated machine tools are either disposed of or their major parts are replaced as otherwise they would witness a significant decrease in product quality and increase in energy consumption. Moreover, piezo-actuated micropositioners play an essential role in vibration-assisted machining, which results in a considerable decrease in consumed energy and an increase in machine life span.

The chapter also introduced the triple major components of a piezo-actuated micro/nanopositioner: actuator, control system and sensor, where the latter does not exist in sensorless micro/nanopositioners. The chapter highlights some shortcomings in this area of technology. As an instance, all methods of both prevalent control approaches, voltage-based and charge-based, suffer from significant drawbacks. Voltage-based methods still struggle to identify and control an intricate nonlinear dynamic system, and charge-based methods are unable to perform in low-frequency operating areas. A hybrid method to appropriately function over the entire operating area is yet to be developed. Due to the increasingly important role of piezo-actuated micro/nanopositioning systems in the preservation of biodiversity and in the sustainability of manufacturing, further investments in technology development of such systems is highly recommended.

References

1. Tang H, Zeng Z, Gao J, Zhang X (2015) A flexible parallel nanopositioner for large-stroke micro/nano machining. In: International conference on manipulation, manufacturing and measurement on the nanoscale (3M-NANO), pp 107–110
2. Li X, Cheah C (2015) Robotic cell manipulation using optical tweezers with unknown trapping stiffness and limited FOV. IEEE/ASME Trans Mechatron 20(4):1624–1632
3. Clayton GM, Tien S, Leang KK, Zou Q, Devasia S (2009) A review of feedforward control approaches in nanopositioning for high-speed SPM. J Dyn Syst Measur Control 131061101
4. Saedi S, Mirbagheri A, Jafari A, Farahmand F (2014) A local hybrid actuator for robotic surgery instruments. Int J Biomech Biomed Robot 3(2):100–105
5. Ghodsi M, Saleem A, Özer A, Bahadur I, Alam K, Al-Yahmadi A et al (2016) Elimination of thermal instability in precise positioning of Galfenol actuators. Behav Mech Multifunct Mater Compos 9800980008
6. Protopopov V (ed) (2014) Beam alignment and positioning techniques in practical opto-electronics. Springer, pp 309–334
7. Mohammadzaheri M, AlQallaf A (2017) Nanopositioning systems with piezoelectric actuators, current state and future perspective. Sci Adv Mater 9(7):1071–1080
8. Bazghaleh M, Grainger S, Mohammadzaheri M, Cazzolato B, Lu T-F (2013) A novel digital charge-based displacement estimator for sensorless control of a grounded-load piezoelectric tube actuator. Sens Actuat A: Phys 198:91–98, 15 August 2013
9. Roshandel N, Soleymanzadeh D, Ghafarirad H, Koupaei AS (2021) A modified sensorless position estimation approach for piezoelectric bending actuators, Mech Syst Sig Process 149107231
10. Lovejoy TE, Hannah L, Wilson EO (2019) Biodiversity and climate change: transforming the biosphere. Yale University Press
11. Supple MA, Shapiro B (2018) Conservation of biodiversity in the genomics era. Genome Biol 19(1):1–12
12. Presicce G (2020) Reproductive technologies in animals. Academic Press
13. Sansinena M (ed) (2020) Assisted reproductive biotechnologies in the horse in reproductive technologies in animals. Elsevier, pp 13–30
14. Yoshida N, Perry AC (2007) Piezo-actuated mouse intracytoplasmic sperm injection (ICSI). Nat Protoc 2(2):296–304
15. Salamone DF, Canel NG, Rodríguez MB (2017) Intracytoplasmic sperm injection in domestic and wild mammals. Reproduction 154(6):F111–F124

16. Zander-Fox D, Lam K, Pacella-Ince L, Tully C, Hamilton H, Hiraoka K et al (2021) PIEZO-ICSI increases fertilization rates compared with standard ICSI-A prospective cohort study, Reproductive BioMedicine Online
17. Iqbal A, Ping J, Ali S, Zhen G, Kang JZ, Yi PZ et al (2021) Conservation of endangered species through somatic cell nuclear transfer (SCNT). Conserv Genet Resourc 1–9
18. Czernik M, Anzalone DA, Palazzese L, Oikawa M, Loi P (2019) Somatic cell nuclear transfer: failures, successes and the challenges ahead. Int J Devel Biol 63(3–4–5):123–130
19. Singh B, Mal G, Verma V, Tiwari R, Khan MI, Mohapatra RK et al (2021) Stem cell therapies and benefaction of somatic cell nuclear transfer cloning in COVID-19 era. Stem Cell Res Ther 12(1):1–16
20. Zhou Q, Yang S, Ding C, He X, Xie Y, Hildebrandt T et al (2006) A comparative approach to somatic cell nuclear transfer in the rhesus monkey. Hum Reprod 21(10):2564–2571
21. Boiani M. Somatic cell nuclear transfer (SCNT) into mouse oocytes using Eppendorf PiezoXpert®
22. Moon J, Jung M, Roh S (2017) Comparison of developmental efficiency of murine somatic cell nuclear transfer protocol. J Embryo Transf 32(3):81–86
23. Uhlmann E, Peukert B, Thom S, Prasol L, Fürstmann P, Sammler F et al (2017) Solut Sustain Mach 139(5)
24. Kianinejad K, Thom S, Kushwaha S, Uhlmann EJPC (2016) Add-on error compensation unit as sustainable solution for outdated milling machines, 40174–178
25. Kianinejad K, Uhlmann E, Peukert B (2015) Investigation into energy efficiency of outdated cutting machine tools and identification of improvement potentials to promote sustainability. Procedia CIRP 26533–26538
26. Uhlmann E, Kianinejad K (2013) Investigation of the upgrading potentials of out-of-date cutting machine tools to promote sustainable and global value creation. In: Proceeding of the 11th global conference on sustainable manufacturing, pp 574–579
27. Singh H, Joshi R (2016) Piezoelectric transducer based devices for development of a sustainable machining system—a review
28. Joshi RS, Singh H (2011) Piezoelectric transducer based devices for development of a sustainable machining system—a review. In: 2011 international symposium on applications of ferroelectrics (ISAF/PFM) and 2011 international symposium on piezoresponse force microscopy and nanoscale phenomena in polar materials, pp 1–4
29. Zheng L, Chen W, Huo D (2020) Review of vibration devices for vibration-assisted machining. Int J Adv Manuf Technol 1081631–108165
30. Minase J, Lu T-F, Cazzolato B, Grainger S (2010) A review, supported by experimental results, of voltage, charge and capacitor insertion method for driving piezoelectric actuators. Precis Eng 34(4):692–700
31. Aggrey P, Salimon A, Korsunsky A (2020) Diatomite inspired nanostructured quartz as a piezoelectric material. Limnol Freshw Biol 828–829
32. Yang X, Li Z, Fei C, Liu Y, Li D, Hou S et al (2020) High frequency needle ultrasonic transducers based on Mn doped piezoelectric single crystal. J Alloys Comp 832154951
33. Sabek W, Al-mana A, Siddiqui AR, Assadi BE, Mohammad-khorasani M, Mohammadzaheri M, Tafreshi R (2015) Experimental investigation of piezoelectric tube actuators dynamics. In: 2nd international conference on robotics and mechatronics, Madrid, Spain, 20–21 July 2015
34. Chopra I (2002) Review of state of art of smart structures and integrated systems. AIAA J 40(11):2145–2187
35. Rios S, Fleming A (2014) Control of piezoelectric benders using a charge drive. Proc Actuat. In: 14th international conference on new actuators, Bremen, Germany, 23–25 June 2014
36. Zhang XL, Tan YH, Su MY, Xie YQ (2010) Neural networks based identification and compensation of rate-dependent hysteresis in piezoelectric actuators. Physica B-Condensed Matter 405(12):2687–2693
37. Hussain F, Khesro A, Lu Z, Wang G, Wang D (2020) Lead free multilayer piezoelectric actuators by economically new approach. Front Mater 7
38. PiezoDrive (2021) Piezoelectric actuators. https://www.piezodrive.com/actuators/

39. Ahmadpour H, Mohammadzaheri M, Emadi M, Ghods V, Mehrabi D, Tafreshi R (2015) Neural modelling of a piezoelectric actuator inspired by the presiach approach. In: International conference on artificial intelligence, energy and manufacturing engineering
40. Wang G, Zhao Z, Tan J, Cui S, Wu H (2020) A novel multifunctional piezoelectric composite device for mechatronics systems by using one single PZT ring. Smart Mater Struct 29(5):055027
41. Ceramics P (2021) Stack actuators. https://www.piceramic.com/en/products/piezoceramic-actuators/stack-actuators/
42. Mohammadzaheri M, Grainger S, Bazghaleh M (2013) A system identification approach to the characterization and control of a piezoelectric tube actuator. Smart Mater Struct 22(10):105022
43. Moheimani SOR, Yong YK (2008) Simultaneous sensing and actuation with a piezoelectric tube scanner. Rev Sci Instr 79(7)
44. Mohammadzaheri M, Grainger S, Kopaei MK, Bazghaleh M (2013) IMC-based feedforward control of a piezoelectric tube actuator. In: IEEE eighth international conference on intelligent sensors, sensor networks and information processing
45. Habibullah H (2020) 30 years of atomic force microscopy: creep, hysteresis, cross-coupling, and vibration problems of piezoelectric tube scanners. Measurement 159107776
46. Abramovitch DY, Andersson SB, Pao LY, Schitter G (2007) A tutorial on the mechanisms, dynamics, and control of atomic force microscopes. Am Control Conf 965–979
47. Moheimani SOR (2008) Invited review article: accurate and fast nanopositioning with piezoelectric tube scanners: emerging trends and future challenges. Rev Sci Instrum 79(7):071101
48. Mohammadzaheri M, Tafreshi R, Mohammad-Khorasani M, Bazghaleh M, Grainger S (2015) Evaluation of the induced voltage in driven electrodes of piezoelectric tube actuators for sensorless nanopositioning. In: IEEE 8th GCC conference and exhibition (GCCCE), pp 1–5
49. Piezo.com (2021) Introduction to Piezos. https://piezo.com/pages/introduction-to-piezos
50. Ceramics P (2021) Piezo bender actuators. https://www.piceramic.com/en/products/piezoceramic-actuators/bender-actuators/
51. Ghodsi M, Modabberifar M, Ueno T (2011) Quality factor, static and dynamic responses of miniature galfenol actuator at wide range of temperature. Int J Phys Sci 6(36):8143–8150
52. Group VP. Strain gage instrumentation, micro-measurements. http://www.vishaypg.com/docs/50002/50002.pdf
53. PiezoDrive. Strain gauge position measurement systems. https://www.piezodrive.com/actuators/
54. Akbari H, Kazerooni A (2018) Improving the coupling errors of a Maltese cross-beams type six-axis force/moment sensor using numerical shape-optimization technique. Measurement 126342–126355
55. Rerkratn A, Luangpol A, Petchmaneelumka W, Riewruja V (2020) Position signal detector for linear variable differential transformer. Energy Rep 6603–6607
56. electronicspecifier, e. https://www.electronicspecifier.com/products/communications/lvdt-range-suitable-for-high-frequency-response
57. OMEGA, O. https://www.omega.com/en-us/control-monitoring/motion-and-position/displacement-transducers/ld500/
58. Keyence K. https://www.keyence.co.uk/products/measure/laser-1d/lk-g5000/
59. Omidi E, Mahmoodi SN (2016) Vibration control of collocated smart structures using H∞ modified positive position and velocity feedback. J Vib Control 22(10):2434–2442
60. Philtec (ed) Fiber optic displacement sensors
61. Ghodsi M, Ueno T, Teshima H, Hirano H, Higuchi T, Summers E (2007) "Zero-power" positioning actuator for cryogenic environments by combining magnetostrictive bimetal and HTS. Sens Actuat, A 135(2):787–791
62. MicroSense. Precision capacitive sensor. http://www.microsense.net/products-position-sensors.htm
63. Mohammadzaheri M, Emadi M, Ghodsi M, Jamshidi E, Bahadur I, Saleem A et al (2019) A variable-resistance digital charge estimator for piezoelectric actuators: an alternative to maximise accuracy and curb voltage drop. J Intell Mater Syst Struct 30(11):1699–1705

64. Mohammadzaheri M, Emadi M, Ghodsi M, Bahadur IM, Zarog M, Saleem AJIJOAI et al. (2020) Development of a charge estimator for piezoelectric actuators: a radial basis function approach. Int J Artif Intell Mach Learn 10(1):31–44
65. Bazghaleh M, Mohammadzaheri M, Grainger S, Cazzolato B, Lu TF (2013) A new hybrid method for sensorless control of piezoelectric actuators, Sens Actuat A: Phys 19425–19430
66. Mousavi Lajimi SA, Friswell MI (2020) Design, analysis, and feedback control of a nonlinear micro-piezoelectric–electrostatic energy harvester. Nonlinear Dyn 1003029–1003042
67. Miri N, Mohammadzaheri M, Chen L (2013) A comparative study of different physics-based approaches to modelling of piezoelectric actuators. In: The IEEE/ASME international conference on advanced intelligent mechatronics
68. Mohammadzaheri M, Ziaeifar H, Bahadur I, Zarog M, Emadi M, Ghodsi M (2019) Data-driven modelling of engineering systems with small data, a comparative study of artificial intelligence techniques. In: 5th Iranian conference on signal processing and intelligent systems (ICSPIS)
69. Mohammadzaheri M, Grainger S, Bazghaleh M (2012) A comparative study on the use of black box modelling for piezoelectric actuators. Int J Adv Manuf Technol 631247–631255
70. Mohammadzaheri M, Grainger S, Bazghaleh M (2012) Fuzzy modeling of a piezoelectric actuator. Int J Precis Eng Manuf 13(5):663–670
71. Mohammadzaheri M, Grainger S, Bazghaleh M, Yaghmaee P (2012) Intelligent modeling of a piezoelectric tube actuator. In: International symposium on innovations in intelligent systems and applications (INISTA)
72. Ahmed K, Yan P (2021) Modeling and identification of rate dependent hysteresis in piezoelectric actuated nano-stage: a gray box neural network based approach. IEEE Access 965440–965448
73. Miri N, Mohammadzaheri M, Chen L, Grainger S, Bazghaleh M (2013) Physics-based modelling of a piezoelectric actuator using genetic algorithm. In: IEEE symposium on industrial electronics and applications (ISIEA)
74. Bazghaleh M, Grainger S, Mohammadzaheri MJJOIMS, Structures (2018) A review of charge methods for driving piezoelectric actuators. J Intell Mater Syst Struct 29(10):2096–22104
75. Mohammadzaheri M, AlSulti S, Ghodsi M, Bahadur I, Emadi M (2021) Assessment of capacitor-based charge estimators for piezoelectric actuators. IEEE Int Conf Mech (ICM) 2021:1–6

Eco Motion Planning for Mechatronic Systems

Paolo Boscariol, Dario Richiedei, and Alberto Trevisani

Abstract Saving energy while retaining productivity is one of the most desired features in the modern production industry, which is currently responsible for a notable amount of the worldwide energy need. Cutting down the costs of an energivorous plant does not just provide economic savings, but it also reduces the carbon footprint of commercial products. In this chapter, the wise use of energy in mechatronic systems is discussed, by focusing on Eco Motion Planning (EMP). The basic idea is that an optimized motion planning, in terms of the selection of the motion time, of the motion profile, of the spatial trajectory and the of motion scheduling, can remarkably reduce energy absorption without affecting the throughput of the mechatronic system. The chapter briefly reviews the most relevant methods that have been proposed in the literature, ranging from simple machines with constant inertia to industrial robots, and provides the most relevant equations and models of the energy consumption that can be used in EMP. A numerical test case is then analyzed to outline a method for designing an energy optimal motion profile and optimal scheduling in a robotic cell made by a system with constant inertia, and a multi-axis robot governed by a nonlinear dynamic model. Despite its conceptual simplicity, the proposed method ensures remarkable energy improvements while retaining productivity and without requiring any physical alteration of the plant and clearly proves the benefit of EMP.

Keywords Energy saving · Sustainability · Motion planning · Robotic cells · Mechatronic systems

P. Boscariol (✉) · D. Richiedei · A. Trevisani
Dipartimento di Tecnica E Gestione Dei Sistemi Industriali, Universitá Degli Studi di Padova, Stradella S. Nicola 3, 36100 Vicenza, Italy
e-mail: paolo.boscariol@unipd.it

D. Richiedei
e-mail: dario.richiedei@unipd.it

A. Trevisani
e-mail: alberto.trevisani@unipd.it

P. Hehenberger et al. (eds.), *EcoMechatronics*,
https://doi.org/10.1007/978-3-031-07555-1_15

Introduction

Improving the energy efficiency of mechatronic systems, while preserving high throughput, is one of the key goals of EcoMechatronics [1] since current trends reveal that global energy demand is expected to grow at a constant rate over the next decades [2]. The relevance of this topic is also testified by the directions set by the European Union policy [3] whose goal is reducing the primary energy consumption of 30% by 2030. A significant amount of the worldwide energy absorption is related to manufacturing processes, where, in turn, a relevant contribution is due to the energy required by automatic machines and robots adopted to perform simple activities with a small added value, such as moving tools or conveying products. A cost-free approach to improving energy efficiency of mechatronic systems is therefore the optimization of these phases by means of the optimization of the planned motion: a wide literature, that will be reviewed in the following, has shown that energy savings up to 30% can be achieved by a wise motion planning focused on the energy absorption, compared to traditional approaches that usually discard this issue. Several techniques have been developed in the literature, thus providing effective solutions for a wide range of machines and robots used in modern industries. Optimization basically involves the selection of the spatial path and of the motion law (often including the duration of motion). If at least two machines are taken into account, also the scheduling of the operation can boost energy efficiency [4]. Indeed, besides optimizing the energy consumption of each axis, a wise scheduling of motion can optimize the energy exchange between accelerating and decelerating axes when regenerative drives are used. The enhancement of the planned motion represents in many cases the most feasible solution for energy saving since it can be applied both when commissioning a new system as well as when retrofitting an existing one, with no physical modifications of the electro-mechanical devices.

The discussion provided in this chapter covers some general issues and their solutions, focusing on methods that are suitable for both researchers and industry practitioners. The first goal of the chapter is to provide a general overview of the methodologies used in modeling and analysis of the energy consumption of mechatronic systems, together with their use for reducing energy absorption. These two issues are the pillars on which the idea of Eco Motion Planning (EMP) is rooted; optimization of the spatial path, of the timing law and of the scheduling are the tools exploited by EMP for improving energy efficiency. Then, the chapter provides and discusses some models of energy absorption, and their use in EMP is proposed through two meaningful examples. The first example consists of the common case of a constant inertia system, as those often adopted in industry, where EMP can be solved through some analytical equations. Then, by considering a multi-axis system with nonlinear dynamic behavior, a different solution technique is proposed by exploiting numerical methods for the solution of the resulting multivariate optimization problem.

Literature Review

Motion design, in its most general definition, is a branch of mechatronics and robotics that deals with the definition of the reference profiles to be executed by an automatic machine. This general problem applies to basic systems as well as to very complex ones, such as robotic cells with several robots and some auxiliary axes. This variety of applications has produced a literature on motion planning that counts for, literally, hundreds of related works. Reviewing all of them is outside the scope of this work: here just a brief overview of the most common applications of EMP is provided. For a general overview of trajectory planning, the reader should refer to the classic titles, such as [5–8]. A detailed list of papers on the problem of boosting energy efficiency by means of motion design is provided also in [9] that quotes also other methods not strictly related to motion design, which are therefore outside the scope of this chapter.

The most investigated problem in EMP is the rest-to-rest motion of a single degree-of-freedom (DOF) system with constant inertia. A meaningful example of a commonly used method can be found in [10], which provides an analytic treatment of the energy consumption and of the optimal motion time of a single servo-driven axis with constant inertia when executing standard motion primitives. This work and several related ones share a similar outline: the energy consumption is found by the time integration of the algebraic equations representing the electric power profile, computed through the system electro-mechanical model that includes the inverse dynamic model of the mechanical part and a simplified model of the motor. The energy consumption therefore takes the form of a function of some parameters, allowing for a simple optimization either through analytical equations, as proposed in [10] and in the more recent paper [11], or through some basic numerical methods based on computing the stationary points of simple cost functions, as proposed in other works, such as [12–14]. Among the analytical solutions of the energy optimal motion planning, the one proposed in [10] results in very simple equations, since the features of the different motion primitives are summarized through a few coefficients that allow for a straightforward calculation and comparison. Simple analytical formulations of the energy consumption can be formulated for a Cartesian robot as well, due to the decoupling between the axis [15].

Some works have followed different approaches: for example, [16–18] solve energy-optimal point-to-point motion within the framework of variational calculus. Hence, the energy minimization problem is approached as an indirect optimization or as an optimal control problem, which usually requires elaborate solution methods [19]. Within a variational problem formulation, the motion profile is generally not constrained to a specific parametrized description of the motion profile (for example, by means of a polynomial function), but it is free to assume any form within the feasible bounds: hence, among the infinite possibilities, a variational problem formulation computes the optimal motion law as the optimal sequence of positions (and consequently speeds and accelerations) that minimizes a given cost function. Variational solutions are virtually effective in finding the optimal solution; however, they might fail to converge, and handling nonlinearities and hard constraints is usually very cumbersome [20]. Additionally, they are hardly appealing for industry practitioners since they require specific solvers.

If mechatronic systems with variable inertia are considered, the use of closed-form solutions is usually impracticable and numerical computation routines should be adopted: the complexity of the nonlinear equations of motions of even a single-DOF system is generally too high to be handled by simple calculations. For example, in [21, 22] numerical methods are adopted to solve the algebraic equations and then to minimize the energy consumption of a toggle mechanism. Difficulties are exacerbated in the case of multi-DOF systems, such as robots. A general overview of the problem is provided in [23], which also sports a well-supported analysis of the various sources of dissipation in industrial robots. A paper to be mentioned is [24], which proposes a commonly used and technically sound approach for EMP in robots: the optimization of a robot task as the result of an optimization procedure. The task is first designed by choosing a set of via-points that ensure a collision-free operation, then an interpolation procedure is used to define the energy-optimal motion, using a meta-heuristic optimization tool. The work [25] deals with the planning of a task that involves at least three via-points, which are to be connected by straight lines and arcs, i.e. by exploiting the most common motion primitives used in industrial robots. The optimization is performed by searching for alternative paths that pass close to the prescribed via-points without affecting the cycle time. A similar approach is proposed in [26] as well, where different motion primitives, such as B-splines, are used to interpolate via-points. In [27, 28], functionally redundant robots are discussed, and the motion of the redundant degrees of freedom is optimally designed to reduce the energy required to perform the task, by exactly passing through some prescribed via-points. Numerical optimization routines are adopted due to the nonlinear dynamic model of the robot.

To handle the complexity of robotic systems, some authors have suggested the use of the so-called "CAD-based simulators", such as in [29, 30], to exploit the capabilities of the multibody dynamics simulators embedded in the most advanced CAD packages that allow running simulations of the dynamics of the device under development without having to explicitly define a dynamic model. These approaches are therefore based on the solution of the differential equations governing the studied systems, by means of repeated simulations to optimize the motion design. These approaches are often denoted as "direct" approaches [10], since optimization often relies on trial-and-error procedures or enumeration and comparison of different solutions. CAD-based simulations are the key elements of the optimization procedures developed in other works as well, among which [31, 32] are worthy of mention, where simulations are exploited to predict and hence optimize energy consumption.

Model-free approaches have been proposed as well, to overcome the difficulties in developing reliable models of energy consumption and to correctly identify their key parameters. One example of such an idea is found in [33], in which some ways to estimate the energy consumption of a robot without having to explicitly develop a dynamic model of the robot, for example by using identification methods applied to simulated data, are suggested. A different technique is suggested in [34], where the data recorded by executing some trajectories with an industrial robot are used to develop a black-box model to be used as the starting point for trajectory optimization. Another model-free approach has been developed in [35, 36]: the method relies on the use of the native robot software and on the optimization of alternative cost functions, which can be based, for example, on pseudo-power (i.e. speed to acceleration product), jerk, or weighted joint accelerations. The authors mention a reduction up to 30% of the overall energy consumption for a given task without altering the execution time or the path.

Finally, a different approach to improve the energetic efficiency of a robot without the explicit use of a dynamic model is the one based on the use of some performance indexes. One example is found in [37] that analyzes the relationship between the motion profile and the location of the task to be executed within the workspace, by taking into consideration a 4-DOF parallel robot. The outcome of this work is the development of energy maps that can guide the practitioners to position, whenever possible, the task in an energy-optimal location. In a related work [38], the approach is generalized by defining two energy and trajectory-related indexes that can be helpful not only during the robot programming phase but also in a previous stage, for example, during the design of the cell or the robot. In this case, the definition of an energetically sound trajectory is achieved by the minimization of a performance index based on inertia ellipsoids, and thus the numerical integration of the inverse dynamics electro-mechanical model of the robot is not needed.

This brief overview of motion-related energy optimization methods has highlighted a rather wide array of solutions and applications: the next section will guide the reader through an example of motion optimization that follows one of the most popular approaches, exploiting nonlinear optimization as the tool to define an energy-optimal motion profile for a robotic cell.

A Paradigmatic Test Case: A Robotic Cell

The aim of this section is to discuss two techniques for EMP that recall the common approaches adopted in the quoted literature, to analyze and optimize the energy efficiency of a robotic cell composed of a SCARA robot and a linear unit. The robotic cell involves two mechatronic systems with different features that are widely adopted in manufacturing plants: the linear unit is a constant inertia single-DOF mechanism, while the robot is a multi-axis system with variable inertia. Discussing this paradigmatic example allows explaining two different approaches to EMP, by clarifying some issues discussed in the Introduction.

The task taken into consideration is split into two main phases to highlight different features: in the first phase the product is conveyed by the linear unit, and the end-effector of the robot is moved from the pallet to the location where the product can be picked up. In the second phase, the product is displaced by the robot, while the linear unit returns to the starting position with no load. The two devices are mechanically decoupled and therefore they can be modeled and analyzed independently from

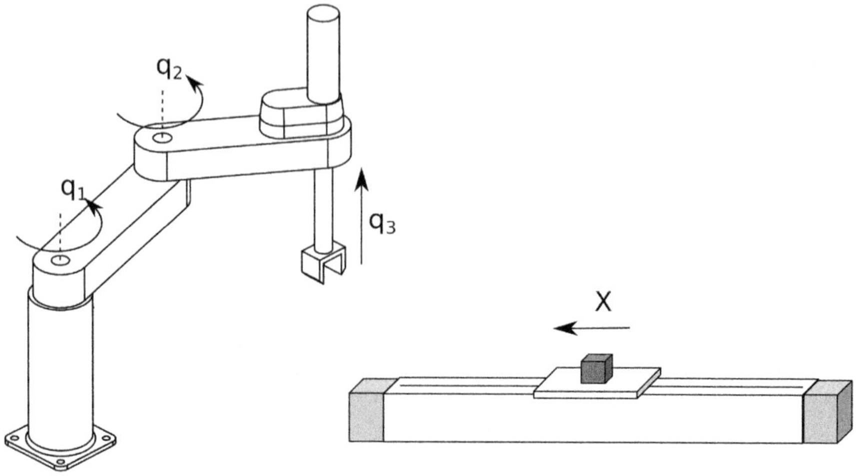

Fig. 1 Layout of the robotic cell: SCARA robot and linear unit

the other. Additionally, it is assumed that no energy regeneration between them is allowed: the energy absorption models are therefore decoupled as well. The system is represented in Fig. 1.

It is assumed that the whole operation has to be optimized for maximum energy efficiency while retaining a fixed total execution time. As an example, it has been assumed that 10 pieces are to be processed in one minute, leading to a cycle time equal to 6 s. The available degrees of freedom are the relative duration of the two phases, and some parameters representing the motion law in time of the linear unit and of the robot during each phase. Some resting phases, if beneficial to reduce the energy consumption, can be included in scheduling the operations.

Energy Optimization

In this section, the energy optimization of the system is developed by exploiting the electro-mechanical models of the two systems. The inverse dynamics, an algebraic model representing the energy consumption of the linear unit, can be integrated both numerically and analytically, while the robot dynamics should be integrated numerically due to the presence of nonlinearities.

Table 1 Linear unit: relevant modeling parameters

Parameter	Value
Pulley radius R_p	0.02626 m
Carriage weight m_c	7 kg
Gear ratio i	5
Gearbox inertia J_r	27×10^{-6} kgm^2
Equivalent inertia J_{eq}	3.8784×10^{-4} kgm^2
Equivalent static friction torque T_{eq}	0.434 Nm
Motor moment of inertia $J_{m,LU}$	1.45×10^{-4} kgm^2
Motor viscous friction coefficient $k_{v,LU}$	1.24×10^{-4} Nms/rad
Motor winding resistance R_{LU}	2.51 Ω
Motor torque constant $k_{t,LU}$	0.74 Nm/A
Motor back-emf constant $K_{b,LU}$	0.45 Nms/rad
Maximum speed v_{max}	3 m/s
Maximum acceleration a_{max}	50 m/s^2

Linear Unit

The linear unit is composed of a belt-driven carriage, supported and guided by ball rails, and its actuation is provided by a brushless motor coupled to the pulley through a gearbox. The main parameters of the linear unit are provided in Table 1, which are extracted from the manufacturer's datasheet available in [39].

The dynamics of the linear unit can be described by the following equation:

$$T_m(t) = J_{LU}\ddot{\vartheta}_m(t) + k_{v,LU}\dot{\vartheta}_m(t) + T_{eq} \tag{1}$$

in which J_{LU} accounts for the reflected moment of inertia of the motor, of the gearbox, of the pulleys and of the mass of the carriage that conveys the product to be displaced. The equivalent static friction torque T_{eq} accounts for the contributions of the motor, of the gearbox and of the linear unit as well. Equation (1) can be used for the evaluation of the instantaneous torque required to track the desired speed and acceleration profiles. Hence, the torque profile $T_m(t)$ is known once the reference motion of the linear unit is defined.

The electric power consumption associated with a point-to-point task can be evaluated by modeling the voltage drop across the motor. According to the common approach adopted that the papers quoted in the introduction, the electro-mechanical model of the motor, are defined by means of its DC-motor equivalent, i.e. by using the well-established Park's direct-quadrature-zero transformation [40]:

$$V(t) = RI(t) + k_b\dot{\vartheta}_m(t) + L\frac{dI(t)}{dt} \tag{2}$$

where the current absorbed by the motor is proportional to the exerted torque T_m:

$$I(t) = \frac{T_m(t)}{k_t} \tag{3}$$

k_t is the motor torque constant, R is its equivalent winding resistance, k_b is the back-emf constant, and L is the motor winding inductance; all these parameters are usually listed in the motor datasheet. The electric power absorption of the motor that drives the linear unit is defined by the voltage-to-current product:

$$W = I(t)V(t) \tag{4}$$

and the energy consumed over the time frame $[0, \tau]$ is the integral of W, as in:

$$E_{LU} = \int_0^{\tau} W(t)dt \tag{5}$$

It should be pointed out that only the positive values of $W(t)$ are to be integrated in Eq. (5) in the proposed example as it is assumed that the motor drive is not regenerative. This model suffices to properly estimate the energy associated with one work cycle, as proved by the literature (see, e.g. [21, 41]). As discussed in the introduction, EMP relies on the analytic computation of the energy, once the motion primitive is chosen or properly written as a function of some meaningful parameters (see, e.g. [10]). In this work, as an example of a widely adopted motion profile, the so-called "S-shaped" motion profile is assumed. This motion law is often used as a smoother alternative to the basic trapezoidal speed profile [7] since it provides continuity up to to the second derivative. It is conveniently described by splitting it into three phases (acceleration, constant speed, deceleration), leading to the following speed profile:

$$\begin{aligned}
\dot{\vartheta}_m(t) &= \frac{v_{max}}{2}\left(1 - \cos(\omega_1 t)\right) & t &\in [0, \tau_a) \\
\dot{\vartheta}_m(t) &= v_{max} & t &\in [\tau_a, \tau_f^{LU} - \tau_a) \\
\dot{\vartheta}_m(t) &= \frac{v_{max}}{2}\left(1 + \cos(\omega_1 t)\right) & t &\in [\tau_f^{LU} - \tau_a, \tau_f^{LU}]
\end{aligned} \tag{6}$$

This motion profile is usually parametrized, in its symmetric form, by the total motion duration τ_f^{LU}, the acceleration (and deceleration) phase duration, τ_a, and by the overall displacement, h. The maximum speed is computed to ensure the proper displacement h, according to the simple relation:

$$v_{max} = \frac{h}{\tau_f^{LU} - \tau_a} \tag{7}$$

while continuity of acceleration is enforced by

$$\omega_1 = \frac{\pi}{\tau_a} \tag{8}$$

This motion profile can also be conveniently formulated through the dimensionless ratio $\lambda = \tau_a / \tau_f^{LU}$, that is, usually assumed as a tuning parameter to trade off between the conflicting requirements of reducing speed and acceleration (and hence torque) [7]. In this work, such a parameter is adopted to optimize the energy consumption, while complying with the specifications of the cycle of the robotic cell. The absorbed energy can be conveniently written as a function of λ and analytic integration of the algebraic equations can be performed, leading to a solution to the following optimization problem that represents EMP for the linear unit, for each of the two phases of the working cycle:

$$\begin{aligned} &\min_{[\lambda]} E_{LU}(\lambda, \tau_f^{LU}) \\ &\text{with: } \tau_f^{LU} \text{ fixed;} \\ &\text{subject to: } \lambda_{min} \leq \lambda \leq \lambda_{max} \end{aligned} \tag{9}$$

The minimum and maximum values of λ are set to ensure the speed (v_{lim}) and acceleration (a_{lim}) limits:

$$\lambda_{min} = \frac{\tau_f^{LU} a_{lim} - \sqrt{a_{lim}\left(2\pi h - a_{lim}\tau_f^{LU^2}\right)}}{2\tau_f^{LU} a_{lim}} \tag{10}$$

$$\lambda_{max} = \min\left\{1 - \frac{h}{\tau_f^{LU} v_{lim}}, 0.5\right\} \tag{11}$$

Other limits could be easily included, such as those stated in [10]. Equation (9) sets a constrained nonlinear optimization problem, which is of fast and straightforward solution using standard numerical optimization routines.

The results of the solution of the optimization problem in Eq. (9) for values of τ_f^{LU} up to 6 s (that is the overall cycle time) are plotted in Fig. 2. Each point of the graph refers to an optimal solution. Figure 2 shows, in the upper plot, the minimum energy required by the linear unit to perform the first phase of the motion to convey the product toward the robot. The total energy is plotted versus the total motion duration: each sample of the energy is evaluated by optimizing the energy consumption for a given value of τ_f^{LU}, and then for each τ_f^{LU} the ratio λ is optimized. The optimal values of λ are reported in the bottom plot of the same figure. τ_f^{LU} is lower limited by the constraints imposed by the maximum allowed speed and torque (and hence acceleration) due to the characteristic curve of the motor.

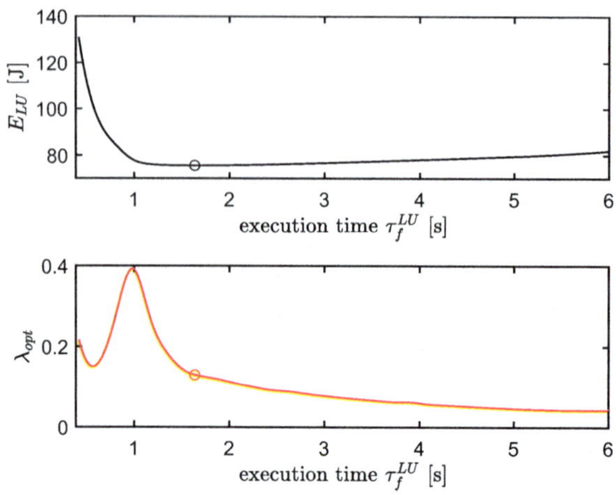

Fig. 2 Linear unit: power consumption E_{LU} and optimal λ value versus execution time τ_f^{LU} of phase I

The first plot in Fig. 2 clearly shows the dependence between total execution time and energy consumption, and reveals that the energy versus time curve can be ideally split into two regions. For execution times smaller than the optimal value, highlighted by the two circles, energy increases steeply as τ_f^{LU} is decreased, while for values of τ_f^{LU} larger than the optimal value, the energy requirement increases almost linearly. This is due to the different balance between two effects [10], i.e. the effect of the terms depending on speed and accelerations, and the effect of the motion-independent terms (such as those related to constant forces). Since the latter are quite pronounced in the test case under consideration, due to high static friction forces T_{eq} in (1), increasing too much the motion time is not beneficial in terms of energy consumption.

These results refer to the first phase of motion, and hence they include the payload $m_L = 3$ kg. The results of the application of the same method to the second phase of motion, when the linear unit does not carry any payload and is returned to the starting position, are very similar to the ones already displayed (and therefore are not shown): the contribution of the payload m_L is scarcely relevant to the dynamics of the linear unit due to the high reduction ratio implemented by the belt drive and the gearbox.

The Eco Motion Planning solution for the linear unit is therefore found by completing the first phase of motion in 1.64 s, while the second one is to be executed in 1.59 s, leading to an energy consumption over a cycle equal to 150.9 J. Both the optimal motion times are consistent with the total execution time which has been set in this example to be equal to 6 s. Two rest times are therefore required to ensure the imposed cycle duration.

Table 2 Electric and mechanical parameters of the SCARA robot

Parameter	Joint 1	Joint 2	Joint 3
Link length	0.45 m	0.35 m	–
Link mass	14 kg	18 kg	2 kg
Gear ratio	1/30	1/30	1/30
Motor inertia	$1 \cdot 10^{-4}$ kgm^2	$1 \cdot 10^{-4}$ kgm^2	$1 \cdot 10^{-4}$ kgm^2
Viscous friction coefficient	0.001 Nms/rad	0.001 Nms/rad	0.001 Nms/rad
Coulomb friction force	$2 \cdot 10^{-2}$ Nm	$2 \cdot 10^{-2}$ Nm	$2 \cdot 10^{-2}$ Nm
Motor winding resistance	3 Ω	3 Ω	3 Ω
Motor back-emf constant	0.375 Vs/rad	0.375 Vs/rad	0.375 Vs/rad
Motor torque constant	0.6 Nm/A	0.6 Nm/A	0.6 Nm/A
Peak motor torque	2.5 N	2.5 N	2.5 N
Peak motor power	75 W	75 W	75 W

SCARA Robot

The classical SCARA robot with three DOFs is assumed, which is commonly used in pick and place operations given its high-speed capabilities and for its good price/performance ratio. The robot is actuated by three brushless motors that are assumed to be controlled by non-regenerative drives. The main parameters used to describe its dynamics are reported in Table 2.

The aim of the EMP is to define an energy optimal motion profile for the robot when executing a pick and place task, from the prescribed initial position to a final one, and then back again to the initial one. In accordance with the payload of the linear unit already discussed, the robot payloads are equal to 0 and 3 kg in the first and second phases, respectively. The same via-points are assumed in the two phases (obviously with reverse order); different motion durations in moving from one point to the next one are instead adopted, since these intervals are exploited to optimize the energy consumption.

The evaluation of the energy consumption starts from the robot inverse dynamic model that allows estimating the torque required to the robot actuators, $\mathbf{T}_j$, as a function of joint reference position, speed and acceleration, which are represented by $\mathbf{q}$ and its time derivatives:

$$\mathbf{M}(\mathbf{q})\ddot{\mathbf{q}}(t) + \mathbf{C}(\mathbf{q}, \dot{\mathbf{q}}) + \mathbf{g}(\mathbf{q}) + \mathbf{f}_v\dot{\mathbf{q}}(t) + \mathbf{F}_c sign(\dot{\mathbf{q}}(t)) = \mathbf{B}\mathbf{T}_j(t) \tag{12}$$

$\mathbf{M}$ is the mass matrix, while $\mathbf{C}$ and $\mathbf{g}$ account for, respectively, the Coriolis acceleration terms and the gravity effects. Friction contributions are split into a speed-dependent viscous term proportional to $\mathbf{f}_v$ and a constant term $\mathbf{F}_c$. Finally, $\mathbf{B}$ is the force distribution matrix. Once the torque to be provided by each motor is computed, the evaluation of energy consumption relies on the electro-mechanical model in Eqs. (2–4), by summing up the energy consumed by the three motors.

The goal of EMP is to minimize the energy consumption E_{robot} associated with the execution of the task, by solving a constrained nonlinear optimization problem. Solving this optimization problem becomes a rather simple task if a proper parametrization of the motion law, and hence of the cost function and of the bounds is defined. Following a common approach, [42–44] trajectory design can be defined through the solution of an interpolation problem. This problem is set up by defining a sequence of N via-points to be visited in sequence by the robot, either defined in the joint space or in the operative space, which are then interpolated according to some trajectory primitives. The choice of the trajectory primitive is a topic which has drawn much attention in the last 30 years [45]; this issue is however outside the scope of this chapter. A standard method, such as the "4-3-4 motion profile", is here used just to provide an example [42]. Without going into details, for which the reader can refer to [28], the motion of the end-effector of the robot is defined by a sequence of polynomial functions: the result is a coordinated motion of the robot that ensures exact interpolation of a set of pre-defined via-points with continuity up to acceleration.

In this case, the energy consumed by the robot, E_{robot}, can be parametrized through the duration of each of the $N-1$ segments that connect the N via-points, τ_i. EMP for each phase is translated into the following optimization problem that is parametric with respect of the total execution time τ_f^R of each phase:

$$\begin{aligned} &\min_{[\boldsymbol{\tau}]} E_{robot} \\ &\text{with: } \sum_{i=1}^{N-1} \tau_i = \tau_f^R \\ &\text{bounded } |\dot{\mathbf{q}}|, |\ddot{\mathbf{q}}|, |\dddot{\mathbf{q}}|; \\ &\text{bounded } |W_{e,j}|, |T_j|, \text{ with } j = 1, 2, 3; \end{aligned} \tag{13}$$

The design variables, i.e. the problem unknowns, are collected in the vector of motion durations $\boldsymbol{\tau}$, i.e. the motion times required for moving from one via-point to another. As the pick and place motion comprises 6 via-points, $\boldsymbol{\tau}$ has 5 entries. The total execution time τ_f^R of each phase is assumed to be fixed, and therefore the optimization problem is parametric with respect to such a variable in both the phases. Additionally, hard bounds are set on the kinematic variables (speed, acceleration), on the motor power absorption $W_{e,k}$ and on motor torques T_j. Constraints on the jerk are set as well, to ensure smooth motion.

Figure 3 separately shows the optimized energy consumption of phases I and II, computed through the solutions of the problem in Eq. (13), by assuming for both the phases the feasible durations. Such a plot reveals that feasibility is ensured for motion durations greater than 1.6 s and 2.6 s for the two phases, respectively, since

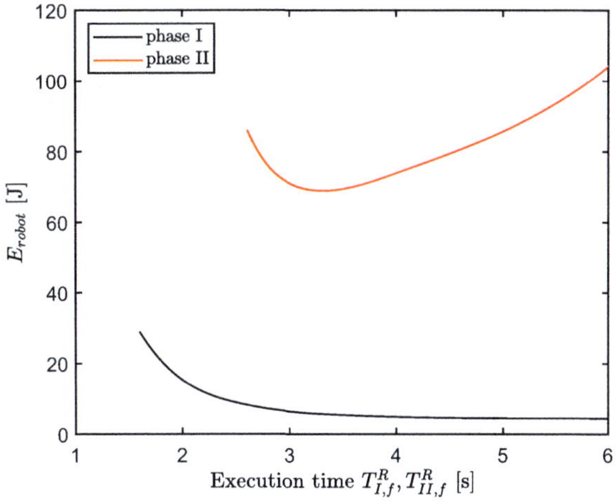

Fig. 3 Optimized energy consumption of the robot versus total execution time: phase I and phase II

faster execution times are incompatible with the operative limits of the robot. Such a Fig. 3 reveals that the increased payload in phase II leads to a convex shape of the energy consumption within the time domain of interest, due to relevant constant loads, such as the ones related to the gravity force. This feature is consistent with the observations provided (and proved analytically with reference to a single-DOF system with constant inertia) in the work [10]. In contrast, the reduction of the energy consumption in phase I is achieved if the motion time is as long as possible. These results suggest that no rest phase should be assumed during the cycle lasting 6 s, and therefore the sum of the durations of the two phases will be 6 s, $\tau^R_{I,f} + \tau^R_{II,f} = 6$ s.

The total energy required by the robot to execute the whole task is plotted in Fig. 4 as a function of the durations of the two phases $\tau^R_{I,f}$ and $\tau^R_{II,f}$ (with the constraint of

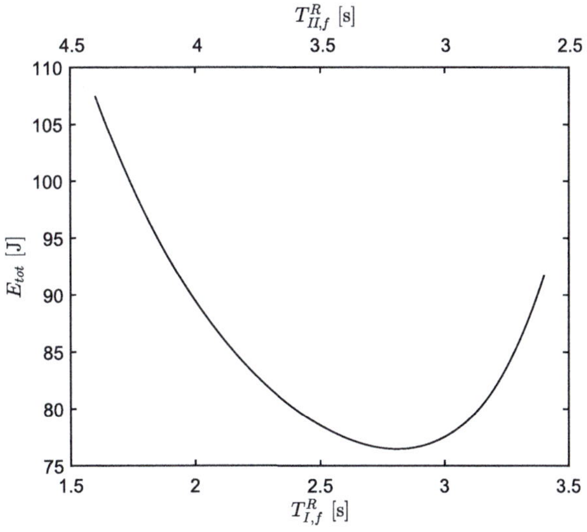

Fig. 4 Robot energy consumption for the whole task versus phase I duration $\tau^R_{I,f}$ and phase I duration $\tau^R_{II,f}$

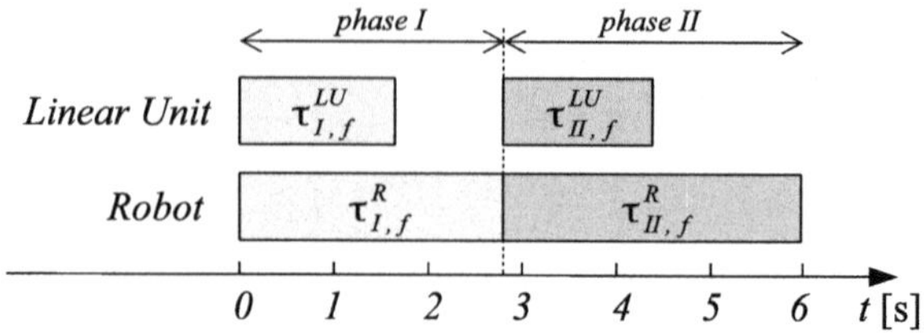

Fig. 5 Eco motion planning solution: scheduling of the motion phases

a cycle time equal to 6 s). The minimum is achieved by splitting the total duration of the task, into a first phase that lasts $\tau^R_{I,f} = 2.81$ s and a second one that lasts $\tau^R_{II,f} = 3.19$ s. This choice sets the overall robot energy consumption to $E_{robot} = 76.4$ J that is the optimal value for the test case under investigation.

Analysis of the Cell

In the light of the optimized motion profiles obtained for both the robot and the linear unit, the sequence of motions of the two cell components during the two phases of motion is optimally scheduled as shown in Fig. 5, which highlights the time frames in which the motion of the linear and of the robot happen, as well as the two rest phases of the linear unit.

Analysis of the Robot Joint Motion

The optimized motion solution is compared with a benchmark motion profile (hereafter denoted also as the "non-optimized" one), obtained by using the standard "chord length distribution" to choose the distribution of the time intervals between two consecutive via-points. As for the distribution between phases I and II, assuming equal motion durations (3 s) for both the phases is not allowed, since the maximum motor torque exerted by the robot would violate the constraints set in Eq. (13) during phase II. The duration of phase II has to be increased up to 4.3 s to obtain a feasible solution, thus the duration of the first one is to be reduced to just 1.7 s. The motion of the linear unit follows the same time frames, and it is not optimized as λ is set equal to 0.25 in both cases, as often done by practitioners, and no rest is assumed. The resulting energy consumption is 75.6 J and 78.6 J for the linear unit, 33.9 J and 78.2 J for the robot, by referring to the first and second phases, respectively. The overall energy consumption over a cycle is therefore equal to 267.3 J. Comparing this result with the optimized motion planning and scheduling reveals that an energy reduction equal to 14.8% is achieved, while ensuring the same productivity of the robotic cell. This is clear evidence of the benefits of using EMP in designing and scheduling motions

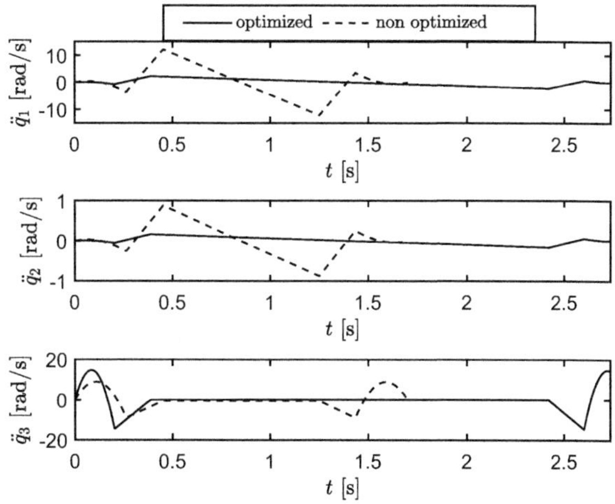

Fig. 6 Robot joint acceleration profiles, phase I: optimized versus non-optimized solution

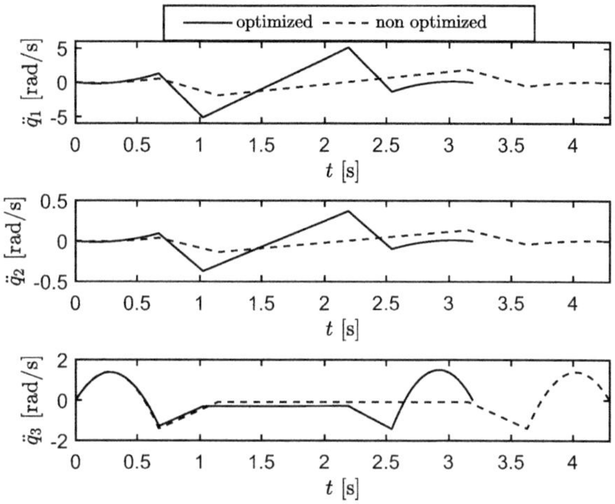

Fig. 7 Robot joint acceleration profiles, phase II: optimized versus non-optimized solution

in a robotic cell. This saving is expected to be even greater on a larger scale system, such as when focusing on a whole production line or even a whole plant.

The obtained motion profiles are shown and compared in figures from Figs. 6, 7, 8 and 9, which show the acceleration profiles of the robot and compare the optimized with the non-optimized solutions. The relationship between joint accelerations and energy consumption is non-obvious, suggesting that solutions based on heuristic approaches that discard the model of the energy consumption are usually hardly capable of leading to an energy-optimal solution.

The instantaneous values of the electric power drawn by each robot actuator are plotted as well in Figs. 8 and 9, which separately refer to each phase. Figure 8 shows that the non-optimized motion is characterized by a pronounced power absorption as a result of its short duration, especially for the first two joints. Looking at Fig. 9, which refers to the second phase, it can be seen that, due to the longer execution

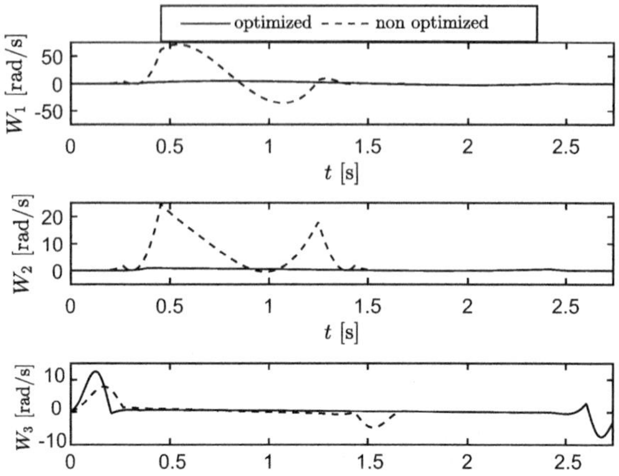

Fig. 8 Robot power absorption, phase I: optimized versus non-optimized solution

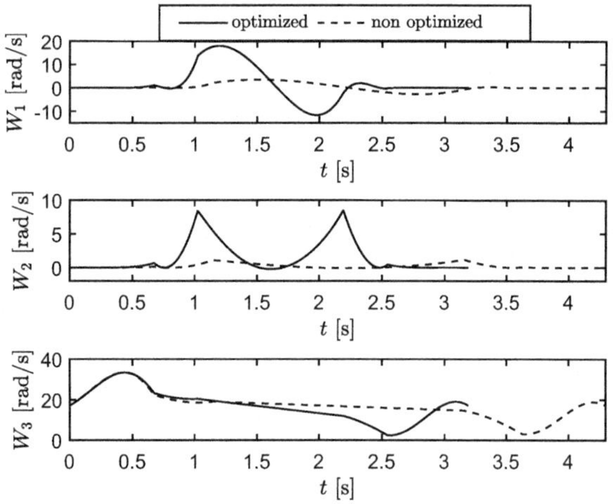

Fig. 9 Robot power absorption, phase II: optimized versus non-optimized solution

time of the non-optimized motion, the power consumption is lower for joints 1 and 2 in comparison with the optimized solution. However, this improvement is made irrelevant by the far worse energetic performance shown by the last joint. As the last joint carries a relevant payload during the last phase, with a relevant effect of gravity force, its duration should be, ideally, shorter.

Conclusion

This chapter proposes, first of all, a literature review of the main methods used to improve energy efficiency by means of motion design, the so-called Eco Motion Planning (EMP) that is one effective tool of EcoMechatronics. To allows the reader to appreciate the benefits and the use of the techniques of EMP, a paradigmatic example is proposed and solved, together with the main models and equations to be adopted. The test case consists of a simple robotic cell, like those often adopted in manufacturing plants, that comprises a belt-driven linear unit, i.e. a single-DOF system with constant inertia, and a SCARA robot, as a representative example of a nonlinear multi-DOFs system. For both the systems involved in the robotic cell under study, an electro-mechanical model is developed and then used to reduce the overall energy consumption when executing two sequential tasks by optimizing some parameters of the motion profiles that are some pillars of EMP: the motion time, the shape of the timing law and the motion scheduling. In particular, the application of the proposed method, that summarizes the features of some common state-of-the-art techniques, allows defining the best distribution of the duration of each phase of motion and the best motion profile for each device. The outcome of the optimization procedure shows that significant improvements in the energy consumption can be obtained, which is found equal to 14.8% in comparison to a standard solution. This improvement is relevant as it can be obtained without any physical alteration of the setup or any modification of the cycle times. The method here discussed, as well as those proposed in the quoted literature, clearly shows how EMP is an essential step toward the development of sustainable mechatronic systems.

References

1. European Parliament and Council (2009) Directive 2009/125/EC of the European Parliament and of the Council of 21 october 2009, establishing a framework for the setting of ecodesign requirements for energy related products (recast). Official J Europ Commun
2. US Energy Information Administration (2019) International energy outlook 2019 with projections to 2050
3. Helm D (2014) The european framework for energy and climate policies. Energy Policy 64:29–35
4. Meike D, Ribickis L (2011) Energy efficient use of robotics in the automobile industry. In 2011 15th international conference on advanced robotics (ICAR), pp 507–511, IEEE
5. Laumond J-P et al (1998) Robot motion planning and control, vol 229. Springer
6. LaValle SM (2006) Planning algorithms. Cambridge university press
7. Biagiotti L, Melchiorri C (2008) Trajectory planning for automatic machines and robots. Springer Science & Business Media
8. Latombe J-C (2012) Robot motion planning, vol 124. Springer Science & Business Media
9. Carabin G, Wehrle E, Vidoni R (2017) A review on energy-saving optimization methods for robotic and automatic systems. Robotics 6(4):39
10. Richiedei D, Trevisani A (2016) Analytical computation of the energy-efficient optimal planning in rest-to-rest motion of constant inertia systems. Mechatronics 39:147–159
11. Carabin G, Vidoni R (2021) Energy-saving optimization method for point-to-point trajectories planned via standard primitives in 1-dof mechatronic systems. Int J Adv Manuf Technol, pp 1–14
12. Yu Z, Han C, Haihua M (2015) A novel approach of tuning trapezoidal velocity profile for energy saving in servomotor systems. In: 2015 34th Chinese control conference (CCC), pp 4412–4417. IEEE

13. Chen H, Mu H, Zhu Y (2016) Real-time generation of trapezoidal velocity profile for minimum energy consumption and zero residual vibration in servomotor systems. In: 2016 American control conference (ACC)
14. Boscariol P, Carabin G, Gasparetto A, Lever N, Vidoni R (2015) Energy-efficient point-to-point trajectory generation for industrial robotic machines. In: Proceedings of the ECCOMAS thematic conference on multibody dynamics, Barcelona, Spain, pp 1425–1433
15. Boscariol P, Richiedei Da (2019) Energy-efficient design of multipoint trajectories for cartesian robots. Int J Adv Manuf Technol 102:1853–1870
16. Inoue K, Ogata K, Kato T (2008) An effcient induction motor drive method with a regenerative power storage system driven by an optimal torque. In: 2008 IEEE power electronics specialists conference, pp 359–364, IEEE
17. Inoue K, Asano Y, Kotera K, Kato T (2014) Optimal energy saving trajectories of induction motor with suppression of sudden acceleration and deceleration. In: 2014 IEEE energy conversion congress and exposition (ECCE), pp 3219–3223. IEEE
18. Asano Y, Inoue K, Kotera K, Kato T (2014) An energy saving drive method of an induction motor with the suppression of sudden acceleration and deceleration. In: 2014 International power electronics conference (IPEC-Hiroshima 2014-ECCE ASIA). IEEE
19. Betts JT (2010) Practical methods for optimal control and estimation using nonlinear programming. SIAM
20. Von Stryk O, Bulirsch R (1992) Direct and indirect methods for trajectory optimization. Annals Operat Res 37(1):357–373
21. Hsu Y-L, Huang M-S, Fung R-F (2013) Energy-saving trajectory planning for a toggle mechanism driven by a PMSM. IFAC Proceed Vol 46(5):628–635
22. Hsu Y-L, Huang M-S, Fung R-F (2014) Energy-saving trajectory planning for a toggle mechanism driven by a PMSM. Mechatronics 24(1):23–31
23. Paryanto, Brossog M, Bornschlegl M, Franke J et al (2015) Reducing the energy consumption of industrial robots in manufacturing systems. Int J Adv Manuf Technol 78(5-8):1315–1328
24. Sengupta A, Chakraborti T, Konar A, Nagar A (2011) Energy efficient trajectory planning by a robot arm using invasive weed optimization technique. In: 2011 Third world congress on nature and biologically inspired computing, pp 311–316. IEEE
25. Fenucci A, Indri M, Romanelli F (2016) An off-line robot motion planning approach for the reduction of the energy consumption. In: 2016 IEEE 21st International conference on emerging technologies and factory automation (ETFA), pp 1–8. IEEE
26. He T, Zhang Y, Sun F, Shi X (2016) Immune optimization based multi-objective six-DOF trajectory planning for industrial robot manipulators. In: 2016 12th World congress on intelligent control and automation (WCICA), pp 2945–2950. IEEE
27. Boscariol P, Caracciolo R, Richiedei D, Trevisani A (2021) Energy optimization of functionally redundant robots through motion design. Appl Sci 10:1–13
28. Boscariol P, Richiedei D (2018) Spline-based energy-optimal trajectory planning for functionally redundant robots. In: 2018 14th IEEE/ASME international conference on mechatronic and embedded systems and applications (MESA), pp 1–6. IEEE
29. Van Oosterwyck N, Vanbecelaere F, Haemers M, Ceulemans D, Stockman K, Derammelaere S (2019) CAD enabled trajectory optimization and accurate motion control for repetitive tasks. In: 2019 IEEE 15th international conference on control and automation (ICCA), pp 387–392. IEEE
30. Van Oosterwyck N, Cuyt A, Derammelaere, S et al (2020) CAD based trajectory optimization of PTP motions using Chebyshev polynomials. In: 2020 IEEE/ASME international conference on advanced intelligent mechatronics (AIM), pp 403–408. IEEE
31. Pellicciari M, Berselli G, Balugani F (2014) On designing optimal trajectories for servo-actuated mechanisms: detailed virtual prototyping and experimental evaluation. IEEE/ASME Trans Mechatron 20(5):2039–2052
32. Berselli G, Balugani F, Pellicciari M, Gadaleta M (2016) Energy-optimal motions for servo-systems: a comparison of spline interpolants and performance indexes using a CAD-based approach. Robot Comput Int Manuf 40:55–65

33. Liu A, Liu H, Yao B, Xu W, Yang M (2018) Energy consumption modeling of industrial robot based on simulated power data and parameter identification. Adv Mech Eng 10(5):1687814018773852
34. Yin S, Ji W, Wang L (2019) A machine learning based energy efficient trajectory planning approach for industrial robots. Procedia CIRP 81:429–434
35. Riazi S, Bengtsson K, Bischoff R, Aurnhammer A, Wigström O, Lennartson B (2016)Energy and peak-power optimization of existing time-optimal robot trajectories. In: 2016 IEEE international conference on automation science and engineering (CASE), pp 321–327. IEEE
36. Riazi S, Wigström O, Bengtsson K, Lennartson B (2017) Energy and peak power optimization of time-bounded robot trajectories. IEEE Trans Automat Sci Eng 14(2):646–657
37. Scalera L, Boscariol P, Carabin G, Vidoni R, Gasparetto A (2020) Enhancing energy efficiency of a 4-DOF parallel robot through task-related analysis. Machines 8(1):10
38. Vidussi F, Boscariol P, Scalera L, Gasparetto A (2021) Local and trajectory-based indexes for task-related energetic performance optimization of robotic manipulators. J Mech Robot 13(2):021018
39. Rexroth B (2021) Compact modules CKK/CKR catalog. published online on www.boschrexroth.com
40. Pillay P, Krishnan R (1988) Modeling of permanent magnet motor drives. IEEE Trans Indust Electron 35(4):537–541
41. Hansen C, Kotlarski J, Ortmaier T (2013) Experimental validation of advanced minimum energy robot trajectory optimization. In: 2013 16th international conference on advanced robotics (ICAR), pp 1–8. IEEE
42. Cook C, Ho C (1984) The application of spline functions to trajectory generation for computer-controlled manipulators. In: Computing techniques for robots, pp 101–110. Springer
43. Petrinec K, Kovacic Z (2007) Trajectory planning algorithm based on the continuity of jerk. In: 2007 mediterranean conference on control & automation, pp 1–5. IEEE
44. Boscariol P, Gasparetto A, Vidoni R (2012) Planning continuous-jerk trajectories for industrial manipulators. In: Engineering systems design and analysis, vol 44861, pp 127–136. American Society of Mechanical Engineers
45. Gasparetto A, Boscariol P, Lanzutti A, Vidoni R (2012) Trajectory planning in robotics. Math Comput Sci 6(3):269–279

Minimization of CO_2-Footprint of Hybrid Propulsion Systems for Mobility and Power-Tool Applications

Stephan Schmidt, Hans-Jürgen Schacht, Alexander Hagenberger, and Dimitrios Vogiatzis

Abstract Research and development activities for propulsion systems follow the demands of users/customers, production companies and legal boundary conditions. While in the past the topics of functionality, producibility and cost in compliance with the actual legal boundary conditions were decisive for R&D, now the minimization of CO_2 footprint tends to become a requirement. In the field of propulsion systems for mobile and power-tool applications, in the past CO_2 emissions during use were a factor of interest in respect of fuel consumption, as traditional fuels are fossil based and emit CO_2 as combustion product. In some sub-fields of these applications, fuel consumption is one part of the legal regulations or a key value which has to be published for customer information. Nowadays, the whole life cycle including production and disposal or recycling of a propulsions system has to be taken into account when calculating CO_2 emissions. In this chapter, the question is addressed, how the development task of minimization of whole life cycle CO_2 balance of mechatronic hybrid propulsion systems in combination with the still valid general demands can be fulfilled by the use of design, simulation and experimental development. A framework of R&D methodologies is described with examples of two hybrid propulsions systems, one for two-wheeler and power sport applications and one designed for hand-held power tools.

Introduction

The CO_2 footprint of a combustion engine propulsion unit is commonly assessed by the CO_2 emitted during its operation ("*tank-to wheel*"). Different methods for life cycle analyses regarding CO_2 and different areas of CO_2 emission exist, whereas the most comprehensive approach is to consider the production, operation and recycling ("*cradle to grave*") [1]. Since the discussion about e-mobility, the inclusion of the powertrain production cost in CO_2 is mandatory for a comparison of different propulsion technologies.

S. Schmidt (✉) · H.-J. Schacht · A. Hagenberger · D. Vogiatzis
Graz University of Technology, Graz, Austria
e-mail: Schmidt@ivt.tugraz.at

P. Hehenberger et al. (eds.), *EcoMechatronics*,
https://doi.org/10.1007/978-3-031-07555-1_16

Up to now, the development of technical systems was driven by the fulfilment of demands like legal regulations, customer needs, marketing aspects, costs and suitability for production. Especially for highly complex hybrid powertrain systems, a balancing between these demands is time-consuming and expensive and the consideration of the CO_2 footprint by a Life Cycle Assessment (LCA) introduces a new parameter into all stages of the development. The task of developing a hybrid propulsion unit taking account all issues including CO_2 footprint can only be solved by integrating CO_2 assessment methods in the development process and into the Computer Aided Technologies (CAT) environment. In the following sections, a framework of R&D methodologies is described using examples of two hybrid propulsions systems with the focus on the production and operation phases.

Mechatronic Powertrain

In today's pure internal combustion engine (ICE) powertrain systems, there is already an interaction taking place between the mechanical and electrical components. This is intensified in hybrid powertrain concepts[1] as a second energy converter and a battery are incorporated into the powertrain. This highly complex mechatronic system can be divided into hardware and software, both existing on component and system levels. A major part of the mechatronic hybrid powertrain is the electronic control of the components and the control of the interaction of the propulsion components with the operation strategy.

Hybrid Powertrain Concepts

In the field of passenger cars, hybrid propulsion units are state of the art, while two-wheeler, power sport vehicles and power tools[2] are still at the beginning of hybrid powertrain introduction. This is due to stringent demands on weight, packaging and cost which leads to simple but satisfying concepts.

Hybrid concepts are used in the automotive sector range from Plug-in-Hybrid, which comprises two fully functional propulsion modes (ICE and electric) and allows for pure ICE, mixed and pure electric operation modes including external battery charging to Full-Hybrid with same functionality as Plug-in Hybrid but without external battery charging and Mild Hybrid where electric propulsion is used as a

---

[1] Hybrid powertrain concepts consist of at least two independent energy sources and/or energy converter. In this context, hybrid means a combination of an internal combustion engine and an electrical machine.

[2] Hand-held power tools are equipment like chainsaw, hedge-trimmer, blower and others.

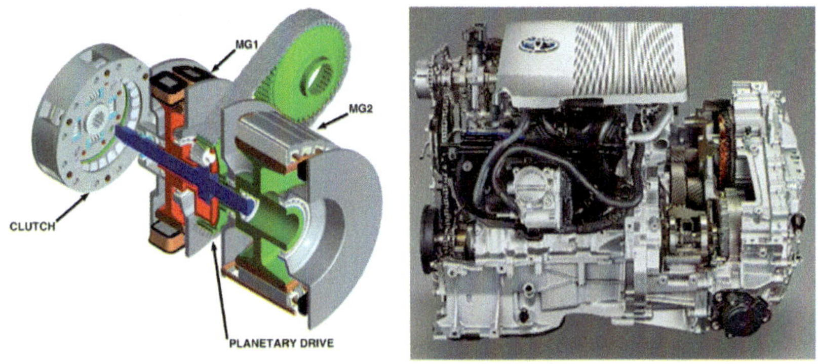

Fig. 1 Toyota prius functional concept schematic [2] and cross-section model [3]

propulsion assist. Another concept is that of a Range Extender, where an internal combustion engine is used as backup for battery charging in an electric propulsion unit.

Three examples of actual hybrid propulsion concepts show the complexity of these systems.

Toyota *Prius* (Power-Split)

One of the first hybrid systems available on the automotive market was the Toyota *Prius,* which was equipped with a highly complex power-split configuration. Figure 1 shows the significantly enlarged powertrain and its additional components consisting of an electric motor, a generator connected to the engine clutch, and a one-speed reduction gearbox driven via a planetary drive. This planetary drive enables control of the power transfer in any direction to represent the different operating modes. The modes range from pure electric drive, electric boost, load point shifting and charging the battery by the ICE during stand-still. It is the most versatile, but also the most complex hybrid configuration. On top of the mechanical and electrical components, a complex operation strategy controls the different operating modes, defining the split of energy coming from the individual sources and torques delivered by the ICE and e-motor. Additionally, the operating strategy ensures the health of all components by monitoring temperatures and system conditions in respect of the operating point decisions.

Range Extender for Small Two-Wheeler Application

Latest figures of motorcycle and moped registrations in Europe [4] show that electric mopeds have a share of over 20% of new registrations, whereas motorcycle have only

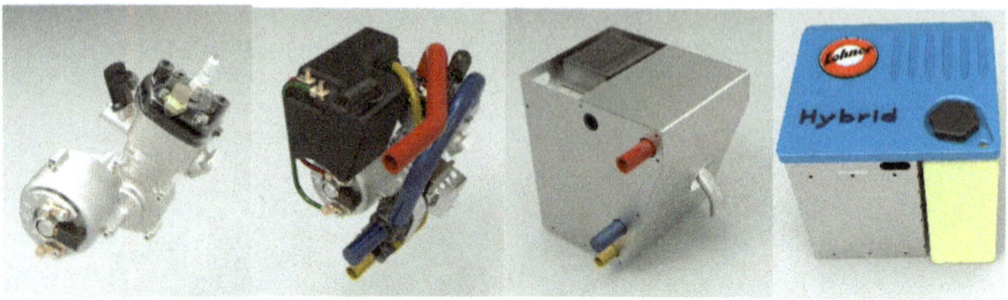

Fig. 2 L1e-Scooter range extender unit developed at FVT[3] for Lohner[4]

Table 1 Technical specifications of the range extender for L1e-category vehicle

Electric power	1.6 kW
Packaging volume	23 ltr
Weight	15 kg
Specific fuel consumption	450 g/KWh
Fuel capacity	4.5 ltr
Engine	50 cm^3 single cylinder 2-stroke internal combustion engine

2% share. This indicates that electric driving is more accepted for low power two-wheelers, which are generally driven in an urban environment and most of the time have less range demands. To overcome range anxiety as main inhibition threshold for acceptance and willingness to buy electric two-wheelers, hybridization by means of a less complex range extender configuration could be a solution. Figure 2 shows a compact and lightweight concept, which can be retrofitted to any existing pure electric scooter as an add-on electric power source, the specifications for which are given in Table 1. This opens up the possibility to connect it whenever greater range is required and omit it when the battery alone covers the distance. The operating logic only starts the engine if a low battery level and/or a high load is detected [5].

Motorcycle Hybrid Powertrain

The L-category, which includes 2, 3 and 4-wheeled vehicles less than 450 kg (for passenger transportation), respectively, and 600 kg (for goods transportation) is lagging behind in terms of hybridization. Although a number of hybrid concepts regularly appear at motor shows, the only hybrid vehicle introduced in the market

---

[3] Forschungsgesellschaft für Verbrennungskraftmaschinen und Thermodynamik mbH, Graz - Austria.

[4] Lohnerwerke GmbH & Co KG.

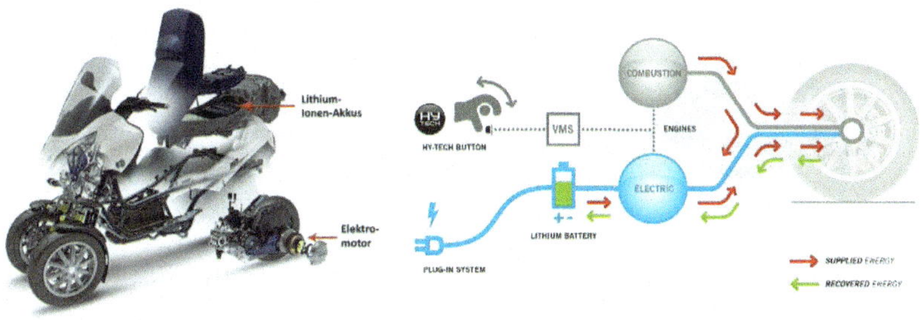

Fig. 3 Piaggio MP3 Hybrid vehicle (left [6]) and powertrain concept (right [7])

was the Piaggi MP3 Hybrid of 2010 (Fig. 3). Two versions were available with 125 cm^3 and 300 cm^3 internal combustion engines, each coupled to a 2.6. kW electric motor installed in the secondary gearbox, driving the rear wheel directly. This corresponds to a P3 parallel hybrid configuration, which allows to drive purely electric (activated by the driver for up to a speed of 30 kph) in restricted urban areas as well as boosting and energy recuperation from braking.

Even though the MP3 is the only homologated hybrid l-category vehicle to date, several manufacturers seem to be preparing hybrid models. According to occasional teasers in the press and patent publications, Kawasaki, Yamaha, Suzuki along with start-ups like Furlon [8] are working in this direction.

Development Process

The process of product development of two-wheelers and powered tools can be divided into the three stages such as "*research and concept development phase*", "*pre-development*" and "*product development*". Several decision gates are contained in the development process, at least one at the end of each stage. At these gate decisions for a certain technology and most importantly, for or against a continuation of the development are made. In all stages, CAx methods are used, starting with CAD for the concept layout, followed by simulations and finally production planning during the final stages of transformation into mass production.

While the research and concept development phase consists largely of non-product-related basic research and concept investigations, the pre-development phase is characterized by detailed concept layouts and concept proofs against the background of a specific product idea. At the stage of product development, all major parameters are defined and the realization of the concept into a mass-production ready product definition takes place.

During the different stages, all aspects and demands on the future product or technology have to be covered and decisions about technical, technological, logistic and commercial topics are made. In order to develop a product with minimal CO_2 emission during production, operation and recycling, the CO_2 issue has to be considered

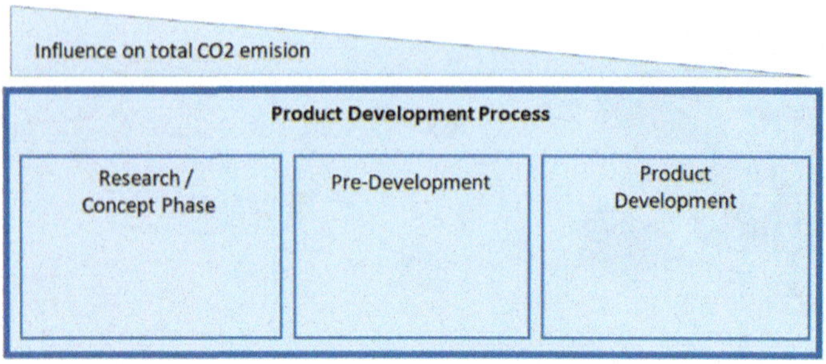

Fig. 4 Influence on carbon footprint in different development phases

in addition to the traditional criteria of each development stage and in respect of each decision.

It is known that within the different development stages, the influence of the decisions on traditional criteria like product key parameters, fuel consumption, cost and so on are very different. A majority of the production costs are defined during the first two steps of the development, while during the third part only a small influence is possible. Considering CO_2, a similar distribution of influence during the development stages can be identified as it is shown in Fig. 4. In the following sections, the possibilities/necessities for a consideration of CO_2 emission during the different development stages are explained.

Research and Concept Development Phase

During the research and concept development phase, a first estimation of the CO_2 contribution during the powertrain operation is made by assessing the vehicle fuel consumption. In parallel, an identification of CO_2-relevant areas in production and recycling can be made. This concerns special CO_2 intensive production technologies or materials, for which recycling is difficult. The CO_2 contribution of various technical and technological solutions also have to be assessed as a rough estimation for comparison, as data for technical and technological decisions are gathered in this phase.

Two examples describe the conflict and frequent trade-off between traditional development targets and reduction of CO_2 footprint. For novel powertrain concepts, the use of alternative fuels is mandatory. These fuels have an impact on CO_2 emissions during operation, as they can be produced in a CO_2 neutral cycle. But as these fuels often have different properties regarding material compatibility, appropriate materials have to be used which might have a worse CO_2 production balance than traditional ones. This can reduce the positive effect of the CO_2 reduction during operation. A hybrid-related example is the use of different types of electric propulsion motors. A decision between an Asynchronous Motor and a Permanent Magnet Synchronous Motor (PMSM) is usually based on criteria of cost and performance in relation to factors such as mass and power density. However, these two options have different CO_2 impacts on the production processes as they use different materials.

Pre-development

During pre-development, all major parameters are defined. Subsequent to the concept phase a proof of concept is performed and data for decisions for or against product development options are generated. The first estimations of the prior phase are verified on the base of more detailed data and, with respect to CO_2 emission, critical areas identified during phase 1 which are evaluated by performing CO_2 life cycle analyses and variant calculations.

During pre-development, novel technologies are investigated for the use in the targeted application. Often, variants are compared with each other regarding the key parameters and suitability for the application. In this comparison, a CO_2 footprint of the technology production has to be included. If, for example, two different cylinder liner coatings are investigated with different influences on the fuel consumption due to different friction behaviour, the difference in CO_2 footprint during the coating process and the materials used (and production sites) have to be taken into account.

Product Development

While in the final development phase the in-use CO_2 emission can be reduced to a certain extent only minor influence on the CO_2 production footprint is possible, as most relevant decisions have been made during the prior phases. One important impact on the production CO_2 emissions is the choice of the production sites. This is due to the circumstances that different boundary conditions in the production site areas lead to a different local CO_2 footprint, and the transport and logistics is an influential parameter for the production CO_2 emissions.

A very important task during product development is the calculation of the total CO_2 footprint using a life cycle analysis based on the detailed available data of the future product. This CO_2 footprint serves as a base for documentation according to legal regulations as well as for marketing and strategic purposes.

Development Framework

In order to cope with the demands of CO_2 footprint consideration in the development process, a framework of tools and methods has to be developed and implemented. Dedicated methods to be used in the different development stages have to be closely coupled to or integrated into the CAx methods to enable a consistent CO_2 footprinting

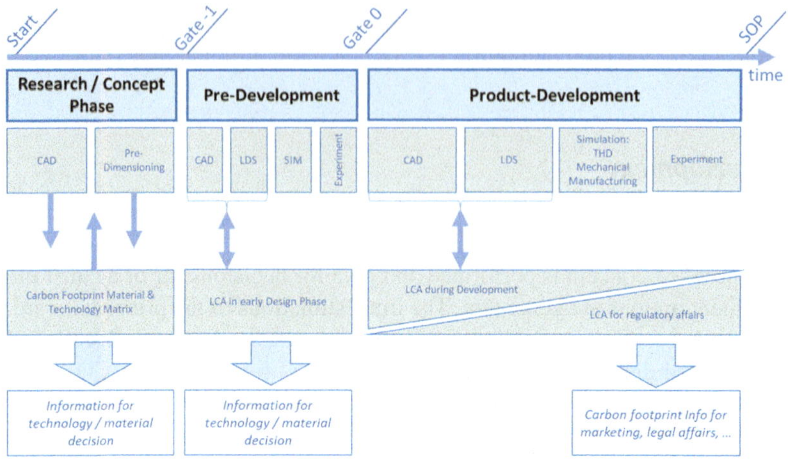

Fig. 5 Development framework for CO_2 footprint consideration

dependent on available information. In Fig. 5, a development framework for the consideration of the CO_2 footprint of a hybrid propulsion powertrain is shown.

In the different development phases, various development tools are used. Some of them deliver information for an assessment of the carbon footprint. In the concept stage, relatively little information is available, and a *material and technology CO_2 matrix* can be used for a first assessment of the hybrid powertrain concept. In the second stage, more detailed information can be gathered from CAD and longitudinal vehicle dynamic simulation (LDS). With this information a "*life cycle analysis in early design stages*" [9] is performed, providing input for a decision for or against a certain technology or material. During the last phase of product development, all detailed parameters of the powertrain are defined, but as most key parameters have been defined during the pre-development, the LCA has a low influence on the system layout. Instead, the complex and entire information set is used to perform a detailed LCA, which is used for legal affairs, marketing and strategic decisions.

In the following sections, tools and methods which can deliver CO_2 footprinting information are presented.

Production Technology and Material Matrix

From studies such as that by Arguillarena et al. [10] the carbon footprint of various materials and production technologies can be extracted and organized in a matrix to compare key parameters including the carbon footprint. This matrix can be used in each step of the development process, but mainly during the concept phase and pre-development together with the LCA in early design stages [2]. This company specific matrix serves as comparison table between different possibilities and does not need to be exact. Nevertheless, a regular check for up-to-datedness as well as enhancement for novel technologies is necessary. If included in the development process as a mandatory tool, the matrix can also raise the awareness and experience regarding CO_2 contribution of different technologies and materials.

Computer-Aided Design Software

Several CAx tools are used during the development of hybrid propulsion systems, with two of them delivering data relevant for carbon footprinting. CAD systems are used in all stages of the development process, from first sketches through packaging analyses to detailed 3D design and manufacturing drawings. For all stages of the CAD design process, CO_2 relevant data can be obtained. These are in the first stages mass and volume of the single parts and later also in respect of production steps and materials. This information can be obtained in an automatic process from the CAD system by user-defined functions implemented into the CAD workflow.

Longitudinal Vehicle Dynamics Simulation LDS

Longitudinal (vehicle) dynamic simulations (LDS) are used to assess the behaviour of hybrid systems and their sub-systems over a period of time for given load profiles or driving cycles. The output regarding CO_2 assessment ranges from direct calculation of CO_2 emission (in terms of fuel consumption and electric energy demand) during operation to indirect assessment of production CO_2 emission by the definition of optimized system architecture, optimal sub-system choice (i.e. e-motor type) and sub-system sizes (i.e. battery size).

The LDS is used for the investigation of driving performance and interaction between the hybrid components, optimization of hybrid structures as well as for the development of the operation strategy of the hybrid system. As the LDS is based on vehicle data such as driving resistance and key parameters of the sub-systems (efficiencies, energy capacities and performance), the significance of the gathered results depend on the depth of detail of these input data and hence on the development stage. Usually, several calculation cycles with increasing accuracy of the input parameters are carried out.

The results with respect to carbon footprint are CO_2 emissions during operation in terms of energy[5] consumption and the choice of sub-systems and sub-system parameters such as the energy content of the battery or power and type of the electric motor. These parameters influence the production carbon footprint, for example, different battery sizes lead to different CO_2 footprints.

[5] Fuel and electric energy.

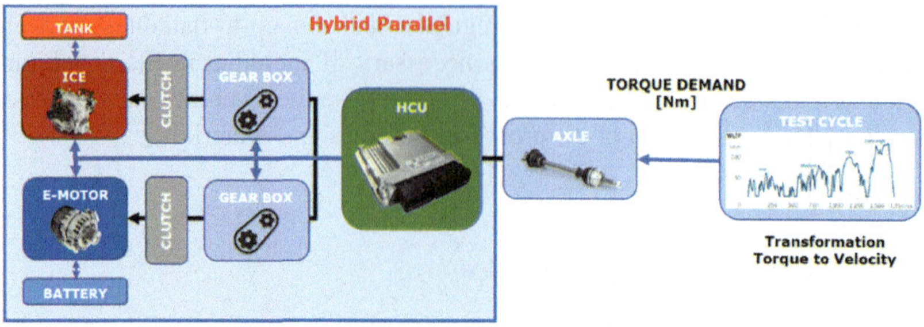

Fig. 6 Backward LDS simulation of a parallel hybrid powertrain

Two main simulation methods of LDS exist: forward and backward simulation. Both methods have advantages and disadvantages. With the help of forward simulation, various control mechanisms are investigated and parametrized and operation strategies can be developed. However, considerably more data about the vehicle and sub-system dynamics is necessary. A backward simulation delivers much faster results, while requiring fewer input parameters, but information about control is not available as there is no embedded feedback control loop. In the following, one example for each simulation method is given.

Backward Simulation

In the backward LDS, the torque demand is calculated at each time step using the target velocity of a driving or load profile and the overall powertrain efficiency. The result is fed back to the hybrid system, and the actual state of the sub-system variables is derived. For instance, the rotational speed of internal combustion engines and e-motor or the trend of state of charge of the battery.

In Fig. 6, a backward simulation of a parallel hybrid powertrain is shown. It consists of two propulsion units, one internal combustion engine and one e-motor, two clutches and gearboxes and a Hybrid Control Unit (HCU). With the backward simulation, an assessment and optimization of the hybrid system and the sub-systems regarding fuel consumption, battery energy content, number of gears and further parameters is possible without consideration of the dynamic behaviour. Adding thermal properties and thermal behaviour to the individual modules allows for an overall thermal assessment of the hybrid system and optimizations of the cooling systems are possible.

Forward Simulation

In contrast to backward simulation, a typical forward simulation (Fig. 7) uses a control module (driver) to compare the actual speed with the desired speed gained

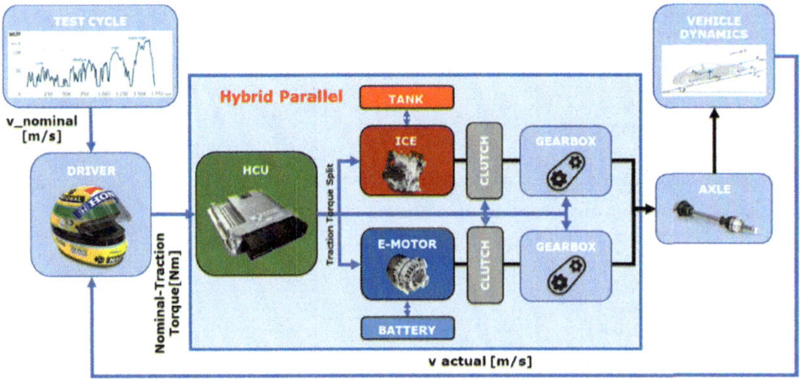

Fig. 7 Forward LDS simulation of a parallel hybrid powertrain

from the test cycle and to adapt the torque demand. The control module is usually a PI-control in a closed-loop control system and needs a parametrization to the dynamic behaviour of the system. The modelling of forward simulation is more time-consuming and needs more input data than a comparable backward simulation, but delivers considerably more information. Special modules like tyre slip or gearbox modules with dynamic parameters allow for an assessment of the dynamic behaviour of the system. The hybrid control unit is used to extract real-time data of the embedded operating strategy and even a software-based parametrization of real-life hardware like the electronic control units of sub-systems is possible.

Examples

Concept and Layout of a Hybrid Propulsion for Power Sport Vehicles

In automotive as well as in power sports electrified powertrain development, there are different ways as to how to integrate the electric motor into the (existing) combustion engine powertrain [11]. In a parallel hybrid configuration not only the internal combustion engine, but also the electric motor is capable of driving the vehicle via its wheels. The fact that in a parallel P2 configuration, the electric motor is connected to the driveshaft between the combustion engine and the driven wheels, allows the integration of an additional (automatic) 2-speed gearbox for the electric motor. In this example, a hybrid powertrain for power sport applications is developed, whereas in a concept layout special focus is put on the introduction of a multi-gear gearbox.

The influence of a 2-speed as compared to a single-speed gearbox for the electric motor is far-reaching as the vehicle driving capabilities, its overall powertrain efficiency, packaging and mass as well as its general powertrain complexity are

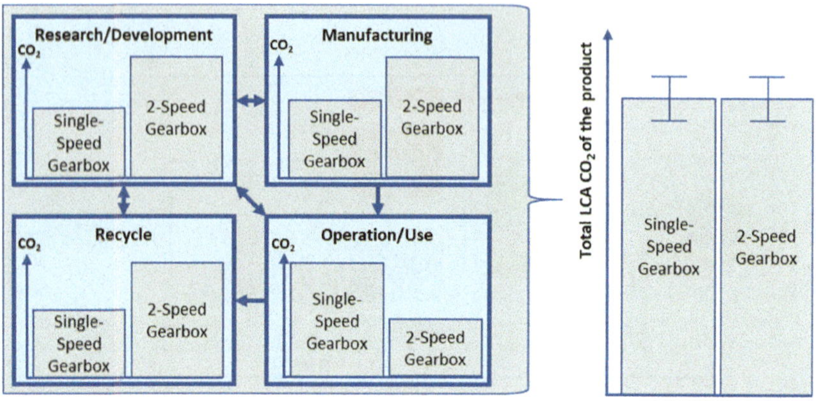

Fig. 8 CO_2 footprint within different life cycle product phases of single-speed and 2-speed gearbox variant for the E-Motor

influenced significantly. These criteria are, furthermore, interacting with the CO_2 assessment of the final product. In Fig. 8, the influences of the two variants in the different CO_2 relevant stages are shown in the box on the left side, resulting in little difference in an overall ranking (right side). While the single-speed variant has obvious benefits in production and recycling CO_2 emission due to less parts, the 2-speed variant shows benefits during operation as the following paragraph describes in detail.

Depending on the vehicle, the typical driving profile (speed and road gradient) and the electric motor characteristics (max. power and torque), an additional second gear can help to increase the vehicle driving capabilities. The benefits prove to be larger for high torque demand scenarios. These scenarios can be characterized by high vehicle masses and steep road gradients or simply pure electric driving, without torque assistance of the ICE. The benefit of two gears has shown to be minor for high power load points, like lower vehicle mass and flat road gradients or while hybrid (ICE + electric motor) driving. The main reason is that one single-speed gear ratio always represents a compromise compared to two separate gear ratios, resulting in limited available torque at the driven wheel.

Figure 9 shows the vehicle speed traces at full-load acceleration starting from a simulation time of 10 s. It illustrates that the performance benefit of a 2-speed over the single-speed gearbox for the electric motor is larger under high torque conditions, which in this case is pure electric mode without ICE support.

In general, the ICE as well as the electric motor show better efficiencies in relatively higher load operation points. Especially the second gear of a 2-speed gearbox with a lower gear ratio can result in higher torque demand at the electric motor, shifting its operational points to regions of better efficiency within the efficiency map. As a consequence of the rise in efficiency, fuel consumption and therefore CO_2 emission decreases proportionally. In the next step, an adequate shifting strategy between the two gears needs to be found. This strategy should, on the one hand,

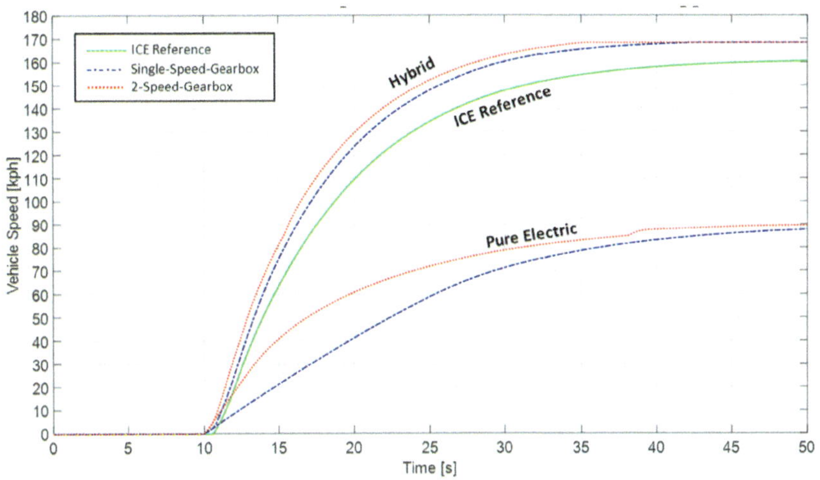

Fig. 9 Acceleration performance of a hybrid vehicle with single-speed and 2-speed gearbox in hybrid and pure e-drive mode compared to the pure ICE reference

satisfy the shifting and driving behaviour according to the driver's throttle input. On the other hand, it needs to be optimized to minimize fuel/energy consumption. The development of this strategy results in higher overall development effort and complexity compared to the single-gear variant. Longitudinal dynamic simulation (in MATLAB/Simulink in this case) is the key tool used for dimensioning and developing such hybrid powertrain concepts and its corresponding operation strategy. LDS though is purely mathematically based and does not consider constructive coherencies within the power- and drivetrain.

Due to the greater mechanical complexity, the number of parts in the 2-speed gearbox is significantly higher compared to the single-speed variant. This not only affects the mass of the gearbox and the complete vehicle, but also presents challenges regarding packaging. In automotive and even more in power sports applications compact packaging and best possible power-to-weight ratio are key criteria. In order to evaluate the change to a 2-speed instead of a single-speed gearbox, a CAD packaging analysis is carried out. The CAD model is capable of assessing the changes in terms of overall mass and the position of the centre of gravity, for instance. These changes are then implemented into the LDS and the results can be evaluated accordingly. Furthermore, the CAD packaging concept allows the drawing of conclusions as to how much resource will be required in course of the production process which influences the production of CO_2 emissions.

With the CO_2 relevant information gathered in the simulation and design processes, the overall CO_2 emissions of the product can be predicted (see Fig. 9). The closed-loop development process mentioned above allows for optimization of product life cycle CO_2 footprint and enables the researcher/developer to make decisions which lead to a minimization of the overall CO_2 emissions of the product.

Concept and Layout of a Hybrid Propulsion for Hand-Held Power Tool

On the basis of hybrid powertrain investigations for CO_2 reduction in hand-held power tools such as brush-cutters or chainsaws, numerous parameters regarding the complete CO_2 footprint of the product but also boundary conditions have to be taken into consideration. Among the parameters of consideration are the hybrid configuration, the number, technology and size of power units to be used, the operational strategy and the technology and size of the battery. As for the boundary conditions, the hybrid tool should respect the overall carbon footprint, life cycle cost, weight and volume limitations of the baseline combustion-engine driven tool. Further, an important constraint is that the energy of the hybrid tool should come only from the fuel and not from the grid, meaning that the State of Charge (SoC) of the battery has to be controlled in order to remain as constant as possible during the whole load cycle. A fully parametric simulation model is thus necessary for identifying the weightiness of each parameter with respect to the aforementioned constraints in order to reduce CO_2 emissions of future hand-held tools [12].

One of the first questions to be addressed for the hybrid tool is the hybrid configuration. Two configurations, the serial and parallel hybrid are possible for this type of application. This decision influences not only CO_2 emissions in operation but also the production carbon footprint, as different parts and units are necessary. In the following, a methodology for determining the optimum hybrid configuration is shown using a simulation model in MATLAB/Simulink (Fig. 10).

A serial hybrid configuration [13] uses the electric motor to drive the tool while the internal combustion engine is coupled with an electric generator which produces the energy needed for driving. In a parallel hybrid 1:1 configuration, the electric motor is attached on the same shaft as the internal combustion engine which enables a utilization of both torques of internal combustion engine and electric motor. The serial hybrid needs three power units and two driveshaft while the parallel hybrid needs two power units and one driveshaft (Fig. 11). With the energy coming only from the fuel, in both cases, the battery is responsible for transmitting energy between the power units by acting as an energy buffer.

Learning from the automotive field, the three main operation modes giving a CO_2 reduction benefit during the operation phase for the hybrid compared with the pure internal combustion engine powertrain are the *idle-off*, *regeneration* and *load-point-shifting* functions.

- The *Idle-off* mode replaces the internal combustion engine idle with e-start. Real-world load cycles of chainsaws, brush-cutters and blowers show, that idling of hand-held tools can reach up to 30% of the duration of the load cycle. Thus, the CO_2 reduction potential of reducing the idle time is not negligible. Comparing the benefit of this mode with the two hybrid configurations, in both the total duration of internal combustion engine idle can be abolished, the frequency of

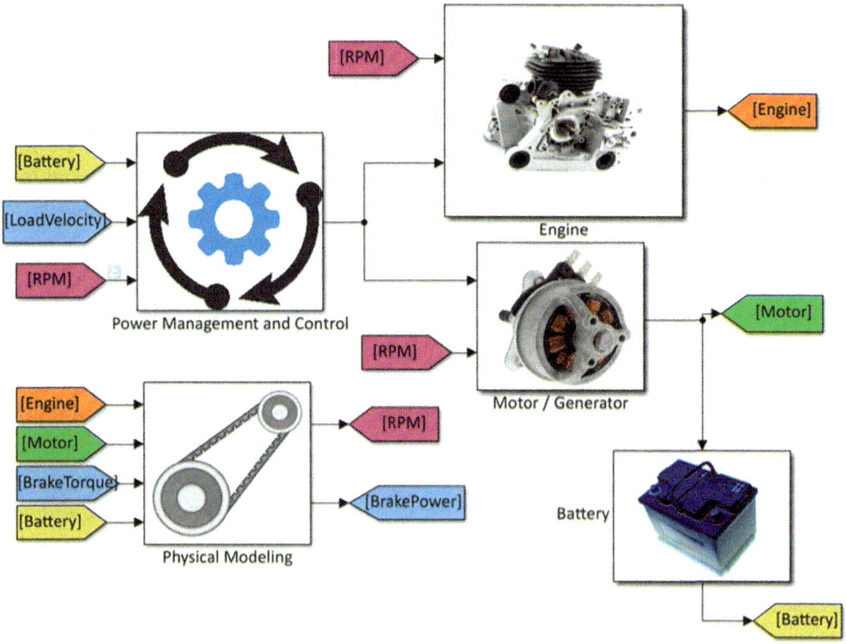

Fig. 10 Simulation model of hand-held hybrid tools

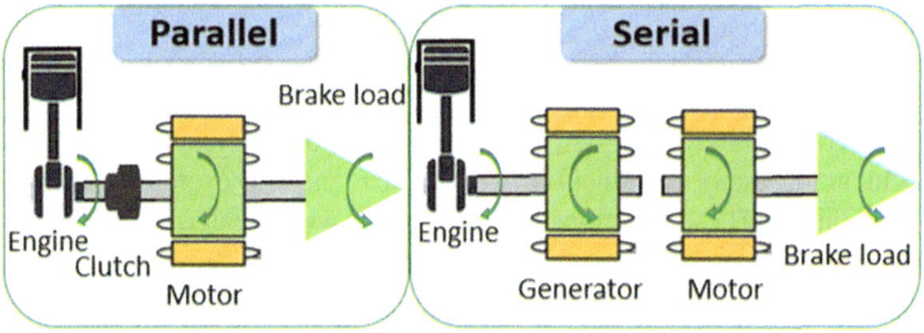

Fig. 11 Parallel versus serial hybrid configuration

start-stops for the parallel hybrid is significantly higher based on the load cycle profile. Further, the number of start-stops for the serial hybrid could be limited in a very small number based on the battery capacity.

- The *Regeneration* mode takes advantage of the load cycle deceleration ramps by braking the tool and storing the energy of the driveshaft inertia to the battery. While the inertias of hand-held tools are small, deceleration ramps of high-speed differences (e.g. from 13,000 rpm down to idle) can give a benefit for this mode of operation. Due to the fact that the dominant inertia of the tool driveshaft is coming from the ICE and not from the tool blade or chain, in this mode the parallel hybrid tool has a benefit over the serial variant.

- The *Load point shifting* mode uses the 1st and 4th quadrant torque of an electric motor to operate the internal combustion engine in the best-efficiency area. While even a small electric motor maintains a high efficiency over almost all regions of the operating map, the small low-complexity internal combustion engine used for hand-held tools optimizes its efficiency only in a small region of its map. Serial hybrids can take the most benefit from this mode by operating the internal combustion engine on the lowest fuel consumption point while parallel hybrid can only operate in the best operational speed line. The best operational speed line for these applications is almost always at the Wide-Open Throttle.

After simulating the load cycle, the accumulative energy consumption differences at the end of the load cycle between the combustion-driven powertrain and the serial hybrid or parallel hybrid powertrain for each of the three operation modii are summarizing the results of the simulation model. As each mode has a well-defined time domain of operation inside the load cycle, the energy consumed or produced during this domain is compared with the energy consumed by the combustion engine-driven tool for the same domain. Here, the domain of idle-off function *(idle-domain)* is below 4000 rpm where the tool[6] is not connected to the drivetrain due to centrifugal clutch disengagement, the domain of regeneration *(dec-domain)* occurs when the tool decelerates from speed larger than 9000 rpm below to 4000 rpm and the domain of load point shifting *(load-domain)* is on load.

Using the simulation model, the overall CO_2 reduction potential (ΔE_c) of different hybrid variants compared to the pure combustion engine version can be calculated (Eq. (1)) from the energy differences of the single operation modii.

$$\Delta E_c = \Delta E_{idle} + \Delta E_{regeneration} + \Delta E_{load} \tag{1}$$

The difference in CO_2 emissions is considered by implementing functions for each influencing factor (production, weight, size and life cycle cost from production to recycling). These functions are normalized with the corresponding values of the baseline combustion-driven tool. So, a cost minimization function (A) of ΔE_c, production-consumed CO_2 (P), weight (W), volume (V) and life cycle cost (LCC) is introduced (Eq. (2)). Based on the design criteria, each one of the five variables is multiplied with a weight factor α.

$$A = \min\left(\begin{array}{l} \left[-a_{\Delta E_c} * \Delta E_c + a_P * P + a_w * W + a_v * V + a_{lcc} * LCC\right]_{SH} \\ , \left[-a_{\Delta E_c} * \Delta E_c + a_P * P + a_w * W + a_v * V + a_{lcc} * LCC\right]_{PH} \end{array} \right) \tag{2}$$

with $SH = Serial\,hybrid\,configuration$ and $PH = Parallel\,hybrid\,configuration$.

The hybrid configuration with the resulting minimum value for A is the optimum for the given tool or load cycle based on a combined criteria matrix.

[6] i.e. cutting tool.

Summary

The consideration of the CO_2 emissions of hybrid powertrain has to cover the stages production, operation and recycling of the product. In order to develop a CO_2 footprint optimized product, measures for an assessment of the CO_2 impact of the powertrain have to be taken in each development phase, starting from the concept phase through pre-development to product development. At the various phases, different depth of information becomes available and different possibilities for technology and material decisions exist, different CO_2 consideration approaches are necessary.

A close collaboration between the traditional development tools like CAD and simulation and the CO_2 life cycle assessment tools and methods is necessary.

Future strategies for the development process have to integrate the LCA methods into the traditional development tools and enable a seamless exchange of data and information. This allows for well-funded decisions during the development process which includes the lifetime CO_2 balance of the powertrain.

References

1. Tagliaferri C, Evangelisti S, Acconcia F, Domenech T, Ekins P, Barletta D, Lettieri P (2016) Life cycle assessment of future electric and hybrid vehicles: a cradle-to-grave systems engineering approach. Chem Eng Res Des 112:298-309. ISSN 0263-8762. https://doi.org/10.1016/j.cherd.2016.07.003
2. http://matteodilallo.tech/il-sistema-hybrid-synergy-drive-di-toyota/ 9.7.2021
3. © Toyota & MotorNews; Der neue Toyota Prius 2016 – Quantensprung in alle Richtungen
4. ACEM "Registrations in the European Union – 2021". https://www.acem.eu/market-data
5. CS-Consulting Patent No. AT 516418 B1
6. https://www.motorradonline.de/elektro/test-piaggio-mp3-der-neue-piaggio-mp3-mit-hybrid antrieb/ 9.7.2021
7. Presentation, Piaggio for eco-enviroment – Hybrid and Electric Technology; Paolo Badalassi; Piaggio&C.s.p.a. 03.04.2014
8. https://furion-motorcycles.com/ loaded on July 19th 2021
9. Merschak S, Hehenberger P, Schmidt S, Kirchberger R (2020) Considerations of life cycle assessment and the estimate of carbon footprint of powertrains. SAE Tech Pap 2020-32-2314. https://doi.org/10.4271/2020-32-2314
10. Arguillarena A, Margallo M, Urtiaga A (2021) Carbon footprint of the hot-dip galvanisation process using a life cycle assessment approach. Clean Eng Technol 2(100041). ISSN 2666-7908. https://doi.org/10.1016/j.clet.2021.100041
11. Singh KV, Bansal HO, Singh D (2019) A comprehensive review on hybrid electric vehicles: architectures and components. J Mod Transp 27:77–107. https://doi.org/10.1007/s40534-019-0184-3
12. Vogiatzis D, Schacht H, Schmidt S, Kirchberger R (2020) A concept investigation simulation model on hybrid powertrains for handheld tools. SAE Tech Pap 2020-32-2316. https://doi.org/10.4271/2020-32-2316.
13. Khajepour A, Fallah S, Goodarzi A (2014) Electric and hybrid vehicles technologies, modelling and control: a mechatronic approach. Wiley Ltd. ISBN 9781118341513

Education

Developing Education in Mechatronics to Support the Challenges for Evolution, Development, and Sustainability

Martin Edin Grimheden and Elias Flening

Abstract In this chapter, we propose a roadmap for the development of education in mechatronics. This development is intended to support future mechatronics engineers in accepting challenges in evolution, development, and sustainability. We explore curricula design, learning strategies, and misalignment between education and professional needs in relation to these challenges. The exploration is based on the fact that the subject mechatronics has now been around for 50 years, and, in particular, relates to the experiences and previous research of the authors, who are based at KTH Royal Institute of Technology where most of the empirical data is from. In this chapter, we argue that the above challenges require new roles, skills, and competencies among mechatronics engineers. With examples from complex product development, we argue that an increased systems perspective with systems architect competencies and experiences can support the embracing of additional systems design factors, which, for example, can enable sustainable development aspects into product design. We explore the interface between mechatronics and systems engineering, via embedded systems, as a possible future development of the subject of mechatronics. Embedded systems can be seen as a related subject, or as a subset of mechatronics, or vice versa. With this chapter, we argue that the subject of embedded systems can provide a shortcut for mechatronics education to approach the role of systems engineering, thereby reaching a level of competence and skills where the above challenges can be approached. The mechatronics engineer can play an important role in sustainable development. With existing curricula, mechatronics engineers are prepared to accept complex problem-solving, utilize solutions from various domains and disciplines, and create domain-independent solutions. These skills are crucial in embracing sustainable development, which puts the mechatronics engineer in a unique position. We also bridge the above discussion with current trends in learning, such as life-long learning and e-learning. The mechanisms that enable the mechatronics engineer to achieve a systems engineering skillset, which,

M. E. Grimheden (✉) · E. Flening
Department of Machine Design, School of Industrial Engineering and Management, KTH Royal Institute of Technology, 100 44, Stockholm, Sweden
e-mail: mjg@kth.se

P. Hehenberger et al. (eds.), *EcoMechatronics*,
https://doi.org/10.1007/978-3-031-07555-1_17

according to our results, needs to be based on engineering experiences, is facilitated both by a professional education system (life-long learning) and by a flexible setup (e-learning).

Introduction

This chapter is based on the following structure:

1. Challenges that relate to sustainable development and eco-design can be approached from a mechatronics perspective, meaning with mechatronics solutions or a mechatronics design thinking perspective.
2. Mechatronics education and mechatronics, in general, provide a suitable base for eco-design with a focus on synergistic integration, non-domain-specific solutions. A mechatronics focus gives a broader solution space.
3. A new type of mechatronics engineer and a new mechatronics program would be required, with more focus on complex problems. Skills in systems engineering and systems architecture would facilitate this.
4. A current shift from mechatronics toward embedded systems design can be a step in this direction. With focus on embedded systems design, skills, and competencies in systems engineering, systems design, and systems architecture can be achieved.
5. The current shift from program-based curricula toward life-long learning and e-learning can support competence development for experienced mechatronics engineers.
6. The new mechatronics engineer will have skills and competencies to accept challenges relating to sustainable development and eco-design.

Background

"We will never be able to solve the problems of tomorrow with the thinking of today".

The famous quote by Albert Einstein implies that we, both as educators and students, always will be a step behind. We develop university programs today. Students will attend these programs, taking their courses in a few years from now. Future employees will hire these students and expect them to contribute to mechatronics engineering, developing new products and engaging in engineering challenges. A decade might pass between the first concepts around program (re)design to when a student from that program get their first job. In current research, we note discrepancies between early engineering professional capabilities in comparison with expectations from employers. There has long been a mismatch between what is expected from a university engineering graduate and the relevant hiring industry. Even so, the Einstein quote takes this an additional step—we must think even further than that, it is not enough with the challenges of today, we need to consider the challenges of tomorrow as well—long after first employment.

Many studies of misalignment; discrepancies between expectations; focus on variations in expectations between engineering skills; and "soft skills", such as teamwork, communication, etc. But studies also show variations within engineering skills, lack of technical depth in certain areas, or the need to include new topics and areas. Such a lack includes the common inability to consciously switch between modes of thinking that are inherent to different engineering disciplines, such as the constituent disciplines of mechatronics. Switching ways of thinking therefore does not only pertain to addressing either "technical" or "non-technical" issues or people, but different aspects of engineering as well.

How should we then train students that will be able to accept and solve these challenges? How do we educate a student to become capable of answering the challenges of climate change? Reduce the carbon footprint? Sustain our quality of life?

Learn from Yesterday—Live for Today

What we do know is that mechatronics, as an interdisciplinary field of engineering science centered on the concept of synergy, provides tools and opportunities to develop new products and new solutions and—in particular—engineers capable of accepting multi-disciplinary challenges. In the early days, around 1970–1980, mechatronics bridged gaps between mechanical engineering and electrical engineering, computer science, and control theory, enabling new technologies and solutions. Today, mechatronics creates bridges also with other areas—embedded systems, systems engineering, cyber-physical systems—creating synergistic products and solutions based on electrically, and software-controlled mechanical parts, for example, distributed via intercommunicating nodes. Looking back at the past 50 years, most breakthroughs can be seen in these interrelations, when ideas from different areas merge, when technological developments in one area are applied to another.

The key factor here is the engineer. It is the engineer who sees the opportunity to apply something from somewhere else, necessitating conscious mode-switching. It is the engineer who can synergistically apply a control method to a mechanical problem. It is the engineer who understands that a computer can control a process, and how this controller can interact with various components. This engineer understands enough—of mechanical engineering, of electrical engineering, of computer science, of control theory—that they can see the opportunity for synergistic integration of these areas. An expert of mechanical engineering or a computer scientist might not be able to see this. Multidisciplinary approaches both require and enable this conscious mode-switching which underlies insights for synergistic integration between disparate areas of engineering knowledge.

In this chapter, our main driver is to promote the idea of learning engineering skills by practice. Learning by training, learning by engineering. How does a person become an elite athlete? The answer is: by conscious and deliberate practice. How does a person become a professional musician? The answer is the same. A professional musician benefits from a lot of theoretical studies as well, but the theoretical studies do not guarantee the musical skills. In this perspective, a university program in engineering does not guarantee all the necessary engineering skills.

Background

A Brief Reflection on Mechatronics History

50 years ago, when the subject of mechatronics emerged, things were different. In hindsight, it can easily be understood that the idea of integrating mechanics and electronics was revolutionary. Mechanical products were controlled by mechanical controllers. The use of electronics in mechanical products presented a radical shift [1].

In 1976, a charismatic professor in machine elements at KTH, Royal Institute of Technology, visited Stanford University during a sabbatical. He brought a microcontroller back to KTH, an Intel 8008, declaring upon his return that this microcontroller, from now on, should be regarded as a machine element. In the years following, the microcontroller indeed radically changed the subject of mechanical engineering and, in particular, machine elements, at KTH. In the 1980s, research projects emerged exploring added functionality in traditional machine elements, such as adaptive vehicle suspension and motion control in general. The 80s and 90s saw some spin-off companies with relation to KTH researchers and research projects, some of which grew into large industries. All were based on radical innovation via technology shifts, basically by the shift from mechanical engineering into mechatronics engineering.

A Brief Reflection of Historic Mechatronics Founders

The professor in machine elements was probably not unique. As a professor, then appointed by the government of Sweden, he had a responsibility to represent and develop the subject of machine elements. It was not uncommon then for professors in applied subjects to be recruited from industry and many professors had the ability to balance both analysis and synthesis, to strive for what we today refer to as innovation and impact. It might have been easier to disregard the microcontroller, to simply state that it's not a machine element, and keep the subject of machine elements narrower. But, with industrial and societal impact in mind, the functionality of machine elements increases drastically if the subject can be expanded to include microcontrollers. When we turn to the role for mechatronics in the current challenges later in this chapter, let's keep this in mind [2, 3].

In one study of the role of the mechatronics engineer in product development [4], several people with previous relationships to the establishment of mechatronics at KTH were interviewed. In one of the interviews, the technical manager of a mid-sized company reflects on the role of their top systems architect. The company develops extremely complex machines that push the limits of both physics (in terms of materials and optical wavelengths) and computer science (in terms of data throughput). Also, the boundaries for mechanics have been pushed in terms of mechanical stability, precision motion control, and vibrations. The technical director of this company, with around 1000 employees, stated in the interview that there is only one person in the organization who truly carries a holistic understanding of the machine: the systems architect [4]. This person understood the limitations of physics, the relation between these limitations and the limitations of applied control, mechanics as well as the capabilities of software systems. A key research focus of this study, and following studies, is the question of the mechanisms that promote this ability. For this chapter, we need to embrace the above, and, at the same time, realize that the challenges of today are even more complex, when including the global sustainability challenges considered in EcoMechatronics.

A Brief Reflection of Mechatronics Education History

In several previous publications [5–7], we have reflected on the first 50 years of mechatronics and the role of universities in the development thereof. As in the example above, most mechatronics programs emerged from mechanical engineering programs and departments.

In one of these studies [8], we discussed future scenarios of the development of the subject of mechatronics. We argued that the subject of mechatronics would cover larger and larger areas, and that there would be a need to reformulate the subject to prevent it from being diminished. If everything becomes mechatronics, and if mechatronics becomes everything, then the identity is lost. In a study 20 years later, we argue that this has in fact happened [5] already.

Current mechatronics educational programs vary greatly in content, level, and scope. At KTH, mechatronics is a master's degree program offered for students with a mechanical engineering background. At Chalmers University of Technology, a 5-year integrated bachelor's and master's program in mechatronics leans toward industrial automation. Chalmers University of Technology also offers a master's program in mechatronics that is based in automatic control. In other universities in Sweden, mechatronics programs aim at either innovation or offer individual courses. In comparison, mechanical engineering, for example, is rather established and quite similar across universities. Even though each university has its own specializations and focus, the general setup is very similar. In comparison, mechatronics seems chaotic.

Outside Sweden, the situation is even more chaotic. In JKU Linz, mechatronics is offered as a 5-year integrated broad bachelor's and master's program that includes subjects such as mechanical engineering. Mechatronics is an umbrella for mechanical engineering, electrical engineering, etc., while at KTH, among several others, mechatronics is a specialization of mechanical engineering. Therefore, a critical feature of the educational field of mechatronics is that it is considerably local and specific, rather than global and general.

Key Competencies Required to Approach the Challenges of the Future

So, what then are the competencies required to tackle the long-term challenges? What kind of training can stimulate the ability to "solve the problems of tomorrow"? What role can mechatronics play in this situation?

Looking Back: Tool Integration, Systems Modeling, and Communication

In one of the earliest mechatronics research projects at KTH, as mentioned above, adaptive vehicle suspension was studied [9]. The implementation required knowledge and skills in vehicle dynamics and control. Modeling tools in the mechanical engineering domain existed that could be used in the design of the controller. The controller could be implemented in a microcontroller-based system with some understandings of sensors and actuators.

In the following years, more and more focus was put on tool integration and integrated product development methods. For many years, many research projects investigated tool development, integration of tools from various domains, integration of tools for mechanical engineering design, and, for example, control systems. Even today the idea of an ultimate tool, the holistic tool that optimizes a design from all technical aspects, is still discussed [10]. A tool that optimizes cost and performance based on variable mechanical design, electrical design, control design, material design, etc. Many have tried to integrate existing tools, and many approaches exist. Even as we speak, spin-off companies emerge focusing on providing, for example, plugins for design tools, enabling a tool to communicate with another tool.

A parallel development relates to work organization. Besides the tools itself, a need to establish new work methods becomes apparent. In the historic setting, a mechanical engineer handed over his design to an electrical engineer, who handed over to a software developer, etc. A product developer handed over drawings to a production engineer. At KTH, the subject of integrated development emerged concurrently with the development of the subject of mechatronics, and showed many similarities. Both these developments showed the need for joint languages, communication, integrated tools, and processes.

A Current Study on Mechatronics Product Development

In a study of a mechatronic product development project at a large Swedish company, we further investigated the role of mechatronics engineers in this integrated development landscape. A researcher followed a development project closely for about 1 year, participating in all project meetings and interviewing relevant team members. The team consisted originally of five people and grew to about ten as product launch neared. The product in focus was a hand-held tool, a product with motors, sensors, gearbox, microcontrollers, and communication modules. The product was intended for professional use, very specific use cases with a narrow range of potential customers.

The product development team originally consisted of a mechanical engineer, an electronics engineer, a project manager, a test engineer, and an embedded systems engineer. In the later stages of the project, several more embedded system engineers were added to support software development.

Only the embedded systems engineer appeared to take a holistic view of the product. The mechanical engineer made many wise design decisions, and was able to make designs based on ergonomic use cases, heat transfer, etc.—some decisions that can be outside the immediate purview of a mechanical designer. But the mechanical designer stayed away from discussions and decisions that related to power, battery design, control systems, communication, etc. Thus, the mechanical design was isolated from the design of the other aspects of the tool.

5 People or 1000 People? Small Project or Large Project?

In the earlier example [4], the technical director of one of the most complex product development projects in Sweden at the time stated that only his top systems architect could fully have a holistic view of the product and project, i.e., only one person out of 1000 had the ability to fully understand the machine. What are the similarities with the above study, in comparison a much simpler product and project, with a team of five people? Is it fair to say that only the embedded systems engineer had a holistic perspective, and only one person (out of five) that could understand the product?

There is one important similarity between these two projects, and it can be illustrated more easily from the smaller project and the individuals there. The embedded systems engineer had an abstract multi-disciplinary mental model of the product. He also had mathematical models of the system behavior, based on the CAD models, the electrical characteristics, and the control models. The embedded systems engineer had a bachelor's degree in mechanical engineering and a master's degree in mechatronics. The embedded systems engineer did not have the detailed understanding of the mechanical characteristics as the mechanical engineer, but he did understand enough. Likewise, he understood enough of the electrical characteristics to be able to design a control system for the product. The embedded systems engineer had sufficient knowledge, skills, and tools to get a holistic view, to synergistically make decisions on allocation of functionality, for example, for making decisions that affect different domains, to optimize the product based on the overall functionality.

Are there any similarities between this embedded systems engineer and the systems architect described earlier? Are there any similarities between the 5-person product development team and the 1000-person product development team? In both cases, one person is seen to have a holistic view of the project. In the smaller project, this emerged from the project meetings and the interviews with the engineers. In the larger project, it is stated by the head of technical development at the company. In the large project, the systems architect is supposed to have this understanding, it's part of his or her role in the company. The systems architect, on the other hand, is not expected to have detailed knowledge in the subdomains. In the case of the smaller project, the embedded systems engineer has detailed knowledge of the subdomains as well.

Shaping Education

Embedded Systems and Systems Engineering

The embedded systems engineer had a master's degree in mechatronics. This is not necessarily a general rule, as an embedded systems engineer today is more likely to be a graduate from computer science or software engineering.

What we now note here is that we can identify a kind of continuum, from mechatronics to systems engineering via embedded systems. Small project: mechatronics and embedded systems engineering. Large project: systems engineering.

In the early years of mechatronics education at KTH, we saw a unique role for mechatronics engineers. They supported radical innovation, enabled new technological shifts by introducing embedded control systems into mainly mechanical engineering organizations. Companies at that time were still organized into mechanical departments and electrical departments. As time passes, we note that the mechatronics engineers move more and more into software. In the beginning of our mechatronics program, we had only one or two programming courses that enabled the mechatronics engineers to understand software. Today, mechatronics engineers from KTH are fully fledged programmers in embedded systems. There is no real competition; there are not so many other programs to recruit from, and the existing computer science programs stay away from the physical hardware, the hardware-oriented aspects of programming. Anyhow, as presented earlier [5], we note that a transition has happened, from mechatronics into embedded systems, which we argue is mainly due to the increased complexity of mechatronics or embedded systems over the last 50 years.

We argue that the transition from mechatronics into embedded systems is a reaction of a weakened identity of the subject of mechatronics, which relates to the broadening of the subject. We also note that the software of a mechatronics product has increased in importance during these 50 years, to encompass most of both the development cost and the value of a product.

Now, let's assume the following. An embedded systems engineer with a degree in mechatronics will have a higher degree of understanding of the physical properties of the product as well as the electrical properties, control aspects, etc. compared to an embedded systems engineer with a more specialized computer science background. If this is the case, how does this come into play when the embedded systems are designed? Can the embedded systems engineer use embedded systems as a tool for creating a holistic view of the product? Are we seeing a transition from embedded systems to embedded systems engineering?

International Council of Systems Engineering, INCOSE

INCOSE, one of the main professional bodies for systems engineering, gathers systems engineers and systems architects. The professional body publishes recommendations and academic guidelines. INCOSE gives a recommendation that is rarely seen in other professional bodies. INCOSE states that systems engineering requires professional experience.

Traditional professional roles have similar requirements. To become an electrician, a plumber, a painter, you traditionally must gather experience first. Educational programs are based on a master-apprentice system where a student works together with an experienced professional for a few years before becoming certified and independent. In medicine, several career steps follow academic graduation and relate to both further theoretical education and professional experiences. In academia, a tenure track system bases career development on professional achievements which directly relate to professional experiences. So, in this perspective, it is not unrealistic to state that a competence in systems engineering relates to professional experience. What is unique though is that, as a consequence, systems engineering, according to INCOSE, cannot be taught at universities the way universities have been running for hundreds of years. There is a huge difference in the comparisons with medicine and academic development in general. In medicine, it's a specialization. In academic tenure track, it's about establishing a subject and growing as a researcher and faculty member. In systems engineering, it's about something else. It's about taking a leap from one area to another, from something into systems engineering, to be able to grasp a wider area, a broadening of understandings, taking a holistic perspective. And to master the tools that support these activities.

ITRL—Moving System Boundaries

The following example is intended to illustrate one of the main pillars of the argumentation, the idea that a shift in the level of abstraction can enable the transition from design to eco-design, from development to sustainable development. The example shows a transition of a research theme, from transport systems into sustainable transport systems. This is not related to the subject of mechatronics but can illustrate how mechatronics could possibly transition into EcoMechatronics or sustainable mechatronics.

This example, from KTH, has been influential in the shaping of the thoughts of this chapter. In 2012, a research center was established based on a substantial donation from a Swedish major truck manufacturer (Scania). Research relating to vehicle technologies had turned out to be fragmented or, to view it more positively, the technological advancements led to several new potential interactions between a truck manufacturer and a university. The new research center, the Integrated Transport Research Lab, ITRL, was established with the purpose of gathering the KTH research that related to transport and allowing the research center to become a bridge builder between various research groups at KTH and various research and development groups at Scania.

Before long, other industries joined, as well as governmental agencies relating to transport. Swedish telecom industry Ericsson saw opportunities in digitalization of transport systems, government agencies saw the need for new legislation. Many research teams at KTH saw opportunities to interact with industrial partners in joint research applications and projects.

In the beginning, the research center explored mainly technical innovations and research, from vehicle dynamics, control systems as well as ongoing research on transport systems. Now, almost 10 years later, a shift from transport systems toward sustainable transport systems has occurred. The introduction of new technologies and the increased focus on sustainable transports drives a paradigm shift, aimed at achieving a sustainable transport system. The business model for the research group has developed into their current mission statement: "…to integrate disciplines, system levels and stakeholders".

In the beginning of ITRL, the core consisted of a selection of KTH researchers, from the various contributing research groups. Now, about 10 years later, the core research team is built based on new recruitments—from various systems engineering programs, researchers with background in social sciences, with design thinking skills, with systems thinking skills. The research group today has competencies in mathematical modeling, systems engineering, and sustainable development. The research group has established itself within the area of "Complex Dynamic Systems". This is an effect of the ambition to approach the grand challenges, and, more specifically, to tackle the issue of sustainable transport systems. The systems dimension was the missing piece of the puzzle that made the group complete. As a result, the group is now able to adopt a system view on complex issues, such as sustainable transport systems, where the system boundary includes models of user behavior, legal issues, and of other stakeholders as well as the technical fundamentals.

We therefore would like to bring the experiences from this development into the discussion on EcoMechatronics, how the subject of mechatronics can support the long-term challenges. We believe there is a middle path here, a balance between depth and width, between the two aspects of the "T" in a T-shaped engineer [11]. An extreme in one direction could lead to pure systems engineering, to include everything in the models, basically to include more areas by raising the level of abstraction. The other extreme could lead to basically putting too much content into a curriculum, to create an ultimate engineer able to solve any challenge. The conclusion of the above is that we base the discussion on a mechatronics engineer, with a specialization in synergistic integration, able to develop mechatronics products. This mechatronics engineer develops into an embedded systems engineer, using embedded systems thinking to apply a systems thinking perspective on the product, project or problem, and that this competence is developed further and further into a more fully fledged systems engineer, with a mechatronics background. We believe that a university can produce these engineers and, if given chances, these engineers can, if supported during a life-time career, be the right people to accept the grand challenges. Challenges of both today and tomorrow.

Proposing a Way Forward

Now we are ready to outline the proposal for future strategy.

We start with a bachelor's program in mechanical engineering. Then we add a master's program in mechatronics. The master's program can have different focus (more about this later). Then, after some time, we add systems engineering as a professional development program. With this setup, we can prepare the more traditional mechatronics engineers, we can provide industry with the embedded systems engineer (as in the smaller project presented above), and we provide opportunities for future systems architects, as in the example with the large project.

The idea of the T-shaped engineer can be applied to this setup. Each T-shaped engineer will have a specialization, a depth, and a competence in a specific area. It can be in mechanical engineering, in electrical engineering, in applied control, or any other subject. The horizontal part of the "T" represents an ability to synergistically interact with other people, other engineers, and with other specializations. These competencies and skills refer not only to teamwork and project management, but also to the technical ability to understand other specializations and interact with them (Fig. 1).

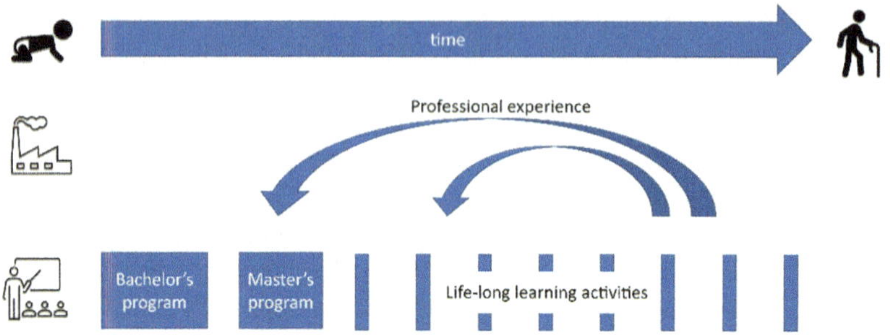

Fig. 1 The proposed way forward: The concept of life-long learning enables professional further education as well as the transfer of experience from senior engineers to more junior engineers. The concept is facilitated by the current move towards more flexible learning processes via e-learning, MOOCs etc.

Bachelor's Degree Level

The bachelor's degree, or part of the academic education, is therefore not crucial in terms of focus. As with the T-shaped engineers, it is more relevant with a heterogeneity than uniformity. It is important, though, that students recruited to the next stage, the master's degree level, have an expertise in a subject that is relevant to mechatronics: mechanical engineering, electrical engineering, etc.

Master's Degree Level

The master's degree level should add both more general competencies and specific skills and competencies. A student with a bachelor's degree in mechanical engineering should be able to broaden their skills with studies in electrical engineering, applied control, etc., and, at the same time, deepen their skills and competencies in mechanical engineering—all with the addition of training in the synergistic integration of these skillsets.

Now, let's relate this to the previous discussion on the importance of embedded systems. We argue that the master's level should include embedded systems education with the purpose of using embedded systems training to create a system perspective on mechatronics. Meaning, embedded systems education should provide both programming skills and architecture skills, as well as introducing students to a systems engineering perspective.

We note many ambitions in favor of the above. Several authors have emphasized the need for a more stringent mathematical approach to mechatronics [12], while others have argued for a deeper focus on computer science and to start investigating the systematic approach [13]. A transition from mechatronics into cyber-physical systems (CPS) is one step in this direction that emphasizes a system perspective.

What we believe is the missing piece here is the ability to integrate other domains into this system perspective. We would like to include what was described earlier as integrated product development; we would like to integrate tool integration ideas, current research on synergistic integration of various domains.

In the first example, we discussed the development of an active suspension. This development project could easily be done by one individual, a mechatronics engineer, at least into a proof of concept, a pre-study. In the second example (the small development project with about five engineers), a mechatronics engineer with competencies in embedded systems could grasp the project, understand all aspects, and integrate various domains. In the third case, the systems architect was an extremely skilled systems engineer.

So, with this perspective, we believe the master's degree level could target these different levels. An individual student could very well reach a T-shaped expertise, with expert knowledge in one area and some skills enabling them to collaborate with others. This person could be heavy in the vertical aspect of the "T", ready to solve intricate challenges. Another person could be more balanced between the horizontal and vertical aspects of the "T", with more emphasis on systems engineering skills, integrative skills, and abilities to collaborate with other domains.

But—to reach the next level requires new thinking.

Life-Long Learning

A combination of e-learning, MOOCs, and the pandemic situation has led to an increased focus on continuous education and re-training of workforces. KTH has announced an ambition that at least 20% of all education within a few years will be offered to non-program students, "outside" the university system. People with existing degrees wishing to learn new things, perhaps people already employed at companies who want to develop their careers. The underlying idea is that university education should no longer be seen as an activity before an industrial career, but rather that university education can support the reality that most people will need to train, re-train, and continue to learn during an entire career.

So the circle is closing. We now have a mechanism in place that can provide the third step, the step that INCOSE mentioned as necessary to develop a systems engineering competence. This third step could be the mechanism where mechatronics and embedded systems engineers train and develop into systems engineers, systems architects, engineers capable of having a holistic view and approaching the challenges of the future. This third step could include more steps, it could be based on the proven method of "master/apprentice" if more experienced systems engineers can share experiences with junior engineers. We therefore believe that the design of this third step, continuous education, life-long learning, is extremely important so that this training can synergistically integrate and utilize both previous educational training periods and experiences from previous industry positions, and include expertise and experiences from senior engineers. Could this be the forum where the systems architect from the 1000-person project trains the embedded systems engineer from the five-person project? And the university, which is the expert on learning and training, facilitates the training?

Conclusions

The underlying reasoning and argument is based on the notion that we cannot solve the problems of tomorrow with the thinking of today. The problems of tomorrow, which are in a way already the problems of today, are directly related to sustainability. All design needs to embrace the concept of eco-design. The thinking of tomorrow, as in the example of the KTH Integrated Transport Research Lab, is that the scope needs to be widened, the level of abstraction needs to be raised, and a system perspective needs to be applied—but without losing the specialist technical domain-specific competencies.

Mechatronics education has for 50 years balanced between breadth and depth, between general and specific knowledge, and between various levels of abstraction. The mechatronics engineer is a specialist in solving problems and creating products that cross disciplines and domains, where the technical solution is optimal regardless of implementation.

With this approach, the mechatronics engineer, with a system perspective, can integrate social aspects, societal aspects, aspects of sustainability, aspects of gender equality—any aspects—as long as the product development team includes enough engineers with the right vertical aspects of the "T"-shape, and so long as enough people in the team get the opportunity to increase the horizontal aspects of their "T"s, for example, by systems engineering training. The example of ITRL shows that these boundaries can be pushed, with the right people and the right training, and that this creates unique possibilities to approach the grand challenges.

An effect of the COVID pandemic is the transition to e-learning, remote education. This transition, together with the emerging need for professional, continuous education, or life-long learning, opens up new possibilities. The main professional body for systems engineering, INCOSE, states that systems engineering can only truly be learnt by experienced engineers. What we see here is the chance to use this opportunity for life-long learning to educate experienced mechatronics and embedded systems engineers in systems engineering, and that this group of new engineers will be sufficiently trained and ready to handle the challenges of sustainable development and eco-design.

References

1. Auslander DM (1996) What is mechatronics? IEEE/ASME Trans Mechatron 1(1):5–9
2. Törngren M, Grimheden M, Adamsson N (2007) Experiences from large embedded systems development projects in education, involving industry and research. ACM SIGBED Rev 4(1):55–63
3. Törngren M, Edin Grimheden M, Gustafsson J, Birk W, Strategies and considerations in shaping cyber-physical systems education. ACM SIGBED Rev 14(1):53–60
4. Ghattas H (2016) Managing complex product development projects: an analytical framework for complex product development (Dissertation), KTH
5. Edin Grimheden M, Flening E, 50 years of mechatronics—what is next? Proceedings of the 20th international conference on research and education in mechatronics
6. Flening E, Edin Grimheden M, Many mechatronics: a discursive model of mechatronics' definitions. Proceedings of the 2018 mechatronics forum, Glasgow
7. Edin M (2013) Can agile methods enhance mechatronics design education? Mechatronics 23:967–973
8. Grimheden M, Hanson M (2005) Mechatronics—the evolution of an academic discipline in engineering education. Mechatronics 15:179–192
9. Lizell M (1985) Micro computer aided suspension. Lic.-avh. Stockholm: Tekn högsk
10. Rahimi F (2021) Towards a concurrent multidisciplinary design optimization of mechatronic systems, Doctoral Thesis, KTH
11. Rogers P, Freuler RJ, The "T-Shaped" engineer paper presented at 2015 ASEE annual conference & exposition, Seattle, Washington. https://doi.org/10.18260/p.24844
12. Lee EA, Seshia SA (2011) Introduction to embedded systems, a cyber-physical systems approach. http://LeeSeshia.org, ISBN 978-0-557-70857-4
13. Marwedel P, Rhode W, Morik K (2013) The TU Dortmund cyber-physical systems program: a step towards multi-disciplinary education. Proceedings of the first workshop on cyber-physical systems education (CPS-Ed 2013) at cyber physical systems week (CPSWeek 2013), Philadelphia, Pennsylvania, USA

Education and Simulation for Electric and Mechatronic Systems in Renewable Energy

P. J. van Duijsen

Abstract Mechatronic systems cover various technical domains, of which electronics and mechanics are the most prominent. But also heat transfer and control are important domains. EcoMechatronic applications for sustainable energy systems are to be found in generation, storage and mobility. Wind power generators, photovoltaic, battery and flywheel storage and especially electric vehicles are the new work-horses in the world of sustainable energy. Most of the applications under discussion use electric machines, and probably the most prominent example of this in EcoMechatronics is the electric vehicle. Therefore, many subjects in this chapter are devoted to the modeling and simulation of electric vehicles and especially how modeling and simulation are applied in the education of EcoMechatronics. In engineering, the role of simulation is undoubtedly strictly woven into the design process, or in other words, modern design without the involvement of simulation is exceptionally rare. Simulation is extensively used to examine the correctness of designs, in addition to its use as a tool to better understand and optimize the performance and/or reliability of a system. In education, the role of simulation is included mainly to teach the use of simulation in the design process. However, also as an educational method, simulation plays a significant role in the form of visualization by means of animation and virtual prototyping. Contrary to the engineering design process where the choice of a simulation method and complexity of the underlying mathematical model is done by the engineer to finally realize a prototype that fulfills expectations, for education the role of simulation is to aid the understanding of the working principles of systems. Engineers seek the highest performance at the lowest cost, whereas for education the goal is to explain the behavior of a system or method, study the application by changing parameters and compare the different results. In this chapter, the various modeling methods are briefly explored and compared. Depending on the domains in which the application is operating, the different modeling methods are discussed. A worked-out example of a sustainable solar-powered electric vehicle demonstrates the use of various simulation methods, along with their mathematical modeling techniques.

P. J. van Duijsen (✉)
The Hague University of Applied Sciences, Delft, The Netherlands
e-mail: p.j.vanduijsen@hhs.nl

P. Hehenberger et al. (eds.), *EcoMechatronics*,
https://doi.org/10.1007/978-3-031-07555-1_18

Introduction

How to apply simulation for education in the field of EcoMechatronics? The simulation should enhance the educational material and not become a subject of its own. Therefore, the use of simulation and the underlying mathematical model should be thoroughly investigated before putting it into practice. In EcoMechatronic systems, sustainability becomes equally important, as is the combination of various technical domains. This means the modeling should also include the special models for sustainable components used in EcoMechatronic systems like solar panels and wind generators, battery and flywheel storage as well as control of electrical machines in electric vehicles. Mechatronics covers a broad spectrum of various disciplines [5, 6]. In this section, we will concentrate on the electrical systems that are used in renewable energy systems. This means that we will focus on motor drives for emobility [7], but not on servo drives. Also, we will discuss solar power generation and storage (Fig. 1).

Overview of Applications

There are many different applications requiring various disciplines. In this section, we will focus on the electrical systems such as

- Electric drives,
- Solar power,
- Battery storage and
- eMobility.

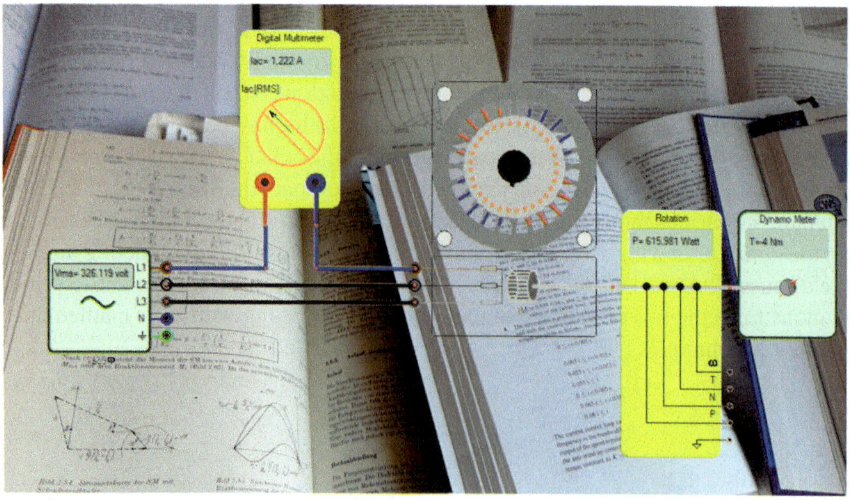

Fig. 1 Education from textbook to virtual prototype

The question is if an application is mechatronic or something else? In reality, various applications are a combination of an actuator with mechanical load, its power supply with an energy source and its control. Varying from a small plunger up to a high-power traction drive, the techniques and tools for designing that system are fairly similar. The physical relations, converted into a mathematical model, are finally solved numerically to obtain static and/or dynamic behavior.

Because of this reason, the methods outlined in the next sections apply to various applications, but because this text is limited in length, we will concentrate on the application of a traction drive for an electric passenger vehicle.

In an electrified traction drive, various domains are coupled, like mechanics, electric, thermal and hydraulic. Control is a combination of basic analog control ICs on the lower level, up to embedded digital control for the complete system at the top level. Using the electric vehicle as an application example throughout this chapter allows us to emphasize on modeling and simulation techniques for a large amount of mechatronic applications.

Educational Methods

There is nothing wrong with a classical printed textbook with static pictures. On the contrary, this source of information is still required to cover the complete topic and explain in detail the physical and mathematical relations. However, adding additional training methods such as visualization and worked-out numerical examples enriches the classical textbook.

Textbook

The textbooks give a good overview of the subject and can detail the evaluation of the mathematical relations that describe the behavior of each mechatronic system. However, practicing those equations requires something more, as will be outlined next.

Numerical Examples

Worked-out numerical examples are valuable since they give insight into the nominal numerical values used in the equations. For example, for an electro-technical student, the numerical value of inertia of a mechanical load in a servo system, or for a large industrial drive like a mill, is not so obvious. The maximum ratings of an operational amplifier regarding output voltage range, current sink capability and slew rate are understandable for him, but a student in the mechanical domain might not be familiar with those values. The equations should not be too complicated, just detailed enough, such that

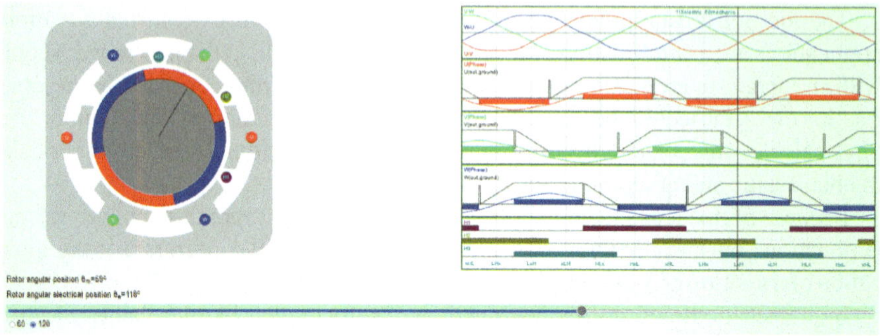

Fig. 2 Animation shows the relation between the angular rotor position, back-emf and outputs from the Hall sensors. The student can choose between 60 and 120° Hall sensor placement and rotate the rotor manually to evaluate the momentaneous back-emf and sensor values [16]

students can follow the approach. For the more detailed numerical examples, the use of simulation (including the explanation of its model and parameters) can give a better understanding. Especially when the student can do parameter variations and can observe the influence of design parameters on the system's behavior.

Animation

Animation can bring the figures from the textbook to life. Manual interaction by setting parameters or shifting sliders immediately shows their influence. Figure 2 gives insight into what Hall signals are generated when the rotor has a certain angular position. Secondly, it gives information on the back-emf of the motor and when the current through each winding should be energized.

Design Tool

The design tool is of help when students encounter an application for the first time and have not yet figured out the size of each component and its parameters. The design tool simplifies and gives default values for dedicated applications and directly provides parameters, graphs and static and dynamic behavior. Changing parameters and/or component size directly reveals their influence, even if the student is not familiar with the operation of the application yet. For example, selecting the values of components in a simulation tool can be challenging if its behavior is unclear and has to be studied. In a design tool, the student has a default value as the starting point. Also, the design tool calculates required parameters after changing the size of a component. For example, the gain and time constant of a proportional-integral controller in a three-phase inverter drive with a permanent magnet synchronous machine [12] are calculated and used as

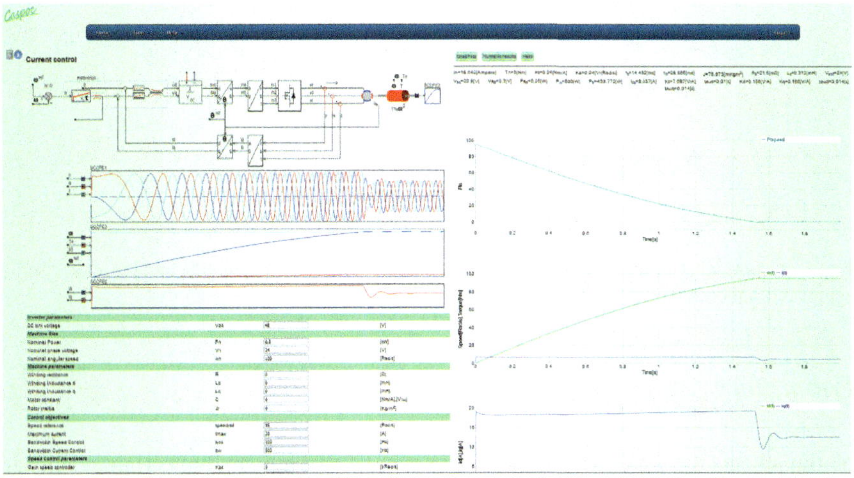

Fig. 3 Design tool directly gives default values for PI controllers and calculates the dynamic behavior in Caspoc [16]

default values for a given electric machine size. The start-up transient response of such a drive is then immediately calculated and displayed in the design tool, as can be seen in Fig. 3.

Simulation and Animation

Simulation shows the dynamic behavior of the overall system. For design and verification, simulation can become complex, but for educational purposes, the simulation and its underlying model and parameters should be as simple as possible. Animation during the simulation reveals visual insight into the application. For example, the flow and amplitude of a current through an electric circuit can be visualized as well as the rotation of a drive train.

Virtual Prototype

Preparing students before going into the laboratory can be done using a virtual prototype. In Fig. 4, an electric drive laboratory exercise can be practiced in advance, and even fault conditions can be studied. As the virtual prototype has the same look and feels as the real setup in the laboratory, students are better prepared when starting the practical assignment in the lab. Also, for explaining basic electronics, virtual breadboards, as shown on the right side in Fig. 4, can be used [16].

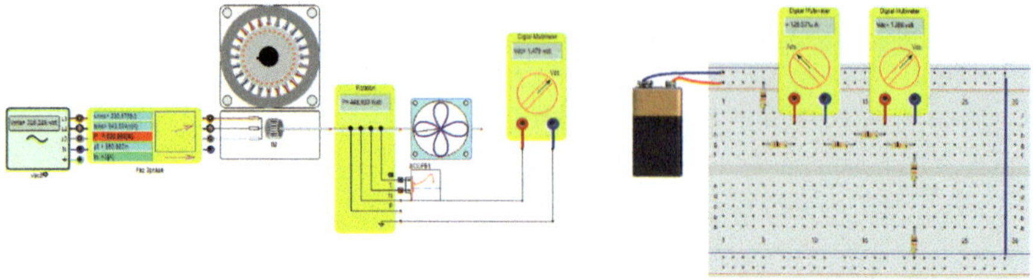

Fig. 4 Virtual laboratory of an induction machine-driven fan on the left side and basic electronics using a virtual breadboard on the right side

Laboratory Assignment

Nothing beats the look and feel of a real practical assignment, but certain results are hard or sometimes even impossible to measure and show, because of safety or difficulty to measure. For example, how to show the magnetic field distribution within an electromagnetic actuator during full operation, or the flow of current in all branches of the circuit at the same time.

However, the combination, where students can prepare the laboratory assignment using design tools, simulation and animation tools and virtual prototyping, helps them to understand what is happening in the laboratory setup on forehand. Secondly, after going through the virtual prototype, the students are prepared for the laboratory setup.

Remote Laboratory

Although multiple students can access a remote laboratory, there is still a limited number of seats available at the same time. Also, the fact that the result is displayed on a computer screen, compared to a real physical measurement, positions this method at the same level as a virtual prototype. Sharing an expensive laboratory setup among several institutes can be arranged via an online booking system.

Verify Laboratory Results with Simulation Results

Verifying the predicted simulation results, with final measurements, can give valuable insight into both the shortcomings of a simulation model on one side and the unexpected behavior of real systems on the other side. Nevertheless, the differences can be explained, in that a measurement is always containing noise and that the simulation result only can be as good as the underlying model and its parameters.

Modeling Domains

The modeling domains, not only for mechatronics, but in general for engineering, include the following [8]:

- Electrical,
- Electromagnetic/Electrostatic,
- Mechanical(Translational and Rotational),
- Hydraulic/Acoustic and
- Thermal.

Also, the signal flow in a control is a modeling domain of its own, but is physically implemented as an electronic or mechanical system and therefore not included in the list above.

The mathematical relations are equivalent in the above domains and are important to understand by the students when moving between the domains [9].

Domain	Across		Through	
Electrical	Voltage	$V[v]$	Current	$I[A]$
Mechanical (Translational)	Speed	$v[m/s]$	Force	$F[N]$
Mechanical (Rotational)	Angular speed	$\omega[rad/s]$	Torque	$T[Nm]$
Electromagnetic	Field intensity	$H[A/m]$	Flux	$B[T]$
Electrostatic	Field intensity	$E[v/m]$	Flux	$D[C/m^2]$
Hydraulic, Acoustic	Pressure	$P[Pa]$	Flow	$Q[m^3/s]$
Thermal	Temperature	$T[C]$	Power	$P[W]$

Before starting going in depth on a specific application, it is worth to introduce the similarities in modeling to the students. This helps in understanding physical properties outside the scope of a student's discipline. For example, students who are profound in electric systems will understand the mechanical relations much easier when the rotating mass or inertia is modeled by a capacitor.

$$I_c = C\frac{dU_c}{dt} \tag{1}$$

$$T = J\frac{d\omega}{dt} \tag{2}$$

By simply replacing current with torque and voltage by angular speed, an electrical engineering student will better understand the numerical value for the inertia J. Secondly, it will help him to set up a multi-domain model as he will be able to translate components from one to another domain. Let us take the example of a permanent magnet dc motor. From an electric viewpoint, the motor has a winding resistance R and inductance L, and there is a back-emf voltage U_{emf}, depending on the angular speed ω of the rotor. From a mechanical viewpoint, there is a rotating mass(the shaft) with inertia J, that is driven by a torque T, originating from the current through the stator winding. Those two models are each given by their own symbols; see Fig. 5.

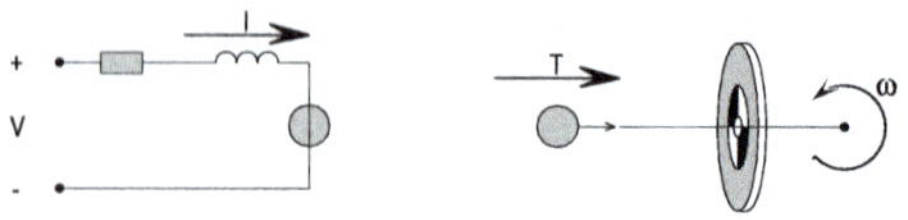

Fig. 5 Electrical and mechanical domains for a permanent magnet dc motor

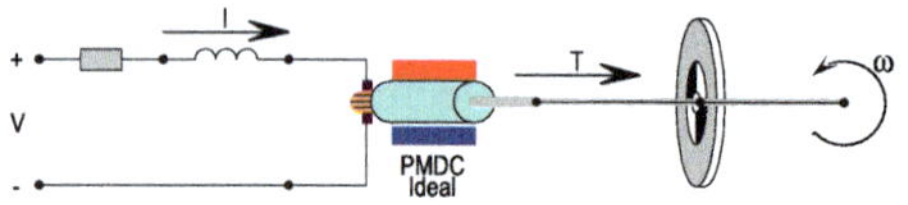

Fig. 6 Coupled electrical and mechanical domain by an ideal model

We can use a dedicated actuator model to couple both domains, and in this case, the ideal dc motor; see Fig. 6.

Combining both domains using a four terminal ideal transformer with gain K, see Fig. 7, couples the electric and mechanical domains as

$$T = K \cdot i \tag{3}$$

$$U_{emf} = K \cdot \omega \tag{4}$$

Because of the wide availability of simulators for electric circuits, the motor is often modeled in the electrical domain where the mechanical domain is translated into the electrical domain; see Fig. 8.

Here, the inertia J as given in (2) is replaced by a capacitor. The capacitor current is related to the torque via (3) and the capacitor voltage is related to the angular speed by 4. An ideal transformer given by (3) and (4) is now modeled by a current-controlled current source, for the current and torque relation (3), and a voltage-controlled voltage source, for the angular rotor speed and the back-emf voltage (4), as shown in Fig. 9.

The main advantage is that now the entire model can be handled by a single simulator for electric circuits. The complexity of the model can even be reduced by transforming the components on the right side of the coupling to the left side, taking into account the gain K of the coupling; see Fig. 10.

The model is now greatly simplified, but the results from the simulation have to be interpreted accordingly. In the simulation, the output torque is measured via the inductor current and is linearly proportional as given by (3). The back-emf U_{emf} is

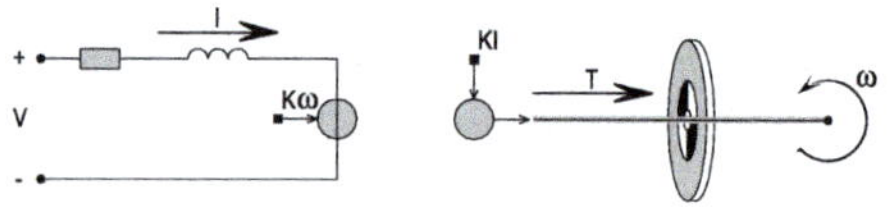

Fig. 7 Coupled electrical and mechanical domain by controlled sources

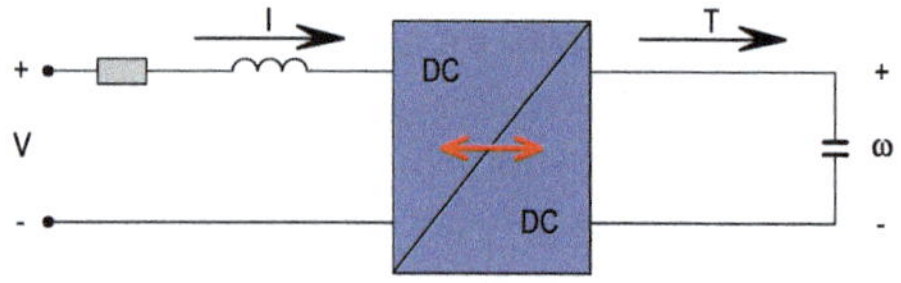

Fig. 8 Coupled electrical and mechanical domain by an ideal transformer and capacitor for modeling the inertia

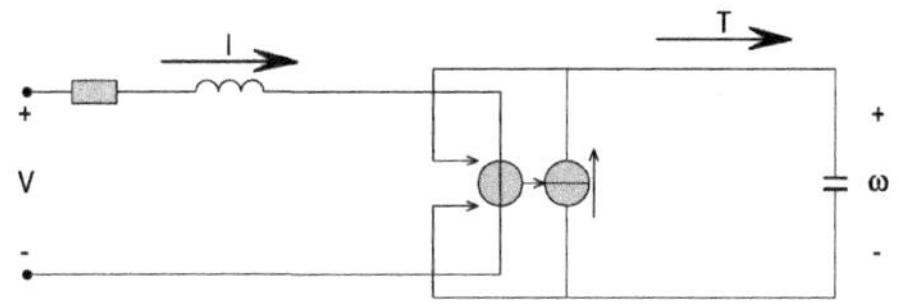

Fig. 9 Coupled electrical and mechanical domain using controlled source and capacitor for modeling the inertia

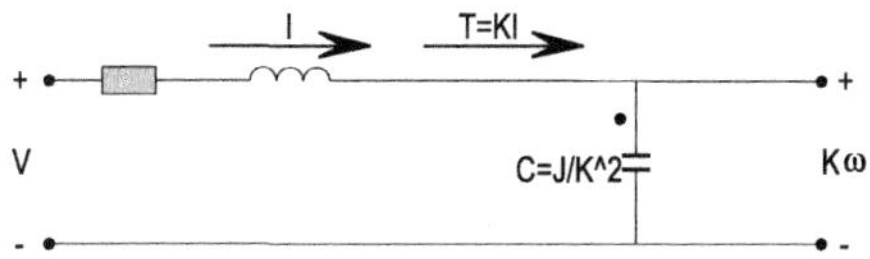

Fig. 10 Coupled electrical and mechanical domain represented by electrical model

measured, which is linearly proportional to the angular speed ω by (4). In Fig. 10, the angular speed ω is now measured by the voltage across the capacitor and has to be scaled according to (4). The value of the capacitor is scaled as given by (2) and (1) as $C = J/K^2$.

In a similar way, thermal models are created and represented by their equivalent circuit model as will be shown in the next section on the underlying mathematics of the model.

Mathematics of Modeling

The mathematical model is used to discover the static or dynamic behavior of the system that is being studied. Depending on the result that is required, first a decision has to be made, on what kind of mathematical model is going to be used. For example, when studying a control system, a block diagram representation of the underlying differential equations is mostly satisfactory, but if one wants to simulate an analog controller build around op amps, a circuit simulation would be preferred.

1D Block Diagram
In a block diagram, a signal flow from input to output is calculated.

2D Circuit model
A circuit model is two-dimensional, describing the mathematical relations between the across and through quantities obeying the two Kirchhoff laws:
Kirchhoff's Current Law: The summation of all currents flowing into a circuit node is zero.
Kirchhoff's Voltage Law: The summation of all voltages in a closed loop is zero.

3D Finite Elements
Thermal, mechanic, magnetic and electric field problems are mainly described by a large number of mathematical non-linear relations and can be described by either 2D or 3D partial differential equations.

Block Diagram

The Block Diagram is in principle a visualization of a differential equation [3, 4]. Therefore, it lends itself to describe dynamic systems as well as control structures. Its unidirectional signal flow can contain feedback loops as long as the loop is broken by inserting memory blocks like an integrator block or delay block.

Naturally, physical dynamic systems can be written using causal mathematical relations, so a block diagram using integrator blocks can be used. However, when non-causal relations are included, a Differential Algebraic Equation [DAE] has to be used [3, 4]. Electric circuit and mechanical systems containing non-linear components almost all have to be modeled by a DAE.

Since modeling of a system in a mathematical model has to be done manually, this type of modeling is not easy to understand in vocational training.

Circuit Model

A circuit model is a two-dimensional representation of a DAE, where the non-causal relations are included easily.

Non-linear relations have to be modeled, but require the use of numerical solvers, like the Newton-Raphson [NR] or Steepest Decent [Gradient] methods; see [1].

Modified Nodal Analysis [2] is applied in most popular circuit simulation programs, because of the straightforward implementation and the inclusion of non-linear mathematical relations. Two-dimensional Finite Difference Methods [FDM], for solving two-dimensional electromagnetic and heat spreading problems, can be seen as circuit models, based on first-order differential equations. For example, heat spreading underneath a semiconductor is modeled by a one-dimensional Foster or Cauer network containing resistors and capacitors. Modeling the thermal coupling between two adjacent semiconductors as a thermal resistor changes the mathematical model from first-order differential equations into two-dimensional partial differential equations. Modeling the connections in a circuit model represents the actual layout of the system and is, therefore, easy to understand.

Finite Elements

Finite Element Method [FEM], Boundary Element Method [BEM], Finite Difference Method [FDM] and Methods of Moment [MoM] are used in two and three-dimensional problems for all disciplines; see [18]. The application in education is limited to the higher grade classes, because of the complexity of the underlying mathematical principles and the understanding of correct meshing. Although applied in mechanical engineering for stress analysis and in heat transfer for fluid and airflow, the user interface hides the complexity of translating the physical application into a mathematical model and the choices made regarding meshing and material properties. Also, there is nearly no insight into the influence of certain parameters of the application on the results, for example, the influence of the thickness of a rod on bending. Only multiple simulations with parameter variations give numerical insight.

Thermal Mathematics Example

The losses and thermal behavior are typical examples of how different types of mathematical equations are used to calculate the behavior. In the first place, a simple linear relation states that power loss through a thermal resistor will give a temperature rise, calculated as

$$T = T_{ambient} + P_{loss} \cdot R_{thermal} \tag{5}$$

This gives a very rude first guess regarding maximum temperature at maximum power loss. A simple block diagram containing a single multiplication would suffice.

However, the temperature rise is a dynamic process, and the temperature-dependent variation of the parameters of semiconductor switches like Mosfets influences the generated losses. The dynamic behavior can be modeled by a differential equation. A circuit model containing resistors and capacitors can model the differential equation [13]; see Fig. 11a.

The power loss from a semiconductor is represented by a flow and modeled by a current source; the thermal resistance is modeled by an electric resistor, and across this resistor, the electric voltage is the temperature rise or fall. Masses that store thermal energy are modeled by a capacitor. The temperature of the mass is represented by the voltage across the capacitor, as shown in Fig. 11a.

This equivalent circuit model from Fig. 11a models the thermal behavior of the Mosfet correctly, as long as the parameters for the equivalent circuit model are chosen correctly.

When the cooling components cannot be modeled by an equivalent electric circuit, because of the complex structure of that component, a partial differential equation has to be used to solve the temperature distribution. A Finite Element or Finite Difference thermal model is then used to calculate the heat spreading on the surface of the printed circuit board; see Fig. 11b.

To model the thermal spreading of the power loss in a printed circuit board, a 2D model would be the first approach to study the heat spreading. A FEM package could be used, but an alternative approach is to use a circuit simulator and model heat spreading by a 2D circuit model containing resistors and capacitors. For modeling the heat spreading in a circuit simulator, 2D and even 3D models are possible, as shown in Fig. 12. A 2D model would contain a single layer from the model in Fig. 12 and only the conducted heat flow is considered. By connecting the 'upper' and 'lower' nodes, the two-dimensional models become three-dimensional.

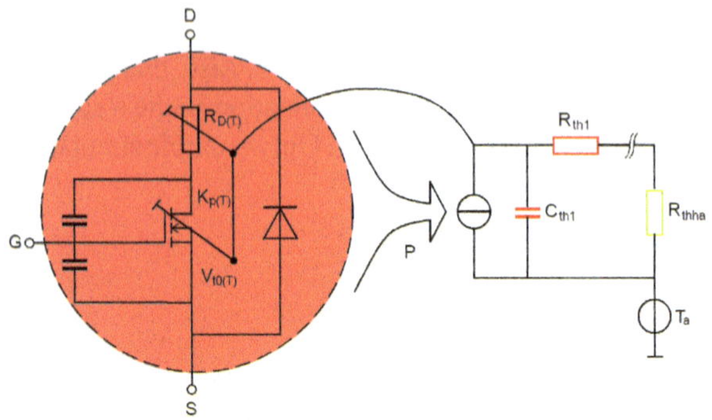

(a) Coupling between the semiconductor model and the thermal model.

(b) Finite Difference model of thermal heat spreading on PCB for a Mosfet.

Fig. 11 Thermal modeling for power electronic losses

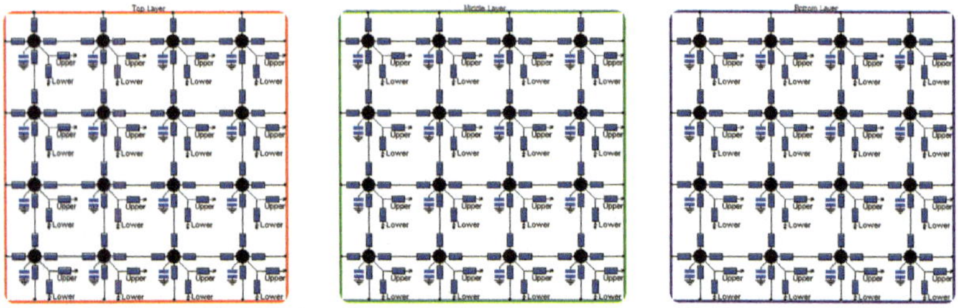

Fig. 12 Thermal modeling of three layers in a printed circuit board [13]

Modeling Detail

Depending on the goal of the outcome of the simulation, one has to choose a level of modeling. Easiness of using the model should be evaluated against the complexity of the model and the number of required parameters; see Fig. 13.

The affordable simulation time is typically what the user expects and what is known as getting simulation results immediately, after a coffee break or overnight. Increasing the model complexity and therewith the number of parameters also increases simulation time. Therefore, only a short time period can be simulated. So, for example, to simulate the detailed transient ringing during on-off switching of the semiconductors taking place on a nanoseconds scale should not be combined with a complete drive cycle simulation for an electric vehicle taking 1200 s for a typical drive cycle. In that case, an averaged model with a maximum simulation time step of, for example, 1 ms to 100 ms should be used.

A complex model can be supported by just a few parameters if internally the detailed parameters of the model can be derived from just a very basic set of parameters from the user. A typical example is the detailed parameters for the electromagnetic relations in a synchronous generator based on a single parameter, being only the power level.

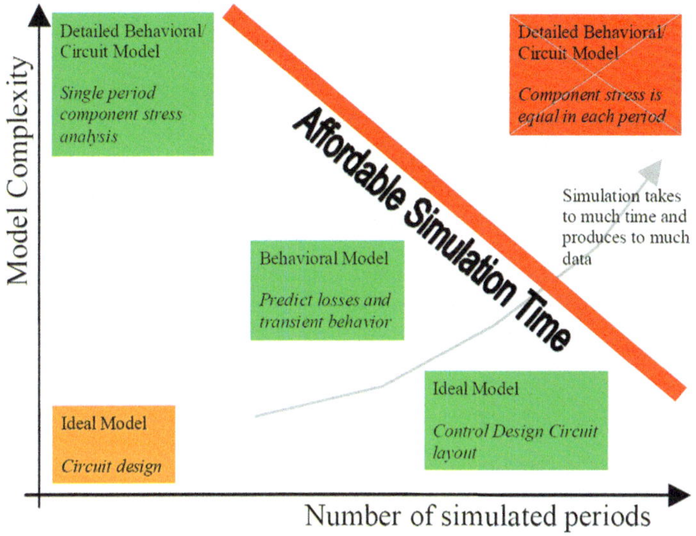

Fig. 13 Model complexity versus simulation time

First Grade Courses

Simulation to show behavior for first grade courses requires a limited set of parameters, but needs internally a complex model to mimic dynamic behavior in detail. Take the example of the solar module for which we would like to see the characteristic current-voltage relation for various solar insolation values; see Fig. 14. To measure these characteristics in a simulation, we would first need a detailed solar module model, and it has to be loaded with a variable load to measure every possible load point; see Fig. 14a. For a first grade course, this would be over-complicating the subject to discuss, namely how do voltage and current change when there is more or less solar insolation. Dedicated solar module models that can be fed with basic parameters like the number of cells in series and the overall size of the module, temperature and a typical efficiency indication give enough information to obtain the characteristics [14, 15]. By including the numerical values for the series wire and parallel leakage resistance, the characteristic can even be tuned to the characteristic given in the datasheet. To see the characteristic for various solar insolation values, this parameter is directly set to step from 100 to 1000 W/m^2 in steps of 200 W/m^2; see Fig. 14b. Each time the parameter is stepped, the complete characteristic was drawn in the scope.

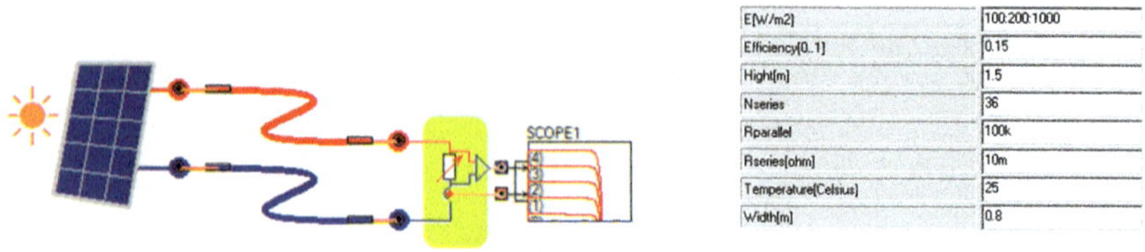

(a) Characteristics for a solar module.

(b) Parameters of the solar module.

Fig. 14 Characteristics for a solar module for various solar insolation values in Caspoc [16], which is varying from 100 to 1000 W/m^2 in steps of 200 W/m^2

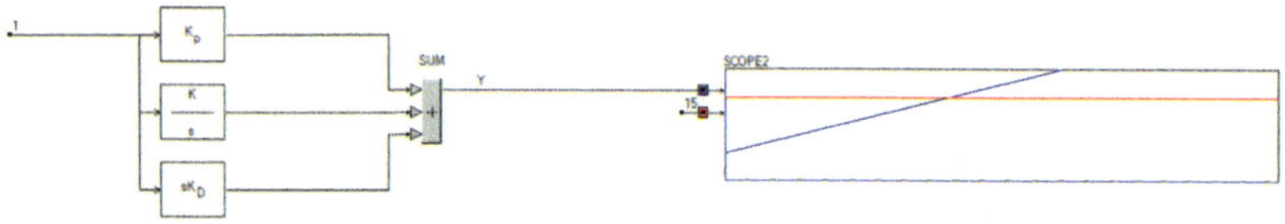

Fig. 15 PID controller modeled by Block Diagram in Caspoc [16]

Detailed Simulation

Detailed models where all parameters are accessible to the user to show the non-linear influence of parameters of components in the overall system sometimes require a different approach. Take the examples of a PID controller as modeled in Fig. 15. Only a single parameter per block is required, being the P, the I and the D value. The simulation will nicely calculate the transient response to an input step function, but will neglect any saturation, bandwidth and slew rate issues known with PID controllers.

If the simulation is modeled by a more detailed circuit model, the Op amp model takes care of those linearities; see Fig. 16. The drawback is that you have to know how the circuit will be designed and what the circuit parameters are going to be. But the simulation will now include the saturation because of the supply voltage and the transient response is now in accordance with the bandwidth and slew rate limitations of the Op amp.

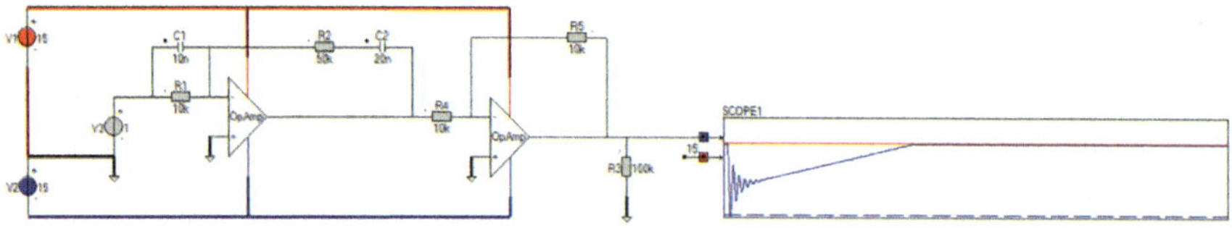

Fig. 16 PID controller modeled by a circuit model with op amps in Caspoc [16]

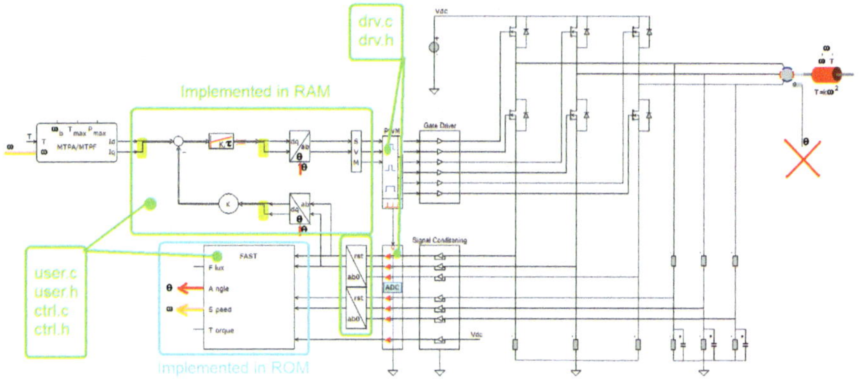

Fig. 17 Speed and current PI controllers modeled by block diagram and embedded into C code in microprocessor [17]

Control Design

Verification of control methods and the influence of parameters on the dynamic behavior is one of the important aspects of simulation. From the control structure, C code for the embedded microcontroller should be possible to derive and to be implemented into the target hardware. If we take the PID control example from Fig. 15, C code can directly be exported. A typical current control in a drive system modeled in Caspoc-Embed [17] is shown in Fig. 17. In this figure, the PID controller from Fig. 15 is part of a larger model that is modeled on different abstract layers. Finally, the C code for the complete control is exported and uploaded into the target hardware and the control system can be tested on the real target hardware.

Example Simulation

Two open-source numerical methods are presented to simulate the acceleration and power consumption of an electric vehicle. The first demonstrates the usage of a spreadsheet, and the second is a numerical integration method implemented using JavaScript in an Internet browser.

Using a simulation program in a spreadsheet, you can see how the drive cycle is used to calculate the amount of power required to drive the vehicle and how much the battery is discharged. Since this is primarily a calculation based on the data from the drive cycle, a simple spreadsheet can be used.

To perform a simulation, a numerical integration method has to be applied. Instead of using a commercial simulation program, the simulation is programmed in JavaScript, and HTML is used to visualize the results. All results can be viewed numerically in a browser.

Drive Cycle

The drive cycle is a look-up table containing the speed of the vehicle against time [10, 11]. From this data, the acceleration is calculated, from which the required power to drive the vehicle is calculated. The spreadsheet from the companion website https://caspoc.com/books/ecomech/ contains the EUDC drive cycle, where in the first column the time in seconds and in the second column the speed of the vehicle in km/h are given. The intermediate speed during two time points can be achieved by interpolation, but is not really required. We are only interested in the acceleration during that time step and the average speed. Both are calculated from the time and speed information as

$$a^i = (\frac{1}{3.6})\frac{v^i - v^{i-1}}{t^i - t^{i-1}} \tag{6}$$

$$v^i_{average} = (\frac{1}{3.6})\frac{v^i + v^{i-1}}{2} \tag{7}$$

where v is the speed of the vehicle in kilometer per hour, $v_0 = 0$ and $v_{average}$ is the average speed in meter per second. The time steps are given by the index i and t is the time in seconds. From this, the amount of traction force $F_t[Nm]$ required to drive the vehicle is calculated. First, the road load force F_r, the slope force F_s and the drag force F_d are calculated as well as the force required to accelerate the vehicle F_a

$$F_s = m \cdot g \cdot sin(\alpha) \tag{8}$$

$$F_r = m \cdot g \cdot c_r \cdot cos(\alpha) \tag{9}$$

$$F_d = \frac{1}{2}\rho \cdot c_d \cdot A \cdot v_{average}{}^2 \tag{10}$$

$$F_a = m \cdot a \tag{11}$$

where m is the total mass of the vehicle in kg, $g \approx 9.8\,\mathrm{m/s^2}$ is the gravitational constant, α is the slope of the road in radians, the road load coefficient $c_r \approx 0.01$, drag coefficient $c_d \approx 3.5$, $\rho = 1.25\,\mathrm{kg/m^3}$ is the air density and the frontal area $A \approx 2.4\,\mathrm{m^2}$. The acceleration a in m/s^2 and the average vehicle speed $v_{average}$ in m/s are the inputs from the columns from the spreadsheet.

The amount of power $P[Watt]$, traction motor torque $T_m[Nm]$ and current $I_{battery}[A]$ required are then calculated from the total traction force $F_t[Nm]$.

$$F_t = F_s + F_r + F_d + F_a \tag{12}$$

$$T_m = \frac{1}{n} r_w \cdot F_t \tag{13}$$

$$\omega_m = n \frac{v_{average}}{r_w} \tag{14}$$

$$P = T_t \cdot \omega_m \tag{15}$$

$$I_{battery} = \frac{P}{V_{battery}} \tag{16}$$

where $V_{battery}[V]$ is the battery voltage, $n \approx 4$ is the gear ratio between the electric machine and the drive axis and $r_w \approx 0.3$ is the wheel radius. The gear ratio is approximately set such that the motor rotates with 314 rad/s while the vehicle runs at 80 km/h to 100 km/h.

For each time period indicated by the index i, the traction force, motor torque and speed are calculated and are displayed in new columns. Using the spreadsheet example, the parameters of the vehicle can be changed to calculate the consumption for various types of vehicles (Fig. 18).

Electric Vehicle Simulation

In the previous paragraph, speed and acceleration were given in a look-up table. But if we are interested in knowing the speed for a given maximum motor power and torque, a numerical simulation in the time domain can be performed. For this, we need the differential equations describing the system, and using a numerical integration method, the dynamic response in the time domain can be simulated. Since vehicle speed and acceleration have to be calculated and are not given in a table like in the previous example, they are calculated in the same way as the maximum torque from the electric machine.

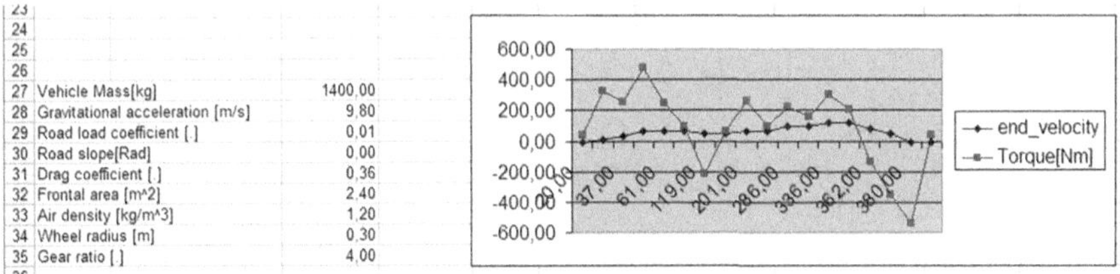

Fig. 18 Parameters of the electric vehicle for the drive cycle simulation in an spreadsheet, with the graphical representation of electric vehicle speed [km/h] and wheel torque [Nm]

$$F_t = \frac{n \cdot T_m}{r_w} \tag{17}$$

$$F_a = F_t - (F_s + F_r + F_d) \tag{18}$$

$$a = F_a / m \tag{19}$$

where F_s, F_r and F_d are calculated as in (8), (9) and (10).

The differential equation $a = \frac{dv}{dt}$ is now rewritten into an integral form

$$v = \int_{time} a\,\mathrm{d}t \tag{20}$$

The numerical integration of the integral in (20) is a good exercise in numerical simulation in the time domain, as students can learn the influence of the simulation time step dt and will understand on how to rewrite a differential equation in the form of an integral equation that can be solved numerically. There are various numerical integration methods that can be taught here, but as a first introduction, Forward Euler [1] is easy to implement,

$$v^{i+1} = v^i + dt \cdot a^i \tag{21}$$

$$t^{i+1} = t^i + dt \tag{22}$$

where i is the index for each numerical integration time step and time t is updated for each numerical integration time step with the integration step size dt.

The listing below shows the implementation of the mathematical relations using JavaScript.

```
t=0;
while(t<=Tsim)
  {
  Ft=n*Tm/rw;                     // Traction force
  Fs=m*g*Math.sin(alpha);         // Road slope force
  Fr=m*g*cr*Math.cos(alpha);      // Road load force
  Fd=0.5*rho*cd*A*v*v;            // Drag force
  a=(Ft-Fs-Fr-Fd)/m;              // Acceleration
  v=v+dt*a;                       // Calculate new speed v
  t=t+dt;                         // Update time with dt
  document.write("Time="+t+"[s]Speed="+v+"[m/s]<br>");
  }
```

Listing 18.1 Javascript source code for calculating vehicle speed where Tsim is the total simulation time

Instead of using a commercial simulation software, Listing 1 shows how to set up such a basic simulation using JavaScript and display results using HTML in a standard

Internet browser. Only a text editor is required to create the HTLM/JavaScript page, and the browser will take care of the interpretation and calculation of the JavaScript.

Worked-Out Example

In this example, the different simulation methods and their underlying mathematical models are described. It is not the intention to derive a single bulky model that includes all aspects, but more an attempt to guide the reader on which simulation method is best suited for a specific component. Instead of focusing on a single mathematical model that includes all behavior, or trying to cover all possible mathematical models that could be used, for some typical components, a single simulation method is demonstrated based on a mathematical model that includes just enough detail to verify the dynamic behavior. Based on the vehicle simulation from the previous section, in the following online sections, typical components are discussed, and how to apply simulation for such a component is also discussed (see Fig. 19).

Companion Website

On the companion website https://caspoc.com/books/ecomech/ you will find the worked-out examples as well as the spreadsheet for the drive cycle simulation and the simulation of the electric vehicle (Fig. 20).

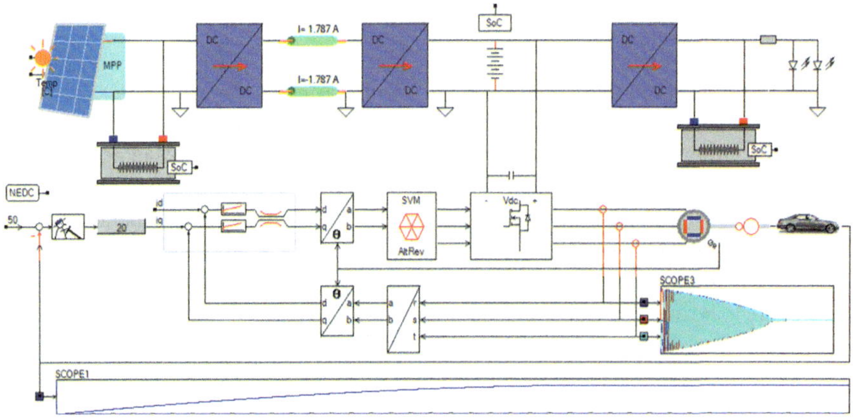

Fig. 19 In a multiphysical simulation in Caspoc [16], the entire electrical model from the solar panel, via batteries, dc-dc converters, inverter and control to the wheel is included

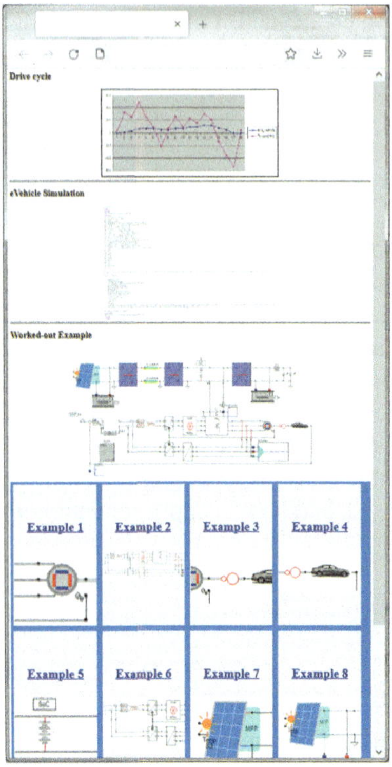

- Interior Permanent Magnet machine
- Current Controller
- Mechanical backlash of the gearbox
- Electric vehicle mechanical load characteristic
- Electric vehicle Li-Ion battery
- Embedded C code export from control
- Solar panel
- Storage of solar power in Lead Acid battery
- Charging the Li-Ion battery in the Electric vehicle
- Losses and heat transfer

Fig. 20 Examples on the https://caspoc.com/books/ecomech/ page

Conclusions

The application of modeling and simulation in EcoMechatronics, for the purpose of education, is various. Teaching the basic principles of mechatronics, simulation and especially animation enriches the process of education. Simulation can be applied in all educational stages, to enhance the understanding, but also to replace laboratory assignments with virtual prototypes. The complexity of the mathematical model, underlying the simulation in EcoMechatronic systems, depends on what kinds of simulation results are required by the user. Depending on this, simple or more detailed models are to be used. The chapter concludes with two basic simulation examples that can be reproduced without the requirement of third-party software. An electric vehicle combines many aspects of EcoMechatronics and is, therefore, used as an application in these simulations. More detailed simulations of a solar power electric vehicle are available on the companion website https://caspoc.com/books/ecomech/.

References

1. Press WH et al (2007) Numerical recipes 3rd Edition: the art of scientific computing. Cambridge University Press; 3rd edn. ISBN-13:978-0521880688
2. Ho CW, Ruehli A, Brennan P (1975) The modified nodal approach to network analysis. IEEE Trans Circuits Syst 22(6):504–509. https://doi.org/10.1109/TCS.1975.1084079
3. Dorf R, Bishop R (2016) Modern control systems, 13th edn. Pearson. ISBN-13:978-0134407623
4. Bishop RH (2017) Mechatronic systems, sensors, and actuators: fundamentals and modeling (Electrical Engineering Handbook) 2nd ed. CRC Press. ISBN-13:978-0849392580
5. Schiessle E, Wolf F, Linser J, Vogt A (2002) Mechatronik 1. Vogel Communications Group GmbH & Co, KG. -13:978-3802318603
6. Schiessle E, Reichert M, Ruf W, Vogt A (2002) Mechatronik 2. Vogel Communications Group GmbH & Co, KG. -13:978-3802319044
7. Emadi A (2005) Handbook of automotive power electronics and motor drives. Taylor & Francis Inc. ISBN:978-0-8247-2361-3
8. Heimann B, Albert A, Ortmaier T, Rissing L (2015) Mechatronik: Komponenten-Methoden-Beispiele. Carl Hanser Verlag GmbH & Co, KG. -13:978-3446444515
9. Lenk A, Pfeifer G, Werthschützky R (2001) Elektromechanische Systeme, Mechanische und akustische Netzwerke, deren Wechselwirkungen und Anwendungen. Springer-Verlag. ISBN:978-3-662-06984-4
10. Guzzella L, Sciarretta A (2015) Vehicle propulsion systems: introduction to modeling and optimization, 3rd edn. Springer. ISBN: 978-3642438479
11. Ehsani M, Gao Y, Longo S, Ebrahimi K (2018) Modern electric, hybrid electric, and fuel cell vehicles, 3rd edn. CRC Press. ISBN: 978-1498761772
12. Schröder D (2015) Elektrische Antriebe-Regelung von Antriebssystemen. Springer Vieweg, 4. Aufl. ISBN:978-3642300950
13. Lutz J, Schlangenotto H, Scheuermann U, De Doncker R (2018) Semiconductor power devices: physics, characteristics, reliability, 2nd edn. Springer. ISBN:978-3319709161
14. Quaschning V (2019) Regenerative energiesysteme: Technologie-Berechnung-Klimaschutz, 10th edn. Carl Hanser Verlag. ISBN: 978-3446461130
15. Mertens K (2020) Photovoltaik: Lehrbuch zu Grundlagen, Technologie und Praxis, 5th edn. Carl Hanser Verlag. ISBN:978-3446464049
16. Caspoc (2021) Simulation research. https://www.caspoc.com/
17. Caspoc-Embed, Caspoc (2022) https://caspoc.com/embed
18. Binns KJ, Lawrenson PJ, Trowbridge CW (1995) The analytical and numerical solution of electric and magnetic fields. Wiley. ISBN: 9780471962427

Robot-Assisted Teaching—The Future of Education?

Robert Rayner, Kristina Kerwin, and Nicholas Valentine

Introduction

One of the most recent and significant developments in education in recent times is the creation of new, innovative teaching aids. These new educational products aim to help teachers provide students of all ages with a more varied and engaging learning experience, with the goal of allowing them to achieve better learning. One such tool, which currently sits at the state of the art, is the creation of a variety of new robotic education devices, each with their own set of features and capabilities. Whilst these devices are no doubt fun and exciting for learners, much debate exists around their pedagogical efficacy. This chapter will seek to contribute to this debate, by providing insights into the world of teaching with robots as experienced from the points of view of three very different individuals, each looking at the topic from a very different angle. Ultimately this chapter will seek to address the question, are robotic teaching aids a passing fad, or are they really the future of education?

In Part I, Kristina Kerwin will consider the theory behind active learning and how it ties with the use of robotic devices in teaching from the research perspective and will look at a study that was carried out to demonstrate how the use of this state-of-the-art technology can provide a more engaging and active learning experience for pupils.

R. Rayner (✉)
University of Hertfordshire, London, UK
e-mail: r.rayner2@herts.ac.uk

K. Kerwin
London South Bank University, London, UK
e-mail: kerwink2@lsbu.ac.uk

N. Valentine
UK Ministry of Defence, London, UK

P. Hehenberger et al. (eds.), *EcoMechatronics*,
https://doi.org/10.1007/978-3-031-07555-1_19

In Part II, Robert Rayner will talk about his experience of using robots in teaching where he has over the years experimented with a variety of robotic systems as teaching tools in his lessons and will share some of his experiences.

Finally, in Part III Nick Valentine, a practising robotics engineer, will then consider the situation from a learner's perspective and will depict how his interaction with robotics in a classroom environment led to him pursuing a career in engineering. A learner who was once inspired by a robot who became an engineer who builds them.

Part I—Kristina Kerwin

The Researcher's View

It is argued [1] that true sign of mastery in the art of teaching is an ability to achieve a state of deep learning in one's students—a high level of sustained engagement that translates into solid understanding of the topic and performance. Success depends on the efficiency of the method by which the teacher delivers subject information to the students. In the past, the dominant teaching methodology was based entirely on directed instruction. More recently, researchers suggested that use of interactive processes in learning has proven to be the most effective and is advocated as one of the best practice methods. This approach is also known as Constructivist Approach, and it is what UK curriculum Key Stages are based upon. It is also often referred to as "*active learning*" or "*hands-on/minds-on approach*".

Every modern-day classroom in the developed world takes the advantage of currently available technologies to enhance student active learning. Thus in 2014, BBC education correspondent Sean Coughlan reported that almost 70% of primary and secondary schools in the UK used tablet computers [2]. Technologies such as overhead projectors, social networking, online teaching, class blogs and wikis, podcasting, interactive whiteboards, tablets, and mobile devices—are now considered as everyday staples in the UK schools. Further, it is predicted that education will continue to improve through integration of new technologies and methods [3].

In recent years, commercially available robotic devices have attracted the attention of Science, Technology, Engineering and Mathematics (STEM) educators. It is perceived that these devices hold a great potential for enhancing active learning and adding interactivity during the lessons. As a consequence, robots are being gradually introduced to the classrooms as teaching aids, particularly when delivering STEM subjects.

For example the *Nao* robot has recently been used successfully to teach Special Educational Needs (SEN) students. Because of its humanoid appearance and lack of distractive emotional facial expressions, *Nao* proved to be efficient in teaching children with autism. The robot is designed to be fitted with sensors for different applications and has an online interface that makes it fairly easy for teachers to customise their teaching sessions [4]. LEGO Robots [5] have been used in teaching basics of control theory at HE and FE level and proved to be a popular tool among university and college level students and staff.

There have been several other projects involving robots for teaching in the recent years and in the UK one of these studies made it to GCSE curriculum as a pilot syllabus aimed at KS3 and KS4 students with an aim to engage students in STEM subjects and inspire them to pursue a career in engineering. The program was designed to help students learn both practical and theoretical STEM skills using FANUK robots in partnership with VEX robotics, teaching students about the installation and maintenance of the robots within the manufacturing environment [6].

Another study, conducted by Faisal and Kapila [7] from the Polytechnic Institute of New York University tested the use of LEGO Mindstorm Robots [5] in an elementary school to determine the advantages of the use of robotic devices over the conventional instructional method of teaching. The study involved students working in groups while conducting an activity that was designed to ensure that every child team member participated. The researchers emphasised that the use of LEGO Robots [5] was viewed by the students as a reward and offered a meaningful minds-on experience as well as accommodating for versatile learning styles and stimulating deep learning.

Interestingly, Hobeika et al. [8], Gerraty et al. [9] and Bunzeck et al. [10] all agree that having meaningful experiences benefits human cognitive development by connecting new information with already existing mental frameworks. This process interconnects the brain networks associated with memory, insight, analogical thinking, awareness, motivation and lastly—**reward**. Hence, there is a good argument that using toy robots to explain abstract STEM theory through play could bring a positive input into the quality of teaching STEM subjects, with children being able to retain more information long term. In addition, the versatility and programmability of robotic toy kits also makes them well-suited for this task.

Nonetheless, although it has been strongly suggested that such devices could be beneficial as teaching aids, little work has been done to quantify their effectiveness. Additionally, little work has been done to identify the best methods of using such devices as teaching aids. In this context, research was carried out to study the efficacy of using LEGO Robots [5] to teach STEM subjects to primary school pupils. The intension was to explore and quantify the effectiveness of using robotic devices in teaching of STEM theory at UK Key Stage 2 (KS2), and to suggest the scheme of best practice methods to do so [11].

The Theory of Learning

To fully evaluate the efficacy of the use of robotic devices in the realm of teaching it was important for me to understand the theory behind deep learning and what triggers the effective information retention in students. Prior to building my experiment I felt it was imperative to study different teaching approaches that involved active learning and identify the most relevant method. The research carried out by Marton and Säljö [12] categorised learning into two types:

Surface learning or short-term memory—All new information we receive is initially stored as a short-term memory. In surface learning students only concentrate on the main facts of given information without understanding of the meaning behind the material. In other words, they are memorising the information in a disengaged manner. This is not an effective way of learning as most effort is spent on memorising rather than understanding information. Obviously, when it comes to recalling the material, significantly less information is remembered [13].

Deep learning is where the real learning begins. Students transform the knowledge they gain by exploring it deeper, aiming to understand the meaning behind the written words in their textbooks. This learning process occurs when students interact by getting real-life examples of the applications of information they learnt. Contrary to memorisation, learners are engaged in critical thought. Because deep learning is more engaging, the information is more likely to be retained long term. It can be argued that the more actively one is engaged with the topic, the better understanding of the topic is achieved leading to one's development. According to McLeod [14] the deeper the information processing level is during the learning process, the easier it would be to recall later. In other words, characteristic of deep learning is a high level of sustained engagement in the topic which then translates into high level of understanding and performance [15]. The ultimate goal in teaching is in-fact to achieve the state of deep learning in students. Marton and Säljö [12] identified 6 conceptions of learning that students experience when studying. Starting from the lowest those 6 conceptions are:

Conception 1—Quantitative increase in knowledge.
Conception 2—Memorisation.
Conception 3—Acquisition of facts for later use.
Conception 4—Abstraction of meaning or making sense.
Conception 5—Process designed to understand reality.
Conception 6—Personal development.

Frederic Reif suggests that knowledge can be split even further into the categories of **usable, effective, efficient and reliable**. Usable knowledge is knowledge deep enough that it can be applied in different spontaneous situations [15].

It appears that in order to achieve the state of deep learning one needs to actively engage the learners in the topic which can be realised through interaction and application of the information using the real-world scenarios, also known as practical tasks. Rosenshine [16] argues that best in their field teaching practitioners always preferred using hands-on activities after basic material was learned, as a most efficient way of ensuring that students acquired, rehearsed, and connected knowledge. He concludes that the most efficient way of supporting deep learning is through reiteration of the information mixed with practical tasks after the end of each step of the lesson. This active learning approach is historically known as or Constructivism. Traditionally, there are two main schools of thought on constructivism:

- Jean Piaget's '*Constructivism*'. Piaget was a pioneer of the theory and the author of the systematic theory of Cognitive Development [16].
- Lev Vygotsky's Cultural-Historical theory and Zone of Proximal Development (ZPD) [18].

Although advocating two different constructivist approaches, both theories agree on the importance of delivering new knowledge to children through active learning approach. Constructivist theory stems from a scientific study based on the observation of human cognitive development. In essence, it states that humans actively create their own new skills, by exploring and assessing what they already know [19].

A constructivist approach considers a number of different teaching practices, but it is narrowed down to encouraging students to use active techniques such as practical experiments and real-world problem solving, and then accessing the learnt information and reflecting on their understanding with some teacher guidance to help build on the knowledge gained.

Piaget argues that children perceive four stages of maturation and experience, that children are active learners and learn through assimilation and accommodation, the key is interaction with physical and social environments. Piaget further states that development occurs in 4 stages [20]:

1. The sensorimotor—age 0–2.
2. Preoperational—age 2–7.
3. Concrete operational—age 7–12.
4. Formal operational period—adolescence through to adulthood.

In 1988 in the UK the Education Reform Act was introduced [22] which brought about the UK National Curriculum, in which constructivist theory is prevalent to this day as is evident from its Key Stage structure. The most recent version of the UK National Curriculum was introduced in September 2014. Influenced by Piaget's theory of Stages of Development, it consists of 4 Key Stages with each stage based around average child's development of that age. The curriculum is designed to address the learning needs within the child's ability [22].

Vygotsky's Cultural-Historical Theory

In his Cultural-Historical theory Lev Vygotsky reasons that the role of culture and social interactions play a major part in human cognitive development. Vygotsky states that any function in the child's cultural development is biplanar: first it appears on the social plane, and then on the psychological plane. Speech is major psychological tool in child's development of thinking. With age and development of child basic speech becomes more complex [23]. Vygotsky's theory is guided by the following basic assumptions:

1. The knowledge derives from culture, community, and society.
2. Speech and thought initially develop separately, then merge with maturation.

3. Conversation with adults assists child development.
4. Children are able to perform more challenging tasks if assisted by advanced individuals—a technique called '*scaffolding*'.
5. Play is accentuated as very important in the learning process since play leads to abstract thought which in turn leads to higher mental functions.
6. The difference between the ability to perform a task assisted, with minimal assistance and independently is called Zone of Proximal Development and that is where the learning occurs.

Although Piaget's Constructivism is the preferred model in the modern-day curriculum, it is interesting to note that Vygotsky's theory found strong support in contemporary models of professional development, where the emphasis is placed on '*social interaction*' as means for faster development of professionals [24]. Furthermore, in recent years, Vygotsky's sociocultural approach has been gaining popularity among the educators and is often used together with standard teaching curriculum methods and in teaching children with SEN [25].

As a matter of fact, there are several contemporary teaching theories stemming from the Constructivist approach that have become popular in the recent decade. The two classics have evolved and merged into new, more refined theories that are applicable in modern-day schooling. For example, Benjamin Bloom's Taxonomy [26], has underlying links to Paige's constructivism. Bloom's 3 main domains are based on Paige's cognitive stages and the constructivist theory that learning development occurs on the basis of building on the previous knowledge. However, it also interlinks with Vygotsky's theory of scaffolding as the means of identifying target skills needed for learning.

To date, there is no single distinctive theory that offers a perfect solution to the problem of maximising the process of deep learning. There is, however, a number of different best practice methods that have been reviewed, modified, and combined over the years, and they all agree that active learning with the guidance of an advanced individual plays a big role in the quality of our deep learning.

Case Study—The Efficacy of Teaching STEM Subjects to Primary School Age Students Using LEGO Robots

A great part of all STEM related subjects is theory-based which is difficult to deliver via active learning. This lack of active learning affects students' potential for deep learning and their perception of STEM subjects in general. With this in mind, the focus was on the delivery of the fundamentals of STEM theory to a young audience because of their innate tendency to learn through play, selecting an abstract STEM topic that would be difficult to convey in an active, tactile manner. The aim was to measure the state of deep learning achieved through the use of robots in class vs conventional teaching methods and to quantify the effectiveness of using robotic toys in delivering STEM related subjects to KS2 students. The chosen age group was Year

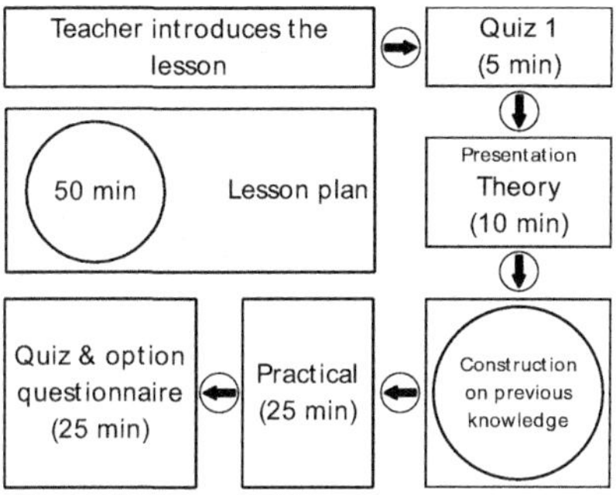

Fig. 1 Systematic procedure in the form of a lesson plan

5 KS2 students (aged between 10 and 11 years). The taught theoretical STEM subject was Speed. All the teaching materials and testing methods developed for the purpose of the study reflected the age of the participants, incorporating constructivist (i.e. Piaget's) argument that children perceive several stages of maturation and experience.

It was important to consider both the quality of deep learning and the quality of children's learning experience. The evaluation of the quality of deep learning was achieved by using a 3-stage knowledge testing process in the form of a quiz and the assessment of the learning experience was achieved using the opinion questionnaire. To help establish the effect of using robotic aids on children's deep learning, the lesson was taught under controlled conditions using the lesson plan sown in Fig. 1. This helped evaluate whether the input variable (of using robots in teaching) had a significant impact on the desired output and manifested in greater information retention in children.

Two groups of children were taught the same topic using the same lesson plan with one group being taught using robotic devices, another (control group) using an alternative classic active learning method. The purpose of the experiment was to compare the two approaches: lesson using the robotic devices versus the lesson using classic teaching methods, to justify the use of robotic devices in teaching STEM theory. A classic active learning method is also referred to as '*bodystorming*' in the study. Bodystorming is a technique often used by product designers in the first stages of design process where subjects represent and act out different ideas using their bodies (for example, when deciding where to put a particular feature on a car seat). In this instance test subjects were using their bodies to solve a number of velocity problems by running, walking, or jumping around the classroom.

Design of Experiment

To effectively carry out the experiment with minimal noise, the input was designed in such a way that it could be applied universally using both models: with robotic aids and without. It was noted that to further reduce the bias and ensure experiment quality, it was favourable for both lesson variations to be delivered by the same

Table 1 Design of experiment

Input		Level	Responses
Lesson on speed			
Controlled Variables	Uncontrolled Variables	Age/KS/Year Group	
Practical using robotic aids	1. Teaching quality 2. Children's cognitive abilities	10 y/o KS2 Year 5	1. Quality of surface learning 2. Quality of deep learning 3. Quality of learning experience
Practical using bodystorming			

teacher, however this was not practically possible in a school setting. It was also essential that the children participating in the experiment were generally not familiar with the topic, of similar age and were not streamed according to their cognitive abilities (see Table 1). The method was designed to be suitable for an unassisted third-party use, i.e. teaching practitioners were able to follow the experiment using provided detailed instructions.

The reason behind choosing KS2 Year 5 s was, because in primary education children are not separated into classes according to their cognitive abilities and are not familiar with the topic, therefore the year group was in line with the parameters outlined above in Table 1 and fell into normal distribution range, since randomisation was necessary to avoid the impact of hidden variables.

The experiment was carried out at a primary school with a total sample of 45 students (two Year 5 groups: 23 ($\pm$1) mixed ability students in class). Each class had its own teacher (a variable that could not be omitted). To assess the quality of deep learning a 3-stage testing procedure was designed and delivered in the form of a quiz which was given to the students at 3 instances: at the beginning of the lesson, immediately after and 4 weeks after the lesson. The tasks were split into three levels of difficulty to give children some feeling of confidence, as difficult testing could result in negative feedback on learning experience which in turn could affect final results. Children were also told that they were expected not to be able to answer the questions in the first test as this is a completely new topic for them. They were encouraged not to guess the answers. All 3 tests were identical in values, however, to avoid any discrepancies in data collection, the test was designed in such a way that it could be delivered to children on three different occasions without the danger of basic memorisation. Hence, the multi-choice answers within the questions had been shuffled while the actual answers remained the same for each question.

The lesson was designed around the questions asked in the Quiz, to ensure the relevant information on the topic of speed could be learned and retained. This, however, did not involve any direct answering or practicing of those questions. Detailed lesson instructions and materials were given to teachers in advance of the lesson. Teachers were also shown how to operate the robots prior to the lesson.

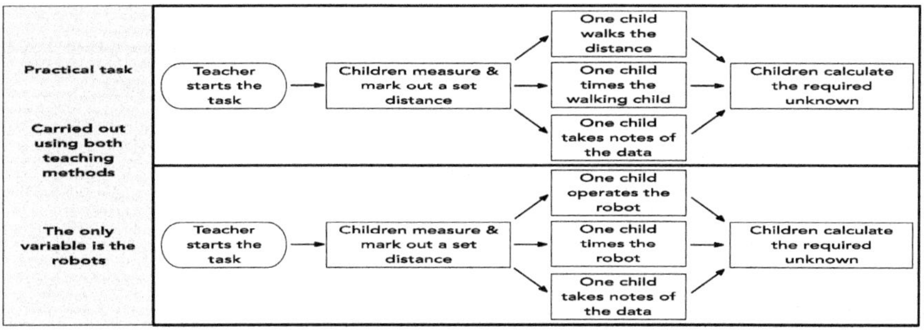

Fig. 2 Task flow diagram

The first test was given to children at the beginning of the lesson to assess their base knowledge level, followed by a short PowerPoint presentation on basic speed theory and a warm-up task to encourage student interaction to construct on children's prior knowledge. After the warm-up task pupils were split into groups and were given a worksheet with instructions and equipment for the practical task. In the group using robots, teacher first demonstrated how to operate the devices for pupils to be able to follow on their own. In the control group the teacher explained about bodystorming. There were 3 tasks in total for the practical part with enough iterations for each child to take part as set out in Fig. 2. Each task reflected a different formula subject: speed, time, and distance.

At the end of the lesson students were given the test again to assess the level of their surface learning, and opinion questionnaire to assess their learning experience. It was observed that children appeared more excited about the final test part as opposed to initial quiz and most of them vocally expressed feeling '*clever*', because they were able to answer the questions.

Results

A total of 45 children participated in the study: 21 pupils in the robotics group and 24 in bodystorming group. The year groups were mixed at a ratio of 5 girls to 6 boys. Results reflecting on the quality of learning experience are shown in Fig. 3. Here, the technology enhanced group (the group using the robots) is represented in light grey with the bodystorming group in black. The results indicate that technology enhanced group enjoyed the experience more. This group rated the first 3 lesson activities higher than the bodystorming group. During this time the robots were introduced to the lesson and used in the practical task. The lesson wrap up was initiated while children completed the second test (the final stage of the lesson) and the drop in the score for that stage of the lesson reflects that the children were no longer exited by the activities of the lesson.

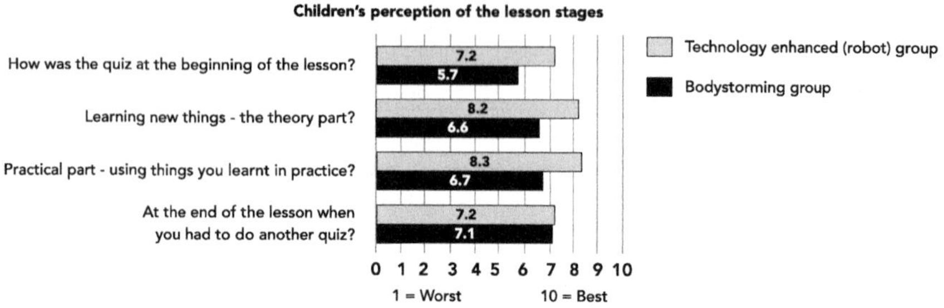

Fig. 3 Quality of learning experience—opinion questionnaire results

Children in the technology enhanced group on average scored higher in the opinion questionnaire, despite initially perceiving STEM subjects with lower confidence that the other group, meaning they had greater quality of learning experience than the bodystorming group, even though both groups were actively engaged in the lesson.

Level of Surface and Deep Learning Data Analysis

The Quiz results are shown in Table 2 and Fig. 4. By taking the mean average scores from the available test results for both groups, it became obvious that the group taught using robotic aids showed a higher percentage increase in subject knowledge when the average scores were compared.

To quantify the degree of surface and deep learning in both groups, percentage change was calculated between tests 1, 2 and 3 using the following metric:

$$\% = \frac{Score\,2 - Score\,1}{Score\,1} \times 100$$

First, by assessing the degree of surface learning, percentage change between Tests 1 and 2 was calculated. The results show that the technology enhanced group benefited slightly more from the lesson on average with 76% vs 63% respectively. This was further confirmed when assessing participant individual scores, the group taught using robots benefited more in general, considering that group on average scored lower in basic knowledge test (Test 1).

Next, percentage change was calculated between Test 2 (Level of Surface Learning) and Test 3 (Level of Deep Learning, given 4 weeks after the lesson). As can be seen from Fig. 4, there is a fair percentage decrease in the non-robotic lesson group. This can be explained by the fact that over time (i.e., after 4 weeks) children naturally forgot some of the information that they had learnt in the lesson, which is in line with currently available literature on cognitive development stating that forgetting rate when using best practice teaching methods can be up to 29% [27].

Table 2 Mean average test scores

Group subject knowledge levels before and after lesson							
	Base knowledge (%)e	Score percentage change Test 1 v Test 2 (%)	Score percentage change Test 1 v Test 3 (%)	Score percentage change Test 2 v Test 3 (%)	Test 1—base knowledge (points out of 9 points)	Test 2—surface learning (points out of 9 points)	Test 3—deep learning (points out of 9 points)
Group 1—Robots	29	76	107	17	2.64	4.65	5.44
Group 2—Bodystorming	43	63	28	−22	3.88	6.32	4.95

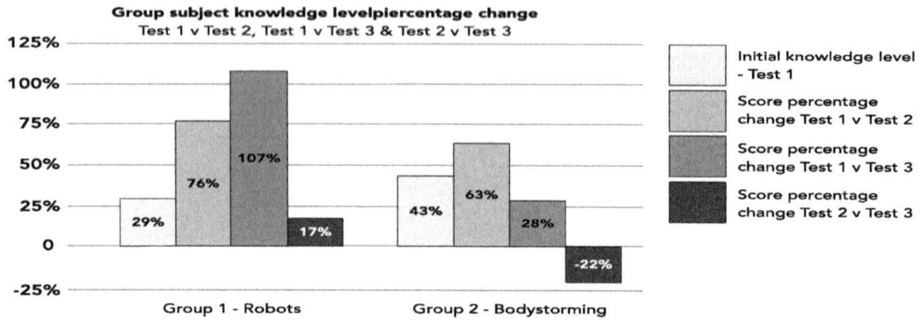

Fig. 4 Subject knowledge level percentage change all tests compared: Group 1—Robots v Group 2—Bodystorming

Overall, technology enhanced group showed a 17% increase in their Test 2 vs Test 3 average score. Group 1, the technology enhanced group scored higher 4 weeks later than straight after the lesson, whereas average final test score 4 weeks after the lesson for Group 2, the group that used bodystorming method, decreased by 22%.

The Quality of Deep Learning score of the technology enhanced group has more than doubled, considering their initial base knowledge score (in Test 1) which was lower than the other group's score. This indicates that the quality of deep learning in the group that had the lesson using robots, is much greater than the group that was taught using conventional best practice methods, although both groups have shown a good degree of information retention long term.

It could be argued that although both groups appeared to have gained and retained information long term in general, the fact that the technology enhanced group's test scores have more than doubled four weeks after the lesson, as opposed to naturally decreasing over time, as was the case with the group that used conventional best practice teaching methods, suggested that using robotic teaching aids significantly impacted on the quality of their surface and deep learning. Furthermore, the evidence from the opinion questionnaire indicated that the quality of learning experience was better for the technology enhanced group, although both lessons were taught in an active manner using hands-on-minds-on approach. It should be noted that several months after the experiment the math tutor involved in the delivery of the lessons mentioned that the students still retained the knowledge on speed.

Conclusion—The Researcher's View

Based on the findings of this research, it is safe to conclude that there is a positive correlation between the levels of learning in test groups, and the presence of the robotic teaching aids during the lesson. This also seems to have had a positive impact on the quality of learning experience, confirming that using robotic aids in teaching can be an effective way of achieving greater information retention levels in learners.

It is clear that there is strong evidence of pedagogical usefulness therefore suggesting that robot-assisted lesson could well be a feature of future teaching.

Part II—Robert Rayner

The Educator's View

An important aspect of being an effective communicator is to have the ability to thoroughly understand the psychology of your audience. With this knowledge, it is then possible to tailor the style and nature of the manner of your delivery to allow you to best connect with their mindset. The best communicators are able to speak in a way that draws in their audiences and captures their attention for just long enough to convey their messages but does not demand so much attention from them as to exhaust or bore them. Good communicators use a variety of tactics to draw in an audience and extend the time that they can hold their attention. Good communicators are also careful to use language, which is natural and easy to understand, avoiding, for example, the use of overly complex technical jargon, which some members of the audience might not understand.

Through my years working as a schoolteacher and university lecturer, practising and honing these fundamental communication skills has been instrumental in increasing my ability as an educator. Thankfully, this aspect of my work has been greatly aided by a large body of research, carried out over many decades, into the ways in which people, and in particular youngsters, think and learn. This research has result in the development of a wide variety of teaching methods and classroom tactics, which have been evidenced to work with the psychology of young minds to increase the chances of achieving the gold standard effect in teaching, deep learning. The nature of these tactics and more specific information on the psychological mechanics of learning is discussed in the first part of the chapter and so will not be repeated here.

A daunting issue for modern educators is that a lot of the fundamental research work into the mechanisms of learning is quite old, with the most significant work by Piaget et al. [18–20, 23–25] having been published over a century ago. A comforting fact for educators of today is that the fundamental psychology of young people has not changed significantly since this work was done and the mechanisms of learning, as identified all those years ago, are just as relevant in today's twenty-first century world as they were when they were first published. Vygotsky, Piaget et al. observed that best learning was achieved when students are emotionally invested and engaged in their lessons. Students have also been observed to learn best when they can draw on their past experiences to reinforce and solidify the concepts being taught. Teaching methods that can achieve these effects have been observed to be effective in drawing in and holding the attention of students, putting their minds in the best possible psychological state to take in and store information. These concepts are practical embodiments of the pedagogical philosophy of constructivism and are embodied in modern-day teaching practice in the theory of active learning.

Whilst the fundamental psychology of students has not changed over the decades, arguably the expectations of students has become more demanding in recent times. Teaching methods and teaching tools that might once have been seen as state of the art may now be seen as dull and old fashioned by modern students. For example, once upon a time, a textbook with coloured, high-resolution images included in its pages might have been a thing of wonder and excitement to students. Today however, such a feature is commonplace and so students are on the lookout for the next big thing to entice and excite them during their learning journey.

In accordance with the teachings of Vygotsky, Piaget et al., significant effort has been placed in recent years to develop learning curricula that are rich in active learning. The rationale for this work is evidential findings that active learning is both more enjoyable for students and is a highly effective way of achieving better deep learning. Developing of such curricula, is a long-standing challenge for teachers, with the development of active learning teaching methods to teach fundamental aspects of STEM subjects being particularly difficult. To help overcome this problem a wide variety of teaching tools have been developed, and made available for purchase by educators, to facilitate a wide range of active learning activities. Particularly in more recent times, some of these tools have been very technically advanced, with a plethora of state-of-the-art products such as computer software and electronic devices being released. Perhaps the current pinnacle of the state of the art of high-tech teaching aids are robots.

A lot of debate currently exists in the teaching community regarding the efficacy of such devices as teaching aids. The first part of this chapter, however, describes research which suggests that robots could be highly effective teaching tools, and can even provide a mechanism for the teaching of fundamental, theoretical STEM topics using active learning methods. The work, however, describes the use of *one* particular type of robot deployed in *one* particular lesson plan. The robot chosen in the work was well understood by the research team, prior to the work being carried out. The capabilities and limitations of the robot were well known, and significant preparatory work was done to identify the best way of using the robot. There are, however, available today a wide range of other educational robotic devices, which can be integrated into lesson plans in numerous different ways.

For educators looking to use robots in a teaching environment, choosing the right robot for a particular purpose can be very daunting. Firstly, it can be difficult to weigh up the strengths and weaknesses of different robotic devices and to be able to identify their key features. It can also be difficult to understand how best to use a given robot in a lesson. With robotic teaching aids being a relatively new development, there is currently only a limited level of understanding of these issues. As a teacher who has had the luxury to teach extensively using robots, I have experimented with numerous robot-assisted lesson plans and observed hundreds of learners interacting with educational robots, with these interactions often leading to some surprising, unexpected but pedagogically useful effects. Perhaps most importantly, I have experienced, first-hand how robots can be powerful teaching tools, drawing looks of amazement and awe from learners and providing them with learning experiences that are unique, varied and highly memorable. In this section of this chapter, I will discuss some of the key observations I have made whilst teaching with robots, observations that will hopefully be useful for other educational professionals considering using them in their lesson plans.

Reactions to Robots

As previously discussed, a particular challenge with modern students is that they are invariably very forward looking and aspirational people. They are constantly looking to the future to visualise what their future world could be like. With the rapidly developing scientific and technological world in which we live today, this mindset is entirely understandable and should be encouraged. The particular challenge that this modern mindset presents educators, however, is that classical, tried, tested and established teaching methods, such as the use of pens, paper and textbooks, can in the presence of a rapidly technically developing world, appear to learners, outdated and dull despite their proven pedagogical benefits. An additional issue with current teaching practice is that, in many classrooms around the world, the use of such classical teaching methods is predominant with little variety in lesson plans being practiced. This lack of variety can, understandably disengage some students from the learning process as the repetitive use of a limited number of teaching methods can over time become dull an uninteresting for them, limiting the quality of their learning.

A particular issue with classical teaching methods is that, when contrasted to some more modern equivalents, they lack the ability to effectively draw on modern learners' experiences and emotions, so limiting the scope for their active engagement in the learning process, again limiting the quality of their learning. In an attempt to combat this, significant work has been done throughout the teaching community to develop new, novel lesson plans which align more closely with the mindset of modern students. The goal of these lesson plans is to make the learning experience more engaging and varied for them, allowing them to immerse themselves more fully in the learning process. One such development, is the use of educational computer programs such as *Kahoot, GeoGebra* and *Mathletics*[1] in teaching plans. Modern students are well used to using computers in their everyday lives and so the use of computers to assist with their learning feels natural and current for them. The use of computer software also provides students with a different learning experience compared to classical teaching methods. The use of computer software assisted lesson plans alongside lessons taught using traditional methods can add variety to a curriculum, so making the learning experience more interesting, enjoyable and engaging.

[1] At www.kahoot.com/, www.geogebra.org & www.mathletics.com respectively.

An additional lesson style that has, to date, been less commonly explored, involves the use of robots as teaching aids. As previously mentioned, I have taught numerous lessons using robots and I have observed students universally enjoying and embracing this form of lesson, no matter what the age of the student. Students understand that robotics is an area of technology which is rapidly developing, with robots becoming more advanced as time goes on. Some people, including me, believe that robots will someday become integral parts of our everyday lives. Indeed, this is a vision of the future that many futuristic, science fiction stories portray. As such students perceive learning with robots as a glimpse of the future world, a world which fascinates and captivates them. From a lesson planning point of view, the inclusion of lessons including robots can allow educators to develop new teaching methods and new lesson plans, which are distinctly different from those already developed. As such, on the most basic level, the use of robots as teaching aids allows for additional lesson variety in taught curricula.

In terms of pedagogical usefulness, much debate currently surrounds robots. In concurrence with the findings of the investigation described in the first section of this chapter, I have found robots to provide a strong positive pedagogical effect. It must be stated however that presently, due to the fact that the use of robots as teaching aids is so rare, robot-enabled lessons currently enjoy a level of exclusivity and novelty and are therefore extremely memorable for students. Indeed, whenever I have taught classes with robots, students have freely commented to me that the class was a unique experience, the like of which they had never experienced before.

This feedback in itself, is evidence of the pedagogical benefit of robots which they generate through their current novel status. The current rarity of lessons taught using robots and the distinct nature of these lessons in contrast to what students routinely experience, result in robot-enabled lessons to be viewed by students as highly memorable, *special events*. Many students are able to remember these lessons long into the future. Indeed, in the next section of this chapter, Nick Valentine, now a grown adult, discusses how he can *still* vividly remember a lesson he was taught involving a robot when he was just a young child. The positive psychological effects of creating highly memorable events can be exploited by educators to achieve excellent learning in students, with the use of robots providing a means of doing so.

Broadly speaking, when humans experience a particularly special event, memories of this event are placed in a portion of the brain responsible for long term memory storage. These memories can be recalled and recollected from memory by individuals, often with great accuracy, long periods of time into the future. By making an educational event a memorable, special event, as information of the particulars of the event is stored in long term memory, so too is information regarding the educational concepts being taught. Information regarding the event as well as information regarding the educational concepts taught are then able to be recalled by individuals long into the future. Since this process has resulted in educational information being stored in long term memory and with this information also being easily retrievable, it can be seen that the highest standard of learning, deep learning, has been achieved.

I have personally, exploited this powerful, novelty-driven, pedagogical effect in a lesson plan to assist with the teaching of a fundamental mathematical concept. On

Fig. 5 The *Nao* robot (© Robert Rayner)

this occasion, I used a humanoid, educational robot called *Nao*, manufactured by Softbank Robotics,[2] to teach an aspect of core algebra to a cohort of 15-year-old boys. An image of this robot can be seen in Fig. 5.

In my lesson plan, I used the *Nao* robot as a robotic *sidekick*. With its ability to generate audible speech and its ability to recreate human-like motion, the robot was used to dictate and act out a story to provide a narrative for the lesson. This story sought to provide an applied context for the mathematical concept being taught. The cohort was enthralled by the robot and freely admitted that the lesson was unlike any they had previously experienced. They also said that they had enjoyed the lesson and found it fun and exciting. Evidence of the quality of learning achieved through that lesson was evident when I taught the same cohort again over a year later. Whilst the cohort struggled to recall and apply some other mathematical concepts taught in other lessons, which had not involved robots, they were able to readily recall the robot-assisted lesson and the mathematical concepts taught in that lesson, with great accuracy. Since so much time had passed between the moment of memory recall and the moment of teaching, there is strong evidence that deep learning had been achieved.

It is important to note that in the robot-assisted lesson, the robot was used purely as an *actor*, with which the students were not able to directly interact. The lesson was very didactic in style with the lesson plan lacking in active learning elements. This is of course a very sub-optimal way of teaching and a very sub-optimal way of using the robot since it does not exploit its active learning enabling features. Despite this however, the students were able to remember the lesson with extreme clarity, an effect, which I suspect is due primarily to the novelty of the lesson. This would

[2] www.softbankrobotics.com.

therefore suggest that currently, due to novelty, robots are able to generate a strong positive pedagogical effect simply by the fact that they are used in a lesson. The effect will occur even if the method of use of the robot is sub-optimal and even if the robot has somewhat inappropriate features for a given lesson plan.

In the short term, educators can certainly exploit this novelty effect to amplify the effectiveness of lessons but in the long term this effect cannot be relied on. Just as students are no longer amazed by high-resolution images in textbooks, should robots become commonplace in educational environments, they too will cease to be viewed as exciting novel devices and this effect will be lost. In the long term what will make robots useful as educational tools will be their designs, their combinations of features and the way they will be integrated into lesson plans.

Robot Design—Creating Immersion and Relatability

A key component to achieving good learning is to create opportunities where learners can directly engage with and immerse themselves in the subject matter and learning process. This of course is a key ingredient for good active learning. A particular benefit of the use of computer software as a learning tool compared with classical pen and paper-based learning, is the fact that using software allows students to be more directly involved in the subject matter and to control their own learning process more fluidly. They are therefore granted an opportunity to experience a more engaging learning experience resulting in better learning.

In my opinion, using robots as teaching aids can provide an even greater sense of engagement and involvement in the learning process, especially if the lesson plan requires students to independently solve problems through the control and manipulation of the robot's actions through direct interaction. I base this opinion, on the observation that whilst a piece of educational software might provide excellent pedagogical results, the level of immersion of the software is limited by the fact that the student has to engage with the experience through a two-dimensional computer screen, with interaction being, typically through a computer keyboard and mouse. These factors can isolate some students somewhat from the learning experience, limiting their quality of learning. In contrast to this, no such barriers exist between students and robots, with robots being physical, real-world, three-dimensional objects which can be seen, touched and directly interacted with, sometimes even using verbal and tactile inputs. This physical presence of a robot and the ability for students to interact with such devices, sometimes using instinctive and simple means, presents a far more immersive and engaging learning experience for students, allowing for better learning to be achieved.

Even more effective active learning can be achieved if a learning activity requires learners to use past life experiences and aspects of pre-knowledge in learning activities. Best learning has also been documented as being achieved when, learners are able to create direct psychological bonds with the learning process. Robots, when well designed, provide an effective mechanism to achieve these very effects through their features and design characteristics.

A good example of a robot that promotes such effects is the previously mentioned *Nao* robot. This robot, through its clever design, exploits many human psychological effects, effects which can create memorable and impactful learning experiences. *Nao* takes the form of a humanoid being and is equipped with a wide variety of sensors allowing it to replicate many of the senses of a human being including sight, hearing and aspects of tactile sensing. The robot can also emit speech and can perform human-like movements with its limbs and has the ability to pick up and grasp objects, walk unaided and can even climb to its feet automatically if it falls over. Whilst clearly robotic in appearance, the physical design and the way in which it presents itself during operation has been cleverly structured to create a strong psychological and emotional bond with learners. In particular, the robot has been designed with the physical appearance of a young child in mind, with the robot have comparable dimensions, being approximately 60 cm tall and being approximately 5 kg in weight.

The ratios of its body elements also provide a strong resemblance to a young child. For example, young children when compared with adults typically have larger heads and eyes when compared to the size of their bodies. These characteristics along with many others are incorporated in the physical design of *Nao*. When in operation, further similarities between the robot and a young child are evident, with the robot having a slightly ungainly walking gait and being set by default to emit synthesised speech in a high-pitched, childlike tone. To people interacting with the robot, these human-like characteristics help create an illusion that it is a young child of some form, prompting students and onlookers alike to form emotional bonds with it, exploiting a subconscious natural tendency of human beings to form caring, emotional bonds with children. With the establishment of psychological bonds between the robot and the students, a scenario is created in which very effective learning can be achieved.

Through my work, I have been lucky enough to observe a wide variety of people of all ages, both old and young, interact with the *Nao* robot and I have always been fascinated by the way this robot captivates, draws in and emotionally influences people. Particularly fascinating is the way that many people interact with the robot as if it genuinely is a young child. For example, when picking up and transporting the robot, it was not uncommon for adults, particularly those who are used to interacting and caring for young children, to handle and carry it in a manner similar to that which might use to carry a young child, namely by placing one hand on the buttock of the robot to support its weight, leaning its body onto the chest of the person involved and then resting the head of the robot on the person's shoulder. I have also observed that when adults provided verbal commands to the robot they often did so in a soft vocal tone and using over-simplified language as if addressing a very young child.

Watching children's interactions with the robot followed a similar pattern, with most treating the robot as an entity which should be cared for, cherished and offered affection. Some treated the robot as if it was a younger sibling whilst other treated the robot as a form of *pet* or *sidekick*. Particularly with the robot's diminutive stature, children were often very careful to treat the robot with care and with a gentle touch.

Often children petted and stroked the robot and when the robot fell over, they were genuinely upset and concerned that the robot had *hurt* itself. They did this despite the fact that the robot has been designed to be sturdy and robust and has design features which suggest that the robot is in some way *armour-plated*. Like adults, many children addressed the robot in over-simplified language as if addressing a pet or a young child.

The key theme running through these patterns of interactions is that as a result of the clever design of the robot, people tended to interact with it as if it was a real-life living entity and not an artificial, man-made machine. The robot reinforces this illusion through it actions when standing idle, by gently swaying its limbs and slowly raising and lowering its chest region, giving the impression that it is *breathing*. With its emotive, childlike features the robot also suggests to people that it is a being of very young age, prompting people to connect and interact with it accordingly. Most strikingly when people speak to the robot, they tend to speak to it in over-simplified language more suited for addressing a young child rather than addressing a state of the art, high powered computer system which the robot actually is!

The fact that people of all ages tend to react in this way to the robot indicates that through its design and operational features, it has prompted people to form an emotional connection and bond with it, a bond which as result of a fundamental human affinity to care and give special attention to children, is particularly strong. Furthermore, despite the robot being an entirely artificial, man-made and emotionless device, people interacting with it, subconsciously impart onto the robot a personality and character. Not only does it indicate that the robot is effective in engaging and holding the attention of people, an important aspect of the learning process, but the fact that people impart onto it a character, shows that they are drawing on past life experiences when interacting with it. Both these aspects are important components of good active learning.

On its most fundamental level, *Nao* succeeds in engaging learners by having design features which people can easily recognise, identify with and relate to and which prompt emotional reactions. Perhaps the most important feature of the robot is its head, which whilst appearing quite alien in appearance has accentuated features which draw people's attention and exploit the human instinct to recognise facial features in images and objects. The design of the head is therefore key to allowing people to relate to the robot and build bonds with it. Indeed, it is interesting to note that in the next section of this chapter, Nick Valentine comments that one of the most memorable features of the robot he worked with, was its cartoon-like face.

I believe that the best teaching robots, particularly for use with younger students, possess design features which allow it to create strong psychological and emotional bonds with users. The *Nao* robot is an excellent example of such a device, with its ability to create such bonds being aided by the fact that it is humanoid in form and so possesses many obvious features to which people can identify and relate. The Nao robot however, despite all its clever design features and dextrous, human-like motion has limits to its performance capabilities. For example, it is not a quick robot and can only propel itself around the floor at a relatively slow speed, slower than typical human walking pace. Should a lesson plan require a robot to be able to move

Fig. 6 The Lego Mindstorm Robot in a wheeled configuration with the ultrasound sensor shown (© Kristina Kerwin)

quickly then the *Nao* robot would be unsuitable for this purpose. For example, the investigation discussed in the previous section of this chapter discusses a lesson about the concept of speed. For this purpose, a far faster wheeled robot, the Lego *Mindstorm* robot, depicted in Fig. 6, is used.

Wheeled robots, whilst they typically have the advantage of being able to move across the ground faster than humanoid, walking robots, lack natural, obvious looking features to which people can relate, such as a head or other limbs, making it more challenging for users to build emotional bonds with such devices. To overcome this, some wheeled robots seek to include such features in their design, including, for example cartoon-like faces on the bodywork or the shaping and positioning of sensors on the robot to resemble humanoid features such as eyes. An example of this can be seen if Fig. 6, where the ultrasonic sensors, on the Lego *Mindstorm* robot, which must usually be placed on the front of a robot to function properly, resemble the *eyes* of the robot. Indeed, having taught lessons using Lego *Mindstorm* robots, constructed in various configurations, far greater levels of student engagement have been observed when robots were designed with such quasi-humanoid features than when constructed without.

Conclusion—The Educator's View

As a professional educator, who has been lucky enough to experiment with teaching lessons using robots, I can honestly say that the experience has been positive and horizon broadening. It has been particularly fascinating, watching how learners of all ages have reacted to such robotic teaching aids, sometimes in very unexpected and surprising but universally positive ways.

Through these observations, and in agreement with the research described in the previous section of this chapter, I strongly believe that robots have the potential to be powerful educational tools, particularly in lesson plans where students are given the opportunity to interact and build bonds with the robot. As with any new device, how to best use educational robots is still the subject of significant research but as time goes on, new and exciting ways will undoubtedly be discovered. Robots too, will continue to develop, with the inclusion of new, exciting features and more refined design.

Currently, a significant aspect of robots' pedagogical benefits come from their novelty and scarcity of use in the classroom, an effect which will certainly diminish over time. Despite this, I strongly believe the usefulness of robots as teaching aids will not be limited by this loss of novelty but will instead be limited only by the skill and imagination of those using and developing them, qualities that teachers, lecturers and engineers alike all exude in abundance. For these reasons I am sure robot-assisted teaching will be the future of education.

Part III—Nick Valentine

A Learner's Perspective

I was recently wondering where my robotics journey began. I knew where my passion for engineering came from, that was easy to explain. From a very early age I loved to design. Be it spaceships, vehicles, sailing ships, town plans, model railways, I was constantly drawing things. Not that I could draw very well, having dyspraxia made fine motor movements particularly challenging. The straight lines necessary for say, the interior decks of a space ship came much more easily to me and my designs never transitioned into what you might call art. The designs were always strictly functional. A spaceship to battle aliens. A sailing ship to fight pirates. A railway to serve a village.

The creativity was always there though which, as it could not be satisfied through drawing, manifested itself in story writing. The stories I wrote were always thoroughly impractical, rarely featured anything technical, and frequently verged on the absurd. This created an interesting contrast with my frequently dismal performances in maths and the sciences. This isn't the conventional story people expect to hear when an engineer recounts their path into the profession. Normally the story goes: engineer meets maths and science, engineer and maths and science fall deeply for one another, engineer and maths and science live happily ever after. I however fall into a tribe defined by the Institution of Mechanical Engineers (IMechE) as the social artist tribe. In the IMechE's report *Five Tribes: Personalising Engineering Education* [28], social artists are described as a large and creative section of the population who seemingly have little affinity with STEM. They are however, the second largest of the five tribes, and a segment of the population that tends to be omitted in STEM outreach. This can mean that the potential skills and talents they bring to engineering and science are largely lost and as such, the diversity of thought and ideas within the industry is reduced.

So, it didn't seem like I was really destined for a career in engineering. How on earth did I end up living a career in engineering, a vocation that I thoroughly love and would recommend? If you need any proof of this commitment, here I am writing a part of a chapter of a book on the subject in my spare time!

Well, the final piece of the puzzle lies in another untapped motivator that STEM outreach frequently neglects, the human capacity for compassion. That need to work as a team, to progress, learn, and grow in order to fulfil our need to empathise and connect with others.

When it comes to engineering endeavour, too often I hear the terms '*because it's cool*' or '*because we can*'. These are normally associated with making something go fast, making it go far, or making it fly high. It is comparatively rare that I hear that the reason for engineering excellence is simply to help others. And whilst there is a certainly a place for the scientific learning gained by doing something because we can, that attitude simply doesn't resonate with sections of the population who are more attuned to the needs and struggles of the world around us. Despite the achievements held up as exemplars of engineering and technology: spaceflight, supersonic cars, or the great feats of civil engineering, at its most fundamental engineering is a way of advancing civilisation and raising our collective standard of living. That's why I ended up studying and pursuing a career in it, it showed me a tangible way that I could take my creativity and apply it in a way that helps others. It wasn't an easy path. As mentioned previously I struggled with maths and science. But that vision of creating and making in order to improve life for others inspired me to persevere. It wasn't a particular love of the sciences that drove me, I had to learn that. No, what I needed was a practical, constructive, and tangible outlet for my creativity. Engineering was to provide that outlet.

That explains my engineering trajectory, but why robotics specifically?

The '*traditional*' view of mine was that my interest in robotics had formed during my industrial placement year at National Grid. My mentor, master of electrical metaphors Carl, introduced me to the different technologies that National Grid were investigating to help manage their equipment. These included Virtual Reality, Artificial Intelligence, mobile robots, and the Internet of Things—technologies which commonly fall into the '*Industry 4.0*' bracket. Whilst working with Carl, I came to realise the potential of these technologies, and particularly the power that could be unlocked by combining them in all kinds of weird and wonderful ways. It was also apparent to me that the energy sector wouldn't be the only one that would be making use of these concepts. And so I firmly decided that I wanted to become an engineer that could take advantage of this potential.

The challenge then, was to find a way of building a career where I could learn and grow in these areas. At the time I was studying for my undergraduate degree and it just so happened that there was a Master's degree course at my institution that included modules in mobile robots, virtual reality, and artificial intelligence. The subject? Robotics and Automation. Robotics was the subject that combined all of the concepts that I was so excited about. Robotics is a discipline that is, by its very nature, a system of systems, and so it made sense that the course would encompass a range of technology areas. However, what particularly drew me to this particular robotics course was its inclusion of a virtual reality module. Robotics and virtual environments are symbiotic in nature, we design, simulate, train, rehearse, and monitor our robotic systems in virtual environments. This, along with modular nature of robots presents a fantastic opportunity for education, outreach and creativity. We'll come back to that idea later.

So that was that, that was definitely one hundred percent why I embarked on a career in robotics. I conducted the placement at National Grid, was inspired by the potential of emerging technologies, undertook an MSc in Robotics and Automation, and just like that I was a roboticist. End of story.

Except that it's not. The reality was that the story begins much, much earlier, in the dark murky days of… primary school.

Whilst thinking about the concepts we talk about in this book, how to teach mechatronics, how to encourage people to engage with it, and what methods could we use to inspire interest in the subject, I was suddenly reminded of an experience I had 25 years ago.

Along with the vague memories of portacabins, school pets, and those plastic balls with all the holes in them that never went in the direction I wanted when I kicked them, I had a distinct memory of playing with a large, grey dome like object with buttons on the top of its surface. By pressing these buttons with arrows and the like on them, the object could be made to move across the polished wooden floor. The arrows would be pressed in a certain order, and the object would then follow this pattern in a route. This was not an activity that was usual in my school. This was a visitor, a special occasion for which we had left our classroom and ventured out into the school corridor, and it was an occasion I distinctly remember. I remember the location, I remember the actions needed to make it move, I remember the feeling, and I remember what it was called, this domelike object. Its name was *Roamer*, and it was a robot.

And this was the thing, I had no concept that this was a robot. *Roamer* was like another, fancier, toy, and I had used my creativity to come up with a route in my head. This was a revelation to me. My creativity could be used, not just to commit ideas to paper, but to have a machine do something I had wanted it to do. This was a way of bringing my ideas off the page, and into the real world. In a very minor way, I had taken ownership and control the real-world actions of something other than myself. I, as a small child, owned the behaviour of something else and I trusted it to perform what I had asked it to do. For that short time *Roamer*, and its actions and behaviours, were mine. This is most likely the very first time that I had such responsibility, and it had been enabled by the use of a robot in a safe, but empowering, environment. How had I achieved this control? I had entered commands into a computer to form a series of instructions. I had had my very first introduction to programming. Not that I knew it at the time, and nor would I be introduced to the concept of programming again for at least another 10 years. This almost certainly presents a missed opportunity, but that nascent spark would not be wasted.

Another element that I find interesting about this memory was that I simply accepted *Roamer*'s existence. Not once did I question *Roamer*'s technological feasibility. Or why I wasn't seeing such technology elsewhere in my day to day existence as a small boy in Stockport. *Roamer* was there. I pressed buttons in a pattern on its exterior, and *Roamer* did what I asked it to. Perhaps at this age we are a lot more accepting and receptive to what is presented to us, this open mindedness makes early years education a critical and often underestimated part of our development. The impact this short activity has had is notable by its endurance in my mind. Even the name has stuck with me, over 25 years later. There aren't many memories of mine that have lasted that amount of time. For some reason, that memory stuck with me. This could simply be down to the novelty of the experience, novelty certainly makes things more memorable, but I also think that interaction with a robot in itself made the experience more memorable. Being able to learn alongside a system expressing its own simple, non-human behaviour deepened the experience and firmly embedded the memory. Although in recent research I have subsequently found that *Roamer* was very much smaller than the image I have in those memories created by a small boy. Memory (and growing up) can be fickle.

When conducting research into *Roamer* (just to be doubly sure that I hadn't made the whole thing up), the images I found often featured *Roamer* with a simple smile and a pair of eyes drawn on its exterior, in an attempt to make *Roamer* more relatable and, for want of a better word, '*cute*'. I don't have a particular memory of the *Roamer* I encountered having a face upon it, but giving educational robots these features, albeit simple ones, can assist in building emotional relationships between us and '*them*' and deepen the interaction we have with them. Having these systems display recognisably organic, friendly, features can be one way of building trust between the children and the machine. Trust between humans and autonomous machines is currently a big topic in robotics, in industry we need to be able to trust autonomous machines to perform their task in a manner that isn't going to cause harm, be it physical or psychological.

In education too, children need to be able to trust the system to behave within the bounds of expected behaviour whilst accompanying them on their learning journey. However, to encourage learning this trust must be balanced with the freedom to explore the unexpected, and importantly the freedom to fail in a safe and forgiving environment. Having a '*companion*' to explore these concepts with, without the behavioural issues you might encounter in another human being, may be helpful when trying to embed learning deeper. This is particularly true with younger children, when social skills might not be fully developed. This interaction with a companion then enables education through stealth. You're completely unaware that you're learning, you're just playing with the cute toy. We don't even need to call it a robot, or engineering, or science. Rather than teaching concepts in an abstract way, we can use ubiquitous, relatively cheap, and accessible technology to build trust and enable education through play. This of course depends on the resources available to a school, but perhaps we can use cheaper mechatronic systems, with their inherent modularity, to unlock creativity and subconsciously teach those concepts.

It seems odd to think about it now, but at the time I encountered *Roamer* I, of course, had no idea that my life would ultimately involve working with mobile robots. This early exposure to the technology has ultimately led to me working on systems that are a lot larger, and in a lot of ways much more frightening than *Roamer*, the benign creature that crawled along the polished wooden floors of a 1990s primary school.

Other later memories from school aren't quite so entertaining or interesting. Following my early experience with *Roamer*, there was then a distinct lack of robots at school. Robots did not feature at all in my high school experience, where I was learning about friendships, relationships, and thinking more seriously about my future. Picture the scene, we've all been here. Tables, plastic chairs, a paper notebook (mine inexplicably '*backed*' with wallpaper) and the board at the front. Lines appear on the board, but the lines don't seem to mean anything to me. They're there, but I can't understand why they're there, what they mean, or what I'm supposed to do with them. '*Turn to page X in your textbooks*'. On page X Farmer Joe (never Farmer Jane) has a problem. He needs to build a fence and so wants to know how much fence he needs for the job. Using maths, figure out how much fence Farmer Joe needs. Answers can be found at the back of the textbook.

In my childhood brain, real-world Farmer Joe was probably out in the sunshine somewhere having great fun whacking in fence poles with a big hammer, and I bet he didn't bother doing any maths to figure out how much fence he needed. He probably just went out with however much fence he had and guessed. Or he just got out some string and measured it like a normal person. All in all I was rather jealous of Farmer Joe and his farm, and his fence. Meanwhile I was stuck in a stuffy classroom, staring at the yellow, worn, pages of a textbook that nobody had ever had any fondness for. I can still remember the smell of the textbooks. And it still gives me a headache.

This contrasted with my experiences at home, where I was engaging more and more with engineering and robotics, not that I knew it in those terms. I had progressed from drawings on paper and was now building things for myself. Using construction toys such as *K'NEX*, which mix electrical and mechanical components, I was expressing my creativity in a constructive way and using the *K'NEX* motors to make my creations move for themselves. This shows that even from a young age, unconstrained by convention, we are able to mix the disciplines that tend to become more separate and more siloed as our technical education continues. This lack of constraints and freedom of imagination in combining concepts is a distinct advantage in robotics. If during our developmental years we can play with educational tools that mix disciplines, those concepts are embedded within us. This is particularly true when that play is coupled with the right learning environment and teaching techniques.

As systems that are themselves a mix of disciplines, robots lend themselves as to teaching these concepts and fortunately, let's be honest, people love robots. They're like us in some ways, and yet they are decidedly not us. Consequently, people connect with them, trust them, and form teams and relationships with them in unique ways. This makes them an obvious choice as educational tools. Not only can we teach technological concepts with them, they can also be used to teach concepts encompassing psychology, sociology, ethical and legal issues. This is because robots are a system of systems and when we interact with them, we ourselves become a part of that system. Interestingly this means that robots can serve as a gateway between engineering and the humanities, not to mention the arts.

Whilst I was able to comfortably mix electronics and mechanical engineering concepts at a young age, one area that was more difficult for me to reach during my childhood was simulation and virtual environments. Today, gaming technology has made simulation and virtual environments far more accessible, and they have become vital within the robotics industry for developing and understanding robot behaviours. The behaviour of a robot is often programmed, simulated and optimised in these virtual worlds, preferably worlds that are representative of the real-world environment that they will be operating in. Once the robot developer is happy with the behaviour of the robot in the simulation, the programming is transferred directly onto the real-world robot and tested in the physical environment.

Some organisations are now controlling their robots directly from the virtual environment, combining the virtual and real worlds in a '*cyber physical system*'. Having their pedigree in the games industry means that contemporary virtual environments are limited only by the imagination of the developer. This almost boundless virtual space to play with, combined with the modular nature of robots, makes for an exciting ecosystem full of educational potential. And children today are much more familiar with digital technologies, they are growing up with them and as such are known as '*digital natives*'. This gives them the mental tools to easily grasp and make use of these technologies, who knows what will come out of the combination of the technology and the skills and creativity of children?

This mixing and matching of professions and disciplines is something that we need more of within engineering. The lines between the disciplines are becoming ever more blurred, with engineering solutions frequently combining mechanical, electrical, and software engineering. Robots are a clear example of this. It is almost inconceivable today that a mechanical system would not include elements of electronics and software. Despite this we still tend to self-categorise ourselves into distinct technical groups. We almost seem to revel in it. To encourage cross pollination of ideas and disciplines, we need ways of further breaking down the self-imposed barriers between our specialisms. There are organisations such as *CHAIN*, which attempt to inspire emerging engineers to work together, learn from each other, and engage with the lessons from across the engineering spectrum. However, by the time we are at the stage where we need to start collaborating and innovating with other disciplines, those divides are already entrenched. This conflicts with the trend for the lines between technologies to become blurred, and as such, we should look to prevent these barriers from even forming at an earlier age, and explore how all of our different professions interact and complement each other to make the world a better place. When we combine this with disciplines that usually fall outside of engineering and technology, this presents us with some powerful and exciting opportunities. By engaging with these disciplines, we can also start to engage with people that wouldn't normally relate with engineering.

Every sector requires diversity of thought and skills. We all need those creatives, those social artists that aren't typically reached by STEM outreach, to help us reach those solutions that we might never conceive. Consequently, we need educational solutions that engage with people in the widest possible terms. That needs flexibility, modularity and novel thinking.

So how do we keep our education of STEM novel? Fortunately, given the pace with which the fields of mechatronics and robotics are progressing, new applications and technologies are constantly emerging and crucially, the price of sensors and processing also falls. Coupled with this, the open source movement has made technology, and the skills to develop it, more accessible than ever, with more and more people constantly uploading new and exciting code for others to build further upon.

Whilst this may create a '*novelty arms rac*e', I believe that we can rely on the creativity and imagination of children to make use of the canvasses that robotics and virtual environments present. A case in point is *Minecraft*, at the time of writing the most popular computer game in the world, with there already being STEM initiatives that rightly make use of it as a powerful education tool.

If we can use mechatronics' inherently modular nature to engage people's creativity, if we can unlock the potential of virtual environments to allow people to explore the limits of their imaginations, if we can unleash the nascent creativity and imagination of those who want to help and do good in the world, and if we can pair this with people and communities who have problems waiting to be solved, then I believe that we'll be a good part of the way towards a future that allows everyone to use their skills to build a more inclusive and fairer future.

References

1. Kerwin K (2019) Studying the efficacy of teaching stem subjects to primary school age students using Lego robots, MSc in engineering projects & systems management Thesis, Kingston University
2. Coughlan S (2014) Tablet computers in '70% of schools', BBC News @ www.bbc.co.uk/news/education-30216408. Accessed 24 Nov 2021
3. Education Technology (2017) How technology will shape the future of education—Education Technology @ edtechnology.co.uk/comments/how-technology-will-shape-the-future-of-education/. Accessed 24 Nov 2021
4. Abilitymagazine (2018) Nao—Robot for Autism @ abilitymagazine.com/Nao-Robot.html. Accessed 24 Nov 2021
5. https://www.lego.com/en-gb/product/lego-mindstorms-ev3-31313
6. Education Technology (2018) First robotics curriculum launched @ edtechnology.co.uk/Article/first-robotics-curriculum-launched. Accessed 24 Nov 2021

7. Faisal A (2012) Using robots to promote learning in elementary grades. Polytechnic Institute of New York University, MSc
8. Hobeika L, Diard-Detoeuf C, Garcin B, Levy R, Volle E (2016) General and specialized brain correlates for analogical reasoning: a meta-analysis of functional imaging studies. Human Brain Mapping 37(5):1953–1969
9. Gerraty R, Davidow J, Wimmer G, Kahn I, Shohamy D (2014) Transfer of learning relates to intrinsic connectivity between hippocampus, ventromedial prefrontal cortex, and large-scale networks. J Neurosci 34(34):11297–11303
10. Bunzeck N, Doeller C, Dolan R, Duzel E (2011) Contextual interaction between novelty and reward processing within the mesolimbic system. Hum Brain Mapp 33(6):1309–1324
11. ukstemconference.files.wordpress.com/2019/06/horizons-conf-19-abstract-booklet-.pdf. Accessed 4 Feb 2022
12. Marton F, Säljö R (1997) Approaches to learning. In: Marton F, Hounsell D, Entwistle ND (eds) The experience of learning. Scottish Academic Press, pp 39–58
13. Mcleod S (2007) Levels of processing | Simply Psychology @ www.simplypsychology.org/levelsofprocessing.html. Accessed 24 Nov 2021
14. Wang M, Derry S, Ge X (2017) Guest editorial: fostering deep learning in problem-solving contexts with the support of technology journal of educational technology & society. Palmerston North 20(4):162–165
15. Reif F (2008) Applying cognitive science to education. The MIT Press
16. Rosenshine B (2012) Principles of instruction: research-based strategies that all teachers should know. Am Educ 36(1):12
17. Ukessays.com (2018) The development of concept through the national curriculum @ www.ukessays.com/essays/education/the-development-of-concept-through-the-national-education-essay.php. Accessed 24 Nov 2021
18. Yasnitsky A, van der Veer R (eds) (2016) Revisionist revolution in Vygotsky studies. Routledge
19. Cook M (1955) The construction of reality in the child. In Piaget J (ed) The construction of reality in the child, Routledge & Kegan Paul, vol 82
20. Teach-nology.com (2018) Piaget's theory on constructivism. @ www.teach-nology.com/currenttrends/constructivism/piaget/. Accessed 24 Nov 2021
21. Legislation.gov.uk. (2018) Education Reform Act 1988 @ www.legislation.gov.uk/ukpga/1988/40/contents. Accessed 24 Nov 2021
22. Department for Education (2018) Next generation of young people gaining the skills Britain needs @ www.gov.uk/government/news/next-generation-of-young-people-gaining-the-skills-britain-needs. Accessed 24 Nov 2021
23. Learning Theories (2018) Social Development Theory @ www.learning-theories.com/vygotskys-social-learning-theory.html. Accessed 24 Nov 2021
24. Shabani K (2016) Applications of Vygotsky's sociocultural approach for teachers' professional development. Cogent Educ 3(1):1252177
25. Daniels H, Hedegaard M (2011) Vygotsky and special needs education. Continuum International Pub. Group
26. Bloom's Taxonomy (2018) Bloom's Taxonomy @ bloomstaxonomy2013.weebly.com. Accessed 24 Nov 2021
27. Thalheimer W (2019) How much do people forget? @ www.worklearning.com/2010/12/14/how-much-do-people-forget/. Accessed 24 Nov 2021
28. IMechE, Five tribes—personalising engineering education @ www.imeche.org/policy-and-press/reports/detail/five-tribes-personalising-engineering-education. Accessed 24 Nov 2021

Conclusion

EcoMechatronics Enabling Technologies, and Future Prospects

Maki K. Habib, Peter Hehenberger, and David Bradley

Abstract EcoMechatronics is an interdisciplinary field that is evolving to have a holistic impact on the development of current and future smart technologies, methodologies, algorithms, techniques, products, processes, and systems. The key features of this evolution are the synergy of the new ideas, designs, and development with the new concepts of eco-design, sustainability, environment friendly, and circular economy supported by clean and energy-saving technologies. The aim is the implementation of Net-Zero EcoMechatronic systems which are the basis of sustainable, environmentally friendly, and cost-effective intelligent products, processes, and systems. This book on EcoMechatronics represents the latest research efforts in the field and it is presented in three sections. The first deals with the underlying issues of sustainability and enabling technologies. The second then looks at a range of its applications, including some, such as the built environment, that have not previously been considered as containing significant mechatronic elements. The final section of the book then focuses on educational concerns and issues. Finally, this chapter highlights some of the enabling and driving technologies contributing to the field of EcoMechatronics and also it introduces the future prospect of the field as projected through a range of open-ended research areas.

M. K. Habib (✉)
The American University in Cairo, New Cairo, Egypt
e-mail: maki@aucegypt.edu

P. Hehenberger
University of Applied Sciences Upper Austria, Wels, Austria

D. Bradley
Abertay University, Dundee, UK

P. Hehenberger et al. (eds.), *EcoMechatronics*,
https://doi.org/10.1007/978-3-031-07555-1_20

Introduction

EcoMechatronics represents a range of new Mechatronics technological development, implementation, and breakthroughs in interdisciplinary innovation. These are rapidly evolving to impact the ways in which to develop and design new smart technologies that can think, learn and adapt, optimize performance, and drive research, leading to the creation of new ideas and effective new and advanced technologies and techniques. Such development aims to improve both the functionality of these new technologies and their reliability, adaptability, and design flexibility while adopting a practical inspiration to managing ecosystems that integrates the latest technological developments with reduced ecological impact. Furthermore, this trend of technological development aims to reduce and manage energy consumption and enhance efficiency and reduce ecological impact while being environmentally friendly while supporting sustainability with effective resource management and being cost-effective with enhanced user comfort. Thus, EcoMechatronics aims to impact every aspect of technological development with an integrative and holistic positive influence on the design process for better planetary ecosystems management that reduces or eliminates waste and controls this through the adoption of a circular economy approach and associated planning.

Natural resources and environmental problems such as air and water pollution are critical issues facing the planet and humanity. At the same time, climate change is a significant challenge that requires joint actions. These issues need to be urgently addressed through effective planning and regulatory policies that reduce the effect of greenhouse gases, particularly carbon dioxide, nitrous oxide, and methane. In addition, there should be a focus on developing and using renewable clean energy while optimizing energy consumption. Further, there is a critical need to develop creative solutions, promote and develop new innovative technologies that limit emissions and help eliminate or reduce the pollutants to the environment by 2050. EcoMechatronics, as an essential interdisciplinary field, has a significant role in designing net-zero energy efficiency technologies at all levels, contributing to the design development process and integrating these within the concepts of a circular economy subject to effective ecosystem management. This design and development approach requires new ways of creative thinking covering ideas and solutions associated with the design and implementation of Net-Zero EcoMechatronics systems as the basis of sustainable devices, products, and systems across a broad range of applications. Developing effective Net-Zero emissions products and systems demands the whole system thinking covering technical, social, and environmental issues. In contrast, lifecycle design covers the design requirements that quantify ecological impacts and assure sustainable design strategies (energy, material, and lifetime).

Mechatronic systems, subsystems, and components form the primary data gathering element of many devices, processes, or systems and also provide the direct action layers of many applications. The conceptual formations of such operations, techniques, systems, and subsystems indicate that Mechatronics design should accommodate functions and technologies featuring smartness, flexibility, and reliability. In addition, achieving an effective technical, social, and environmental balance during the design process should be given significant and detailed consideration.

Thus, many of the actions of EcoMechatronics systems described by smartness are associated with computational capabilities and experiential learning that utilizes digitized data with Artificial Intelligence (AI) and Machine learning (ML) techniques. These techniques raise the question of individual privacy, cybersecurity, and ethical issues that need to be adequately addressed and effectively treated. Of particular concern is the increasing use of autonomous systems, ranging from real-time healthcare to military applications, where the actions of the AI component, and hence the response of the Mechatronics component, can be a threat to life.

EcoMechatronics Development and Enabling Technologies

The driving and enabling technologies, techniques, and algorithms that support evolving EcoMechatronics devices, processes, and systems aim to assure sustainable design strategies to facilitate the design and development that focuses on energy, materials, and lifetime. Some of these technologies evolved from integrating traditional technologies with IT-based tools and technology, while the other trends in the development are purely new innovative design development. This enabling development includes integrating different layers of the technologies and the subsequent integration of vital concepts, such as ecology management and the Green Circular Economic. This demands skilled personnel and an effective education system to play a critical role in establishing and maintaining an understanding of what is involved in EcoMechatronics and the evolution of the associated technologies. Table 1 presents some of the enabling technologies supporting the growth of EcoMechatronics.

EcoMechatronics: Issues and Future Prospects

Ecodesign and sustainable design strategies are critical issues facing the development of EcoMechatronics technologies, materials, processes, intelligence, systems, and future evolution. The evolution of EcoMechatronics needs to focus on developments in respect of a wide range of open-ended research areas, including, but not exclusive to:

- Advanced, distributed computing
- Agile systems
- Artificial Intelligence (AI) and Machine Learning including for Education
- Bio-hybrid robots
- Biosensors and sensor networking
- Clean energy and energy harvesting

Table 1 EcoMechatronics enabling technologies

Domain	Technologies
Computing	• Distributed and Optical Computing • DNA Data Storage • Miniaturized Computing Technologies • Molecular Computers • Neuromorphic Computers • Quantum Computers
Digitalisation	• Artificial Intelligence • Blockchain • Cyber Security and Privacy Protection • Digital Twins for Real-Time Insight, Optimization, and Decision-Making • Human/Robot Interaction • Smart Cities and Urban Mobility • Virtual and Augmented Reality • Wearable Devices
Intelligence	• Artificial Intelligence, Machine Learning, Real-time Learning, and Deep Learning • Autonomous Systems, Autonomous Robots, Vehicles, and Navigation • Big Data, Data Analytics, and Data Fusion • Connected devices, Internet of Things (IoT), and Cyber-Physical Systems (CPS) • Real-Time Interactions • Robotics • Smart Grids • Smart Factories
Nanotechnology	• Biophysics • Micro and Nano Manufacturing • Nanocomputation • Nanomedicine • Nanorobotics • Nano Smart Materials and Green Technologies
Robotics	• Additive Manufacturing and Rapid Prototyping • Autonomous Aerial, Maritime, Road, and Underwater Vehicles • Heterogeneous Multi-Robot Systems • Industrial, Service a, and Personal Robotics • Intelligent Robot Mechanisms • Medical Robots • Robot Agility
Sensors and actuators	• Actuator Design and Technologies • Biosensors and Bio-Actuators • Nano-Sensors and Nano-Actuators • Sensor Design and Integration • Wireless Sensor Networks • Sensors, Actuators, and Smart Materials

- Distributed data, computing, analysis, and decision-making
- Ecodesign and circular economy
- Edge computing
- Effective and fast networking to support distributed real-time requirements in different domains
- Evolution of eco-design and circular economies
- Fully autonomous systems
- Green technologies and eco-friendly solutions
- Human adaptive and friendly technologies
- Integrated intelligence and digitalization
- Integrative multi-dimensional computational intelligence
- Intelligent adaptive behaviors
- Intelligent and adaptive mechanisms
- Machine intelligence, intelligent systems, and techniques
- Nanotechnologies leading to effective nanorobotics, nano-devices, and nano-systems
- New and novel actuator designs involving bioinspiration
- Smart manufacturing systems and nanomanufacturingSmart materials and bio-materials
- Virtuality, augmented reality, and mixed realities

It is necessary to recognize that this list of future research and development topics and areas is intended to be indicative rather than comprehensive. As the understanding of the impact of technologies and methods in relation to sustainability goals develops, additional new and novel areas of activity, interest, and research will emerge. In this context, consideration of the short, medium, and long term actions and responses to EcoMechatronics and the meetings of climate change goals suggests the following.

Short Term Responses and Actions—These should be based on the use of existing technologies, and it is therefore synchronized to a degree reflected in the background to EcoMechatronics as expressed in *Mechatronics in Action, Mechatronic Futures,* and *Reinventing Mechatronics* along with studies by organizations such as the World Economic Forum (WEF) on the impact of technologies on society and employment and the work of Conference Of the Parties (COP26) in association with the 2021 United Nations Climate Change Conference.

Medium Term Responses and Actions—This represents implementing systems that are reaching the final stages of their research and development immediately prior to their deployment. The methodologies, technologies, algorithms, techniques, and systems described in EcoMechatronics could be considered to fall into this category with implementation horizons of 5–10 years.

Long Term Responses and Actions—These represent open-ended potential new motivations, concepts, and ideas supporting the evolution of EcoMechatronics in the early stages of research or perhaps may not even have reached the initial stages of concept development. Some of the contributed materials presented in *EcoMechatronics* hint at some of these future development prospects that are difficult to isolate and worth mentioning.

The manufacturer's authorised representative in the EU is Springer Nature Customer Service Centre GmbH, Europaplatz 3, 69115 Heidelberg, Germany. If you have any concerns regarding our products, please contact ProductSafety@springernature.com

Printed and bound by CPI Group (UK) Ltd, Croydon, CR0 4YY

01/10/2026

02225872-0012